AUDIO/RADIO
HANDBOOK

Technical Editor & Contributing Author:
Martin Giles
Manager - Consumer Linear Applications

Contributors:

Dennis Bohn	Tim D. Isbell	Don Sauer
K. H. Chiu	Kerry Lacanette	Jim Sherwin
Gene Garrison	John Maxwell	Tim Skovmand
William Gross	Thomas B. Mills	John Wright
Steve Hobrecht	Ron Page	Milt Wilcox
Wong Hee	Tim Regan	

National Semiconductor Corporation • 2900 Semiconductor Drive • Santa Clara, CA 95051
© 1980 National Semiconductor Corp.

Reprinted with permission by Audio Amateur Inc.

Note: "The United States Army and Air Force do not endorse the publication and do not derive profit from the sale of this publication."

W.S. Werner, GS-15
Chief, Officer for Security Review
Secretary of the Air Force
Office of Public Affairs

© **National Semiconductor Corporation**
2900 Semiconductor Drive, Santa Clara, California 95051
(408) 737-5000/TWX (910) 339-9240
National does not assume any responsibility for use of any circuitry described; no circuit patent licenses are implied, and National reserves the right, at any time without notice, to change said circuitry.

Section Edge Index

Introduction 1
Preamplifiers 2
AM, FM and FM Stereo 3
Power Amplifiers 4
Floobydust 5
Appendices 6
Index 7

Table of Contents

1.0 Introduction
- 1.1 Scope of Handbook 1-1
- 1.2 IC Parameters Applied to Audio 1-1

2.0 Preamplifiers
- 2.1 Feedback — To Invert or Non-Invert 2-1
- 2.2 Design Tips on Layout, Ground Loops and Supply Bypassing 2-1
- 2.3 Noise 2-3
- 2.4 Audio Rectification — or, "How Come My Phono Detects AM?" 2-11
- 2.5 Dual Preamplifier Selection 2-12
- 2.6 LM381 2-13
- 2.7 LM381A 2-16
- 2.8 LM387 or LM387A 2-17
- 2.9 LM382 2-18
- 2.10 LM1303 2-22
- 2.11 Phono Preamps and RIAA Equalization 2-23
- 2.12 Tape Preamps and NAB Equalization 2-28
- 2.13 Mic Preamps 2-43
- 2.14 Tone Controls — Passive and Active 2-46
- 2.15 Scratch, Rumble and Speech Filters 2-55
- 2.16 Bandpass Active Filters 2-58
- 2.17 Octave Equalizers 2-59
- 2.18 Mixers 2-65
- 2.19 Driving Low Impedance Lines 2-67
- 2.20 Noiseless Audio Switching 2-68

3.0 AM, FM and FM Stereo
- 3.1 AM Radio 3-1
- 3.2 LM3820 3-4
- 3.3 FM-IF Amplifiers/Detectors 3-7
- 3.4 LM3089 — Today's Most Popular FM-IF System 3-8
- 3.5 LM3189 3-13
- 3.6 FM Stereo Multiplex — LM1310/1800 3-14
- 3.7 Stereo Blend — LM4500A/LM1870 3-18

4.0 Power Amplifiers
- 4.1 Inside Power Integrated Circuits 4-1
- 4.2 Design Tips on Layout, Ground Loops and Supply Bypassing 4-5
- 4.3 Power Amplifier Selection 4-5
- 4.4 LM1877/378/379/1896/2887/2896 4-7
- 4.5 LM380 4-22
- 4.6 LM384 4-29
- 4.7 LM386 4-31
- 4.8 LM389 4-36
- 4.9 LM388 4-41
- 4.10 LM390 4-45
- 4.11 LM383 4-46
- 4.12 Power Dissipation 4-48
- 4.13 Boosted Power Amps/LM391/LM2000 4-50
- 4.14 Heatsinking 4-64

Table of Contents
(continued)

5.0 Floobydust*

5.1	Biamplification	5-1
5.2	Active Crossover Networks	5-1
5.3	Reverb	5-7
5.4	Phase Shifter	5-10
5.5	Fuzz	5-11
5.6	Tremolo	5-11
5.7	Acoustic Pickup Preamp	5-12
5.8	Non-Complementary Noise Reduction	5-13

6.0 Appendices

A1	Power Supply Design	6-1
A2	Decibel Conversion	6-11
A3	Wye-Delta Transformation	6-11
A4	Standard Building Block Circuits	6-12
A5	Magnetic Phono Cartridge Noise Analysis	6-13
A6	General Purpose Op Amps Useful for Audio	6-16
A7	Feedback Resistors and Amplifier Noise	6-17
A8	Reliability	6-18
A9	Audio-Radio Glossary	6-19

7.0 Index

*"Floobydust" is a contemporary term derived from the archaic Latin *miscellaneus*, whose disputed history probably springs from Greek origins (influenced, of course, by Egyptian linguists) — meaning here "a mixed bag."

Device Index

Device	Page	Device	Page
LM378	4-7	LM1310	3-14
LM379	4-7	LM1800	3-14
LM380	4-22	LM1818	2-37
LM381	2-13	LM1870	3-20
LM381A	2-16	LM1877	4-7
LM382	2-18	LM1896	4-7
LM383	4-46	LM2000	4-62
LM384	4-29	LM2001	4-62
LM386	4-31	LM2887	4-7
LM387	2-17	LM2896	4-7
LM387A	2-17	LM3089	3-8
LM388	4-41	LM3189	3-13
LM389	4-36	LM3820	3-4
LM390	4-45	LM4500A	3-19
LM391	4-52	MM5837	2-62
LM1011	2-42	LM13600	5-13
LM1303	2-22		

1.0 Introduction

In just a few years time, National Semiconductor Corporation has emerged as a leader — indeed, if not *the* leader in all areas of integrated circuit products. National's well-known linear and digital ICs have become industry standards in all areas of design. This handbook exists to acquaint those involved in audio systems design with National Semiconductor's broad selection of integrated circuits specifically designed to meet the stringent requirements of accurate audio reproduction. Far from just a collection of data sheets, this manual contains detailed discussions, including complete design particulars, covering many areas of audio. Thorough explanations, complete with real-world design examples, make clear several audio areas never before available to the general public.

1.1 SCOPE OF HANDBOOK

Between the hobbyist and the engineer, the amateur and the professional, the casual experimenter and the serious product designer there exists a chaotic space filled with Laplace transforms, Fourier analysis, complex calculus, Maxwell's equations, solid-state physics, wave mechanics, holes, electrons, about four miles of effete mysticism, and, maybe, one inch of compassion. This audio handbook attempts to disperse some of the mist. Its contents cover many of the multidimensional fields of audio, with emphasis placed on intuition rather than rigor, favoring the practical over the theoretical. Each area is treated at the minimum depth felt necessary for adequate comprehension. Mathematics is not avoided — only reserved for just those areas demanding it. Some areas are more "cookbook" than others, the choice being dictated by the material and Mother Nature.

General concepts receive the same thorough treatment as do specific devices, based upon the belief that the more informed integrated circuit user has fewer problems using integrated circuits. Scanning the Table of Contents will indicate the diversity and relevance of what is inside. Within the broad scope of audio, only a few areas could be covered in a book this size; those omitted tend to be ones not requiring active devices for implementation (e.g., loudspeakers, microphones, transformers, styli, etc.).

Have fun.

1.2 IC PARAMETERS APPLIED TO AUDIO

Audio circuits place unique requirements upon IC parameters which, if understood, make proper selection of a specific device easier. Most linear integrated circuits fall into the "operational amplifier" category where design emphasis has traditionally been placed upon perfecting those parameters most applicable to DC performance. But what about AC performance? Specifically, what about audio performance?

Audio is really a rather specialized area, and its requirements upon an integrated circuit may be stated quite concisely: The IC must process complex AC signals comprised of frequencies ranging from 20 hertz to 20k hertz, whose amplitudes vary from a few hundred microvolts to several volts, with a transient nature characterized by steep, compound wavefronts separated by unknown periods of absolute silence. This must be done without adding distortion of any sort, either harmonic, amplitude, or phase; and it must be done noiselessly — in the sun, and in the snow — forever.

Unfortunately, this IC doesn't exist; we're working on it, but it's not ready for immediate release. Meanwhile, the problem remains of how to choose from what is available. For the most part, DC parameters such as offset voltages and currents, input bias currents and drift rates may be ignored. Capacitively coupling for bandwidth control and single supply operation negates the need for concern about DC characteristics. Among the various specifications applicable to AC operation, perhaps slew rate is the most important.

1.2.1 Slew Rate

The slew rate limit is the maximum rate of change of the amplifier's output voltage and is due to the fact that the compensation capacitor inside the amplifier only has finite currents[1] available for charging and discharging (see Section 4.1.2). A sinusoidal output signal will cease being small signal when its maximum rate of change equals the slew rate limit S_r of the amplifier. The maximum rate of change for a sine wave occurs at the zero crossing and may be derived as follows:

$$v_o = V_p \sin 2\pi \, ft \qquad (1.2.1)$$

$$\frac{dv_o}{dt} = 2\pi f V_p \cos 2\pi \, ft \qquad (1.2.2)$$

$$\left.\frac{dv_o}{dt}\right|_{t=0} = 2\pi f V_p \qquad (1.2.3)$$

$$S_r = 2\pi f_{max} V_p \qquad (1.2.4)$$

where: v_o = output voltage

V_p = peak output voltage

S_r = maximum $\frac{dv_o}{dt}$

The maximum sine wave frequency an amplifier with a given slew rate will sustain without causing the output to take on a triangular shape is therefore a function of the peak amplitude of the output and is expressed as:

$$f_{max} = \frac{S_r}{2\pi V_p} \qquad (1.2.5)$$

Equation (1.2.5) demonstrates that the borderline between small signal response and slew rate limited response is not just a function of the peak output signal but that by trading off either frequency or peak amplitude one can continue to have a distortion free output. Figure (1.2.1) shows a quick reference graphical presentation of Equation (1.2.5) with the area above any V_{PEAK} line representing an undistorted small signal response and the area below a given V_{PEAK} line representing a distorted sine wave response due to slew rate limiting.

As a matter of convenience, amplifier manufacturers often give a "full-power bandwidth" or "large signal response" on their specification sheets.

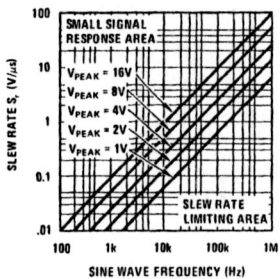

FIGURE 1.2.1 Sine Wave Response

This frequency can be derived by inserting the amplifier slew rate and peak rated output voltage into Equation (1.2.5). The bandwidth from DC to the resulting f_{max} is the full-power bandwidth or "large signal response" of the amplifier. For example, the full-power bandwidth of the LM741 with a $0.5 V/\mu s$ S_r is approximately 6 kHz while the full-power bandwidth of the LF356 with a S_r of $12 V/\mu s$ is approximately 160 kHz.

1.2.2 Open Loop Gain

Since virtually all of an amplifier's closed loop performance depends heavily upon the amount of loop-gain available, open loop gain becomes very important. Input impedance, output impedance, harmonic distortion and frequency response all are determined by the difference between open loop gain and closed loop gain, i.e., the loop gain (in dB). Details of this relationship are covered in Section 2.1. What is desired is high open loop gain — the higher the better.

1.2.3 Bandwidth and Gain-Bandwidth

Closely related to the slew rate capabilities of an amplifier is its unity gain bandwidth, or just "bandwidth." The "bandwidth" is defined as the frequency where the open loop gain crosses unity. High slew rate devices will exhibit wide bandwidths.

Because the size of the capacitor required for internally compensated devices determines the slew rate — hence, the bandwidth — one method used to design faster amplifiers is to simply make the capacitor smaller. This creates a faster IC but at the expense of unity-gain stability. Known as a *decompensated* (as opposed to *uncompensated* — no capacitor) amplifier, it is ideal for most audio applications requiring gain.

The term *gain-bandwidth* is used frequently in place of "unity gain bandwidth." The two terms are equal numerically but convey slightly different information. Gain-bandwidth, or gain-bandwidth product, is a combined measure of open loop gain and frequency response — being the product of the available gain at any frequency times that frequency. For example, an LM381 with gain of around 2000 V/V at 10 kHz yields a GBW equal to 20 MHz. The GBW requirement for accurate audio reproduction may be derived for general use by requiring a minimum loop-gain of 40 dB (for distortion reduction) at 20 kHz for an amplifier with a closed loop gain of 20 dB. This means a minimum open loop gain of 60 dB (1000 V/V) at 20 kHz, or a GBW equal to 20 MHz. Requirements for lo-fi and mid-fi designs, where reduced frequency response and higher distortion are allowable, would, of course, be less.

1.2.4 Noise

The importance of noise performance from an integrated circuit used to process audio is obvious and needs little discussion. Noise specifications normally appear as "Total Equivalent Input Noise Voltage," stated for a certain source impedance and bandwidth. This is the most useful number, since it is what gets amplified by the closed loop gain of the amplifier. For high source impedances, noise current becomes important and must be considered, but most driving impedances are less than 600Ω, so knowledge of noise voltage is sufficient.

1.2.5 Total Harmonic Distortion

Need for low total harmonic distortion (THD) is also obvious and need not be belabored. THD performance for preamplifier ICs will state the closed loop gain and frequency at which it was measured, while audio power amplifiers will also include the power output.

1.2.6 Supply Voltage

Consideration of supply voltage limits may be more important than casual thought would indicate. For preamplifier ICs and general purpose op amps, attention needs to be directed to supply voltage from a dynamic range, or "headroom," standpoint. Much of audio processing requires headroom on the order of 20-40 dB if transient clipping is to be avoided. For a design needing 26 dB dynamic range with a nominal input of 50 mV and operating at a closed loop gain of 20 dB, a supply voltage of at least 30 V would be required. It is important, therefore, to be sure the IC has a supply voltage rating adequate to handle the worst case conditions. These occur for high power line cases and low current drain, requiring the IC user to check the "absolute maximum" ratings for supply voltage to be sure there are no conditions under which they will be exceeded. Remember, "absolute maximum" means just that — it is not the largest supply you can apply; it is the value which, if exceeded, causes all bets to be cancelled. This problem is more acute for audio power devices since their supplies tend to sag greatly, i.e., the difference between no power out and full power out can cause variations in power supply level of several volts.

1.2.7 Ripple Rejection

An integrated circuit's ability to reject supply ripple is important in audio applications. The reason has to do with minimizing hum within the system — high ripple rejection means low ripple bleedthrough to the output, where it adds to the signal as hum. Relaxed power supply design (i.e., ability to tolerate large amounts of ripple) is allowed with high ripple rejection parts.

Supply ripple rejection specifications cite the amount of rejection to be expected at a particular frequency (normally 120 Hz), or over a frequency band, and is usually stated in dB. The figure may be "input referred" or "output referred." If input referred, then it is analogous to input referred noise and this amount of ripple will be multiplied by the gain of the amplifier. If output referred, then it is the amount of ripple expected at the output for the given conditions.

REFERENCES

1. Solomon, J. E., Davis, W. R., and Lee, P. L., "A Self-Compensated Monolithic Operational Amplifier with Low Input Current and High Slew Rate," *ISSCC Digest Tech. Papers,* February 1969, pp. 14-15.

Section 2.0
Preamplifiers

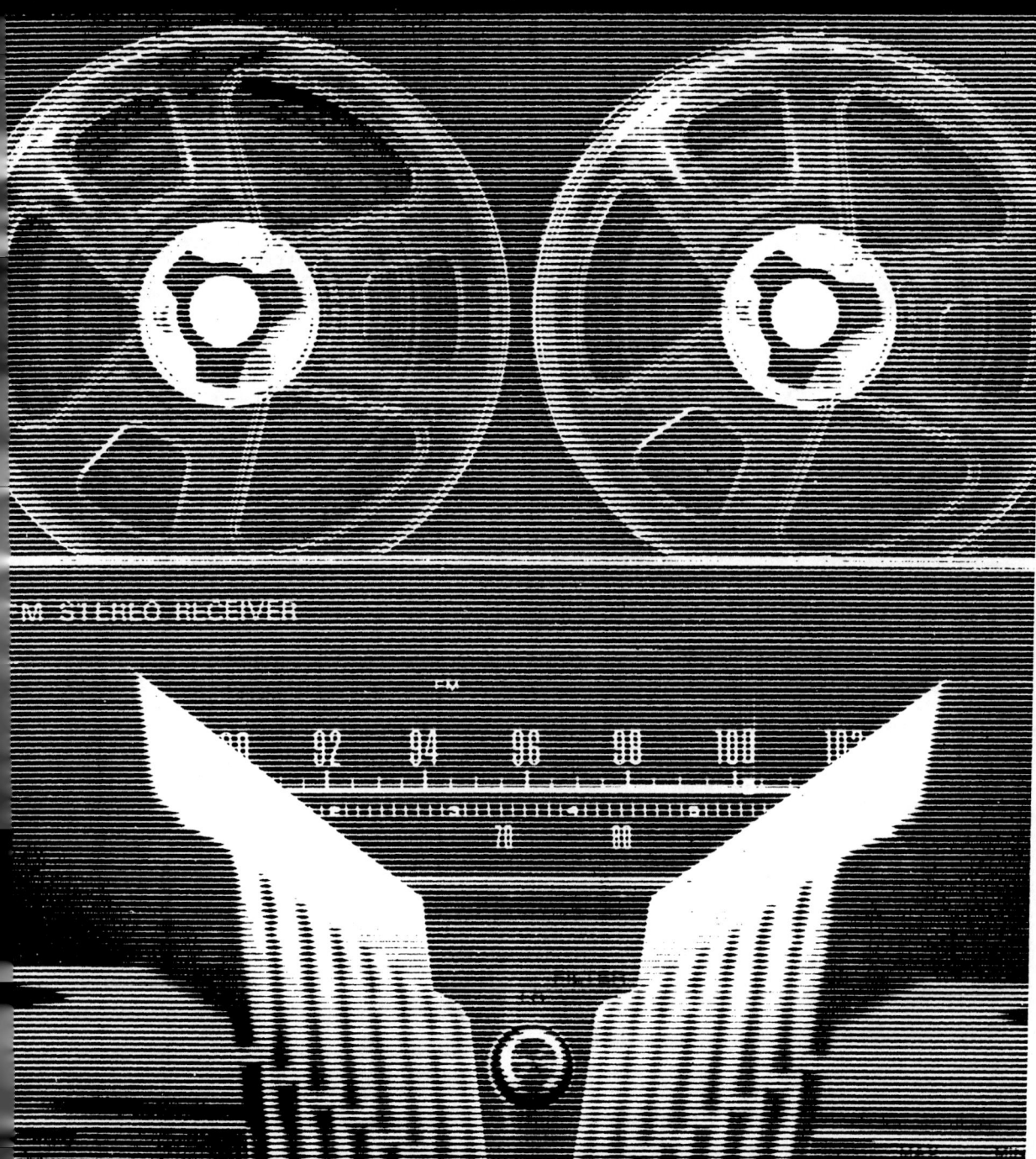

Real-world ground leads possess finite resistance, and the currents running through them will cause finite voltage drops. If two ground return lines tie into the same path at different points there will be a voltage drop between them. Figure 2.2.1a shows a common-ground example where the positive input ground and the load ground are returned to the supply ground point via the same wire. The addition of the finite wire resistance (Figure 2.2.1b) results in a voltage difference between the two points as shown.

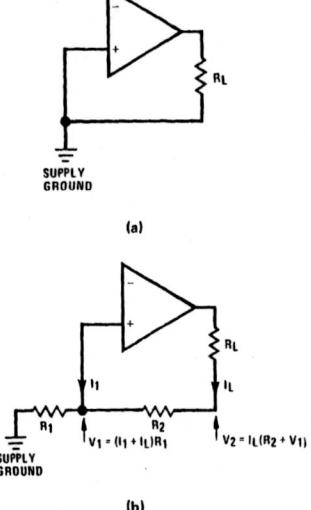

FIGURE 2.2.1 Ground Loop Example

Load current I_L will be much larger than input bias current I_1, thus V_1 will follow the output voltage directly, i.e., in phase. Therefore the voltage appearing at the non-inverting input is effectively positive feedback and the circuit may oscillate. If there were only one device to worry about then the values of R1 and R2 would probably be small enough to be ignored; however, several devices normally comprise a total system. Any ground return of a separate device, whose output is in phase, can feedback in a similar manner and cause instabilities. Out of phase ground loops also are troublesome, causing unexpected gain and phase errors.

The solution to this and other ground loop problems is to *always use a single-point ground system*. Figure 2.2.2 shows a single-point ground system applied to the example of Figure 2.2.1. The load current now returns directly to the supply ground without inducing a feedback voltage as before.

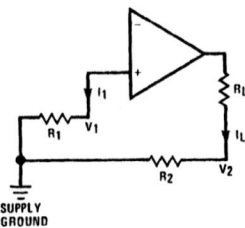

FIGURE 2.2.2 Single-Point Ground System

The single-point ground concept should be applied rigorously to all components and all circuits. Violations of single-point grounding are most common among printed circuit board designs. Since the circuit is surrounded by large ground areas the temptation to run a device to the closest ground spot is high. This temptation must be avoided if stable circuits are to result.

A final rule is to make all ground returns low resistance and low inductance by using large wire and wide traces.

2.2.3 Supply Bypassing

Many IC circuits appearing in print (including many in this handbook) do not show the power supply connections or the associated bypass capacitors for reasons of circuit clarity. *Shown or not, bypass capacitors are always required.* Ceramic disc capacitors (0.1µF) or solid tantalum (1µF) with short leads, and located close (within one inch) to the integrated circuit are usually necessary to prevent interstage coupling through the power supply internal impedance. Inadequate bypassing will manifest itself by a low frequency oscillation called "motorboating" or by high frequency instabilities. Occasionally multiple bypassing is required where a 10µF (or larger) capacitor is used to absorb low frequency variations and a smaller 0.1µF disc is paralleled across it to prevent any high frequency feedback through the power supply lines.

In general, audio ICs are wide bandwidth (~ 10MHz) devices and decoupling of each device is required. Some applications and layouts will allow one set of supply bypassing capacitors to be used common to several ICs. This condition cannot be assumed, but must be checked out prior to acceptance of the layout. Motorboating will be audible, while high frequency oscillations must be observed with an oscilloscope.

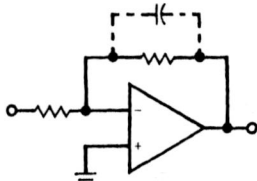

(a) Unity-Gain Stable Device

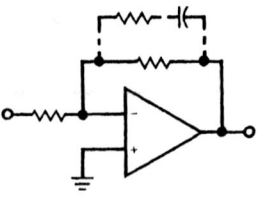

(b) Decompensated Device

FIGURE 2.2.3 Addition of Feedback Capacitor

2.2.4 Additional Stabilizing Tips

If all of the previous rules are followed closely, no instabilities should occur within the circuit; however, Murphy being the way he is, some circuits defy these rules and oscillate anyway. Several additional techniques may be required when persistent oscillations plague a circuit:

- Reduce high impedance positive inputs to the minimum allowable value (e.g., replace 1 Meg biasing resistors with 47k ohm, etc.).
- Add small (< 100 pF) capacitors across feedback resistors to reduce amplifier gain at high frequencies (Figure 2.2.3). **Caution: this assumes the amplifier is unity-gain stable.** If not, addition of this capacitor will *guarantee* oscillations. (For amplifiers that are not unity-gain stable, place a resistor in series with the capacitor such that the gain does not drop below where it is stable.)
- Add a small capacitor (size is a function of source resistance) at the positive input to reduce the impedance to high frequencies and effectively shunt them to ground.

2.3 NOISE

2.3.1 Introduction

The noise performance of IC amplifiers is determined by four primary noise sources: thermal noise, shot noise, 1/f, and popcorn noise. These four sources of noise are briefly discussed. Their contribution to overall noise performance is represented by equivalent input generators. In addition to these equivalent input generators, the effects of feedback and frequency compensation on noise are also examined. The noise behavior of the differential amplifier is noted since most op amps today use a differential pair. Finally noise measurement techniques are presented.

2.3.2 Thermal Noise

Thermal noise is generated by any passive resistive element. This noise is "white," meaning it has a constant spectral density. Thermal noise can be represented by a mean-square voltage generator $\overline{e_R^2}$ in series with a noiseless resistor, where $\overline{e_R^2}$ is given by Equation (2.3.1).

$$\overline{e_R^2} = 4k\,TRB \text{ (volts)}^2$$

where: T = temperature in °K

R = resistor value in ohms

B = noise bandwidth in Hz

k = Boltzmann's constant (1.38×10^{-23} W-sec/°K)

The RMS value of Equation (2.3.1) is plotted in Figure 2.3.1 for a one Hz bandwidth. If the bandwidth is increased, the plot is still valid so long as e_R is multiplied by $\sqrt{B}$.

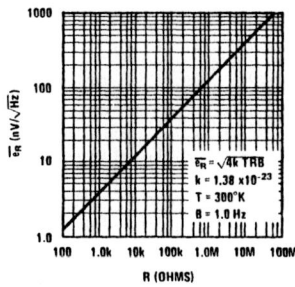

FIGURE 2.3.1 Thermal Noise of Resistor

Actual resistor noise measurements may have more noise than shown in Figure 2.3.1. This additional noise component is known as *excess noise*. Excess noise has a 1/f spectral response, and is proportional to the voltage drop across the resistor. It is convenient to define a *noise index* when referring to excess noise in resistors. *The noise index is the RMS value in µV of noise in the resistor per volt of DC drop across the resistor in a decade of frequency.* Noise index expressed in dB is:

$$NI = 20 \log \left(\frac{E_{ex}}{V_{DC}} \times 10^6 \right) dB \quad (2.3.1)$$

where: E_{ex} = resistor excess noise in µV per frequency decade.

V_{DC} = DC voltage drop across the resistor.

Excess noise in carbon composition resistors corresponds to a large noise index of +10 dB to −20 dB. Carbon film resistors have a noise index of −10 dB to −25 dB. Metal film and wire wound resistors show the least amount of excess noise, with a noise index figure of −15 dB to −40 dB. For a complete discussion of excess noise see Reference 2.

2.3.3 Noise Bandwidth

Noise bandwidth is not the same as the common amplifier or transfer function −3 dB bandwidth. Instead, noise bandwidth has a "brick-wall" filter response. The maximum power gain of a transfer function $T(j\omega)$ multiplied by the noise bandwidth must equal the total noise which passes through the transfer function. Since the transfer function power gain is related to the square of its voltage gain we have:

$$(T_{MAX}^2)B = \int_0^\infty |T(j\omega)|^2 d\omega \quad (2.3.2)$$

where: T_{MAX} = maximum value of $T(j\omega)$

$T(j\omega)$ = transfer function voltage gain

B = noise bandwidth in Hz

For a single RC roll-off, the noise bandwidth B is $\pi/2\,f_{-3dB}$, and for higher order maximally flat filters, see Table 2.3.1.

TABLE 2.3.1 Noise Bandwidth Filter Order

Filter Order	Noise Bandwidth B
1	$1.57\,f_{-3dB}$
2	$1.11\,f_{-3dB}$
3	$1.05\,f_{-3dB}$
4	$1.025\,f_{-3dB}$
"Brick-wall"	$1.00\,f_{-3dB}$

2.3.4 Shot Noise

Shot noise is generated by charge crossing a potential barrier. It is the dominant noise mechanism in transistors and op amps at medium and high frequencies. The mean square value of shot noise is given by:

$$\overline{i_s^2} = 2q\,I_{DC}\,B \text{ (amps)}^2 \quad (2.3.3)$$

where: q = charge of an electron in coulombs

I_{DC} = direct current in amps

B = noise bandwidth in Hz

Like thermal noise, shot noise has a constant spectral density.

2.3.5 1/f Noise

1/f or flicker noise is similar to shot noise and thermal noise since its amplitude is random. Unlike thermal and shot noise, 1/f noise has a 1/f spectral density. This means that the noise increases at low frequencies. 1/f noise is caused by material and manufacturing imperfections, and is usually associated with a direct current:

$$\overline{I_f^2} = K \frac{(I_{DC})^a}{f} B \text{ (amps)}^2 \qquad (2.3.4)$$

where: I_{DC} = direct current in amps
K and a = constants
f = frequency in Hz
B = noise bandwidth in Hz

2.3.6 Popcorn Noise (PCN)

Popcorn noise derives its name from the popcorn-like sound made when connected to a loudspeaker. It is characterized by a sudden change in output DC level, lasting from milliseconds to seconds, recurring randomly. Although there is no clear explanation of PCN to date, it is usually reduced by cleaner processing (see Reference 5). Extensive testing techniques are used to screen for PCN units.

2.3.7 Modelling

Every element in an amplifier is a potential source of noise. Each transistor, for instance, shows all three of the above mentioned noise sources. The net effect is that noise sources are distributed throughout the amplifier, making analysis of amplifier noise extremely difficult. Consequently, amplifier noise is completely specified by a noise voltage and a noise current generator at the input of a noiseless amplifier. Such a model is shown in Figure 2.3.2. Correlation between generators is neglected unless otherwise noted.

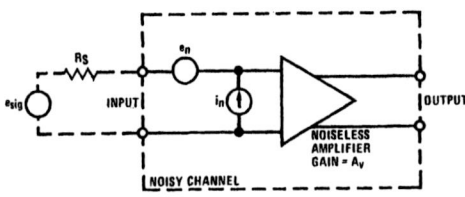

FIGURE 2.3.2 Noise Characterization of Amplifier

Noise voltage e_n, or more properly, equivalent short-circuit input RMS noise voltage, is simply that noise voltage which would appear to originate at the input of the noiseless amplifier if the input terminals were shorted. It is expressed in "nanovolts per root Hertz" ($nV/\sqrt{Hz}$) at a specified frequency, or in microvolts for a given frequency band. It is measured by shorting the input terminals, measuring the output RMS noise, dividing by amplifier gain, and referencing to the input — hence the term "equivalent input noise voltage." An output bandpass filter of known characteristic is used in measurements, and the measured value is divided by the square root of the bandwidth if data are to be expressed per unit bandwidth.

Figure 2.3.3 shows e_n of a typical op amp. For this amplifier, the region above 1 kHz is the shot noise region, and below 1 kHz is the amplifier's 1/f region.

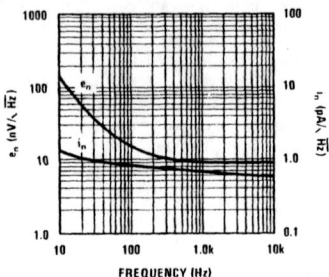

FIGURE 2.3.3 Noise Voltage and Current for an Op Amp

Noise Current, i_n, or more properly, equivalent open-circuit RMS noise current, is that noise which occurs apparently at the input of the noiseless amplifier due only to noise currents. It is expressed in "picoamps per root Hertz" ($pA/\sqrt{Hz}$) at a specified frequency or in nanoamps in a given frequency band. It is measured by shunting a capacitor or resistor across the input terminals such that the noise current will give rise to an additional noise voltage which is $i_n \times R_{in}$ (or X_{Cin}). The output is measured, divided by amplifier gain, and that contribution known to be due to e_n and resistor noise is appropriately subtracted from the total measured noise. If a capacitor is used at the input, there is only e_n and $i_n X_{Cin}$. The i_n is measured with a bandpass filter and converted to $pA/\sqrt{Hz}$ if appropriate. Again, note the 1/f and shot noise regions of Figure 2.3.3.

Now we can examine the relationship between e_n and i_n at the amplifier input. When the signal source is connected, the e_n appears in series with the e_{sig} and e_R. The i_n flows through R_s, thus producing another noise voltage of value $i_n \times R_s$. This noise voltage is clearly dependent upon the value of R_s. All of these noise voltages add at the input of Figure 2.3.2 in RMS fashion, that is, as the square root of the sum of the squares. Thus, neglecting possible correlation between e_n and i_n, the total input noise is:

$$\overline{e_N^2} = \overline{e_n^2} + \overline{e_R^2} + \overline{i_n^2} R_s^2 \qquad (2.3.5)$$

2.3.8 Effects of Ideal Feedback on Noise

Extensive use of voltage and current feedback are common in op amp circuits today. Figures 2.3.4a and 2.3.4b can be used to show the effect of voltage feedback on the noise performance of an op amp.

Figure 2.3.4a shows application of negative feedback to an op amp with generators $\overline{e_n^2}$ and $\overline{i_n^2}$. Figure 2.3.4b shows that the noise generators can be moved outside the feedback loop. This operation is possible since shorting both amplifiers' inputs results in the same noise voltage at the outputs. Likewise, opening both inputs gives the same noise currents at the outputs. For current feedback, the same result can be found. This is seen in Figure 2.3.5a and Figure 2.3.5b.

The significance of the above result is that the equivalent input noise generators completely specify circuit noise. *The application of ideal negative feedback does not alter the noise performance of the circuit.* Feedback reduces the output noise, but it also reduces the output signal. *In other words, with ideal feedback, the equivalent input noise is independent of gain.*

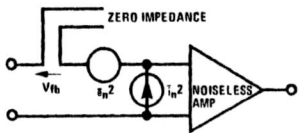
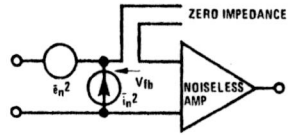

(a) Feedback Applied to Op Amp with Noise Generators   (b) Noise Generators Outside Feedback Loop

FIGURE 2.3.4

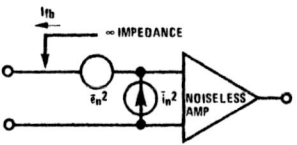
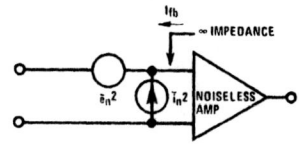

(a) Current Feedback Applied to Op Amp   (b) Noise Generators Moved Outside Feedback Loop

FIGURE 2.3.5

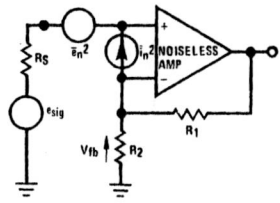
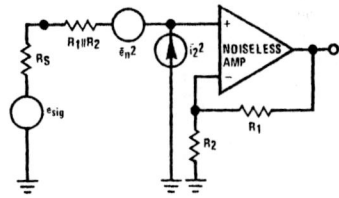

(a) Practical Voltage Feedback Amplifier   (b) Voltage Feedback with Noise Generators Moved Outside Feedback Loop

FIGURE 2.3.6

2.3.9 Effects of Practical Feedback on Noise

Voltage feedback is implemented by series-shunt feedback as shown in Figure 2.3.6a.

The noise generators can be moved outside the feedback loop as shown in Figure 2.3.6b if the thermal noise of $R_1 \| R_2$ is included in $\overline{e_N^2}$. In addition, the noise generated by $i_n \times (R_1 \| R_2)$ must be added even though the (−) input is a virtual ground (see Appendix 7). The above effects can be easily included if $R_1 \| R_2$ is considered to be in series with R_s.

$$\overline{e_N^2} = \overline{e_n^2} + 4kT(R_s + R_1\|R_2) + \overline{i_n^2}(R_s + R_1\|R_2)^2$$

$$\overline{i_2^2} = \overline{i_n^2}$$

Example 2.3.1

Determine the total equivalent input noise per unit bandwidth for the amplifier of Figure 2.3.6a operating at 1 kHz from a source resistance of 1 kΩ. R_1 and R_2 are 100 kΩ and 1 kΩ respectively.

Solution:

Use data from Figure 2.3.1 and Figure 2.3.3.

1. Thermal noise from $R_s + R_1 \| R_2 \approx 2k$ is $5.65 \text{nV}/\sqrt{\text{Hz}}$.
2. Read e_n from Figure 2.3.3 at 1 kHz; this value is $9.5 \text{nV}/\sqrt{\text{Hz}}$.
3. Read i_n from Figure 2.3.3 at 1 kHz; this value is $0.68 \text{pA}/\sqrt{\text{Hz}}$. Multiply this noise current by $R_s + R_1 \| R_2$ to obtain $1.36 \text{nV}/\sqrt{\text{Hz}}$.
4. Square each term and enter into Equation (2.3.5).

$$e_N = \sqrt{\overline{e_n^2} + 4kT(R_s + R_1\|R_2) + \overline{i_n^2}(R_s + R_1\|R_2)^2}$$

$$e_N = \sqrt{(9.5)^2 + (5.65)^2 + (1.36)^2}$$

$$e_N = 11.1 \text{nV}/\sqrt{\text{Hz}}$$

This is total RMS noise at the input in one Hertz bandwidth at 1 kHz. If total noise in a given bandwidth is desired, one must integrate the noise over a bandwidth as specified. This is most easily done in a noise measurement set-up, but may be approximated as follows:

1. If the frequency range of interest is in the flat band, i.e., between 1 kHz and 10 kHz in Figure 2.3.3, it is simply a matter of multiplying e_N by the square root of the noise bandwidth. Then, in the 1 kHz-10 kHz band, total noise is:

$$e_N = 11.1\sqrt{9000}$$

$$= 1.05\mu V$$

2. If the frequency band of interest is not in the flat band of Figure 2.3.3, one must break the band into sections, calculating average noise in each section, squaring, multiplying by section bandwidth, summing all sections, and finally taking square root of the sum as follows:

$$e_N = \sqrt{\overline{e_R}^2 B + \sum_1^i (\overline{e_N}^2 + \overline{i_n}^2 (R_s + R_1 \| R_2)^2)_i B_i}$$

(2.3.6)

where: i is the total number of sub-blocks

For details and examples of this type of calculation, see application note AN-104, "Noise Specs Confusing?"

Current feedback is accomplished by shunt-shunt feedback as shown in Figure 2.3.7a.

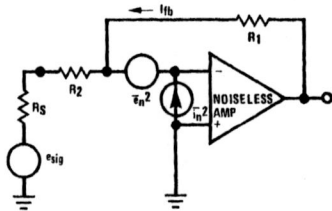

(a) Practical Current Feedback Amplifier

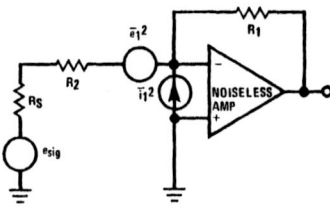

(b) Intermediate Move of Noise Generators

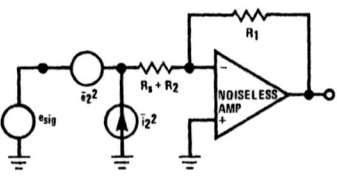

(c) Current Feedback with Noise Generators Moved Outside Feedback Loop

FIGURE 2.3.7

$\overline{e_n}^2$ and $\overline{i_n}^2$ can be moved outside the feedback loop if the noise generated by R_1 and R_2 are taken into account.

First, move the noise generators outside feedback R_1. To do this, represent the thermal noise generated by R_1 as a noise current source (Figure 2.3.7b):

$$\overline{i_{R_1}}^2 = 4kT\frac{1}{R_1}$$

so: $\overline{e_1}^2 = \overline{e_n}^2$

and: $\overline{i_1}^2 = \overline{i_n}^2 + 4kT\frac{1}{R_1}$

Now move these noise generators outside $R_s + R_2$ as shown in Figure 2.3.7c to obtain $\overline{e_2}^2$ and $\overline{i_2}^2$:

$$\overline{e_2}^2 = \overline{e_n}^2 + 4kT(R_s + R_2) \quad (2.3.7)$$

$$\overline{i_2}^2 = \overline{i_n}^2 + 4kT\frac{1}{R_1} \quad (2.3.8)$$

$\overline{e_2}^2$ and $\overline{i_2}^2$ are the equivalent input generators with feedback applied. The total equivalent input noise, e_N, is the sum of the noise produced with the input shorted, and the noise produced with the input opened. With the input of Figure 2.3.7c shorted, the input referred noise is $\overline{e_2}^2$. With the input opened, the input referred noise is:

$$\left(\frac{i_2 R_1}{A_V}\right)^2 = \overline{i_2}^2 (R_s + R_2)^2$$

The total equivalent input noise is:

$$e_N = \sqrt{\overline{e_2}^2 + \overline{i_2}^2 (R_s + R_2)^2}$$

Example 2.3.2

Determine the total equivalent input noise per unit bandwidth for the amp of Figure 2.3.7a operating at 1 kHz from a 1 kΩ source. Assume R_1 is 100 kΩ and R_2 is 9 kΩ.

Solution

Use data from Figures 2.3.1 and 2.3.3.

1. Thermal noise from $R_s + R_2$ is $12.7\,nV/\sqrt{Hz}$.
2. Read e_n from figure 2.3.3 at 1 kHz; this value is $9.5\,nV/\sqrt{Hz}$. Enter these values into Equation (2.3.7).
3. Determine the thermal noise current contributed by R_1:

$$i_{R_1} = \sqrt{4kT\frac{1}{R_1}B} = \sqrt{\frac{1.61 \times 10^{-20}}{100k}} = 0.401\,pA/\sqrt{Hz}$$

4. Read i_n from Figure 2.3.3 at 1 kHz; this value is $0.68\,pA/\sqrt{Hz}$. Enter these values into Equation (2.3.7).

$$e_N = \sqrt{\overline{e_n}^2 + (R_s + R_2)^2 (\overline{i_n}^2 + 4kT\frac{1}{R_1}) + 4kT(R_s + R_2)}$$

$$e_N = \sqrt{(9.5)^2 + (10k)^2(0.68^2 + 0.401^2) + (12.7)^2}\,nV/\sqrt{Hz}$$

$$e_N = 17.7\,nV/\sqrt{Hz}$$

For the noise in the bandwidth from 1 kHz to 10 kHz, $e_N = 17.7\,nV\sqrt{9000} = 1.68\mu V$. If the noise is not constant with frequency, the method shown in Equation (2.3.6) should be used.

TABLE 2.3.2 Equivalent Input Noise Comparison

NON-INVERTING AMPLIFIER					INVERTING AMPLIFIER				
A_V	R_s	R_1	R_2	e_N (nV $\sqrt{Hz}$)	A_V	R_s	R_1	R_2	e_N (nV $\sqrt{Hz}$)
101	1k	100k	1k	11.1	100	1k	100k	0	10.3
11	1k	100k	10k	17.3	10	1k	100k	9k	17.7
2	1k	100k	100k	46.0	2	1k	100k	49k	49.5
1	1k	100k	∞	80.2	1	1k	100k	99k	89.1

Example 2.3.3

Compare the noise performance of the non-inverting amplifier of Figure 2.3.6a to the inverting amplifier of Figure 2.3.7a.

Solution:

The best way to proceed here is to make a table and compare the noise performance with various gains.

Table 2.3.2 shows only a small difference in equivalent input noise for the two amplifiers. There is, however, a large difference in the flexibility of the two amplifiers. The gain of the inverting amplifier is a function of its input resistance, R_2. Thus, for a given gain and input resistance, R_1 is fixed. This is not the case for the non-inverting amplifier. The designer is free to pick R_1 and R_2 independent of the amplifier's input impedance. Thus in the case of unity gain, where $R_2 = \infty$, R_1 can be zero ohms. The equivalent input noise is:

$$e_N = \sqrt{e_n^2 + 4kTR_s + i_r^2 R_s^2}$$

$$e_N = 10.3 \text{nV}/\sqrt{Hz}$$

There is now a large difference in the noise performance of the two amplifiers. Table 2.3.2 also shows that *the equivalent input noise for practical feedback can change as a function of closed loop gain A_V. This result is somewhat different from the case of ideal feedback.*

Example 2.3.4

Determine the signal-to-noise ratio for the amplifier of Example 2.3.2 if e_{SIG} has a nominal value of 100mV.

Solution:

Signal to noise ratio is defined as:

$$S/N = 20 \log \frac{e_{SIG}}{e_N} \quad (2.3.9)$$

$$= 20 \log \frac{100 \text{mV}}{1.68 \mu V} = 95.5 \text{dB}$$

2.3.10 RF Precautions

A source of potential RF interference that needs to be considered in AM radio applications lies in the radiated wideband noise voltage developed at the speaker terminals. The method of amplifier compensation (Figure 2.3.8a) fixes the point of unity gain cross at approximately 10MHz (Figure 2.3.8b). A wideband design is essential in achieving low distortion performance at high audio frequencies, since it allows adequate loop-gain to reduce THD. (Figure 2.3.8b shows that for a closed-loop gain of 34dB there still exists 26dB of loop-gain at 10kHz.)

The undesirable consequence of a single-pole roll-off, wideband design is the excess gain beyond audio frequencies, which includes the AM band; hence, noise of this frequency is amplified and delivered to the load where it can radiate back to the AM (magnetic) antenna and sensitive RF circuits. A simple and economical remedy is shown in Figure 2.3.8c, where a ferrite bead, or small RF choke is added in series with the output lead. Experiments have demonstrated that this is an effective method in suppressing the unwanted RF signals.

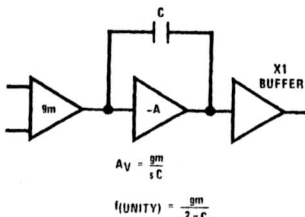

$$A_V = \frac{g_m}{sC}$$

$$f_{(UNITY)} = \frac{g_m}{2 \pi C}$$

(a) Typical Compensation

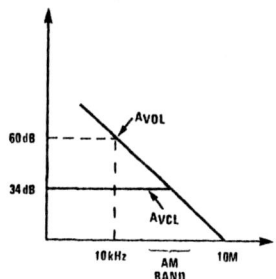

(b) Source of RF Interference

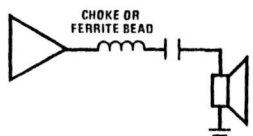

(c) Reduction of RF Interference

FIGURE 2.3.8

2.3.11 Noise in the Differential Pair

Figure 2.3.9a shows a differential amplifier with noise generators e_{n1}, i_{n1}, e_{n2}, and i_{n2}.

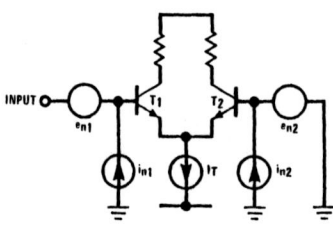

(a) Differential Pair with Noise Generators

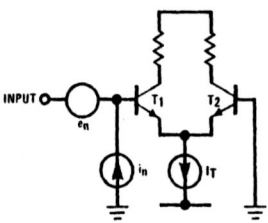

(b) Differential Pair with Generators Input Referred

FIGURE 2.3.9

To see the intrinsic noise of the pair, short the base of T_2 to ground, and refer the four generators to an input noise voltage and noise current as shown in Figure 2.3.9b. To determine e_n, short the input of 9(a) and 9(b) to ground. e_n is then the series combination of e_{n1} and e_{n2}. These add in an RMS fashion, so:

$$e_n = \sqrt{e_{n1}^2 + e_{n2}^2}$$

Both generators contribute the same noise, since the transistors are similar and operate at the same current; thus, $e_n = \sqrt{2}\, e_{n1}$, i.e., 3dB more noise than a single ended amplifier. This can be significant in critical noise applications.

In order to find the input noise current generator, i_n, open the input and equate the output noise from Figure 2.3.9a and Figure 2.3.9b. The result of this operation is $i_n = i_{n1}$. Thus, from a high impedance source, the differential pair gives similar noise current as a single transistor.

2.3.12 Noise Measurement Techniques

This section presents techniques for measuring e_n, i_n, and e_N. The method can be used to determine the spectral density of noise, or the noise in a given bandwidth. The circuit for measuring the noise of an LM387 is shown in Figure 2.3.10.

The system gain, V_{OUT}/e_n, of the circuit in Figure 2.3.10 is large — 80dB. This large gain is required since we are trying to measure *input referred noise generators* on the order of $5nV/\sqrt{Hz}$, which corresponds to $50\mu V/\sqrt{Hz}$ at the output. R_1 and R_2 form a 100:1 attenuator to provide a low input signal for measuring the system gain. The gain should be measured in both the e_n and i_n positions, since LM387 has a 250k bias resistor which is between input and ground. The LM387 of Figure 2.3.10 has a closed loop gain of 40dB which is set by feedback elements R_5 and R_6. 40dB provides adequate gain for the input referred generators of the LM387. The output noise of the LM387 is large compared to the input referred generators of the LM381; consequently, noise at the output of the LM381 will be due to the LM387. To measure the noise voltage e_n, and noise current $i_n \times R_3$, a wave analyzer or noise filter set is connected. In addition the noise in a given bandwidth can be measured by using a bandpass filter and an RMS voltmeter. If a true RMS voltmeter is not available, an average responding meter works well. When using an average responding meter, the measured noise must be multiplied by 1.13 since the meter is calibrated to measure RMS *sine waves*. The meter used for measuring noise should have a crest factor (ratio of peak to RMS value) from 3 to 5, as the peak to RMS ratio of noise is on that order. Thus, if an average responding meter measures 1mV of noise, the RMS value would be $1.13mV_{RMS}$, and the peak-to-peak value observed on an oscilloscope could be as high as 11.3mV ($1.13mV \times 2 \times 5$).

Some construction tips for the circuit of Figure 2.3.10 are as follows:

1. R_4 and R_6 should be metal film resistors, as they exhibit lower excess noise than carbon film resistors.

2. C_1 should be large, to provide low capacitive reactance at low frequency, in order to accurately observe the 1/f noise in e_n.

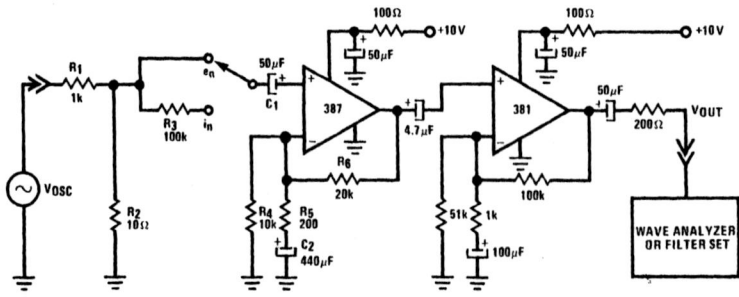

FIGURE 2.3.10 Noise Test Setup for Measuring e_n and i_n of an LM387

3. C_2 should be large to maintain the gain of 80 dB down to low frequencies for accurate 1/f measurements.
4. The circuit should be built in a small grounded metal box to eliminate hum and noise pick-up, especially in i_n.
5. The LM387 and LM381 should be separated by a metal divider within the metal box. This is to prevent output to input oscillations.

Typical LM387 noise voltage and noise current are plotted in Figure 2.3.11.

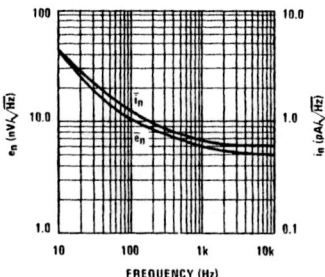

FIGURE 2.3.11 LM387 Noise Voltage and Noise Current

Many times we do not care about the actual spectral distribution of noise, rather we want to know the noise voltage in a given bandwidth for comparison purposes. For audio frequencies, we are interested only in a 20 kHz bandwidth. The noise voltage is often the dominant noise source since many systems use a low impedance voltage drive as the signal. For this common case we use a test set-up as shown in Figure 2.3.12.

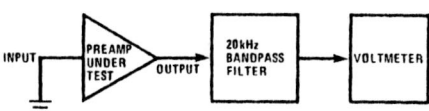

FIGURE 2.3.12 Test Setup for Measuring Equivalent Input Noise for a 20 kHz Bandwidth

Example 2.3.5

Determine the equivalent input noise voltage for the preamp of Figure 2.3.12. The gain, A_V, of the preamp is 40 dB and the voltmeter reads 0.2 mV. Assume the voltmeter is average responding and the 20 kHz low-pass filter has a single R-C roll-off.

Solution:

Since the voltmeter is average responding, the RMS voltage is V_{RMS} = 0.2 mV × 1.13 = 0.226 mV. Using an average responding meter causes only a 13% error. The filter has a single R-C roll-off, so the noise bandwidth is $\pi/2$ × 20 kHz = 31.4 kHz, i.e., the true noise bandwidth is 31.4 kHz and not 20 kHz. Since RMS noise is related to the square root of the noise bandwidth, we can correct for this difference:

$$V_{OUT} = \sqrt{\frac{0.226}{\pi/2}} = 0.18 \text{ mV}$$

The equivalent input noise is:

$$\frac{V_{OUT}}{A_V} = \frac{0.18 \text{ mV}}{100} = 1.8 \mu V \text{ in a 20 kHz bandwidth.}$$

If this preamp had RIAA playback equalization, the output noise, V_{OUT}, would have been divided by the gain at 1 kHz.

Typical values of noise, measured by the technique of Figure 2.3.12, are shown in Table 2.3.3. For this data, B = 10 kHz and R_s = 600 Ω.

TABLE 2.3.3 Typical Flat Band Equivalent Input Noise

Type	e_N (μV)
LM381	0.70
LM381A	0.50
LM382	0.80
LM387	0.80
LM387A	0.65

2.3.13 Noise Measurement for Consumer Audio Equipment — The Use of Weighting Filters

The previous discussion of noise and its measurement has been mainly concerned with obtaining a noise voltage "number" over a given frequency bandwidth in order to provide a S/N ratio for signals that can occupy all or part of the same bandwidth. The usefulness of this is restricted by the fact that there is no indication from this "number" of the subjective annoyance of noise spectra present within this bandwidth of interest. For example, two systems with measured identical signal/noise ratios can sound very different because one may have a uniform distribution of noise spectra whereas the other may have most of the noise concentrated in one particular portion of the frequency band. The total noise voltage is the same in each case but the audible effect is that one system sounds "noisier" than the other.

To understand why this should be, we need to investigate in a little more detail the relative sensitivity of the human ear and the effects of auditory masking phenomena. Readers familiar with the Fletcher-Munson equal loudness contours (Section 2.14.7) and the more recent work by Robinson and Dadson[6] will already know that the ear is not uniformly sensitive to all frequencies in the audio band, an effect that is emphasized at extremely low sound levels. Further, in a steady state condition, the threshold of hearing for a given tone is changed by the presence of another tone (the masker). The amount of change is dependent on the relative pitch and loudness of the masker and the maskee. Noise will also raise the threshold of hearing for tones — i.e. the tone has to be louder to be heard if noise is also present in some part of the frequency band. Figure 2.12.23 is a plot of the hearing threshold of acute ears for noise in a typical home environment (noise spectra below this curve are inaudible). Below 200 Hz and above 6 kHz the shape of this curve is caused by the hearing mechanism, and between 200 Hz and 6 kHz is caused by the masking effect of room noise. This means that if noise is just audible at 1 kHz, the amplitude of noise at 100 Hz has to be 30 db higher to be equally audible. A further complication is that the audibility of the noise is not necessarily indicative of its obtrusiveness or annoyance.

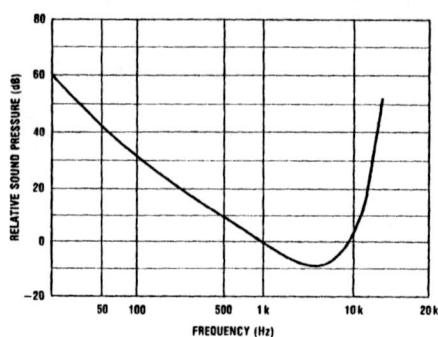

Figure 2.12.23 Threshold of Hearing for Noise in the Home Environment

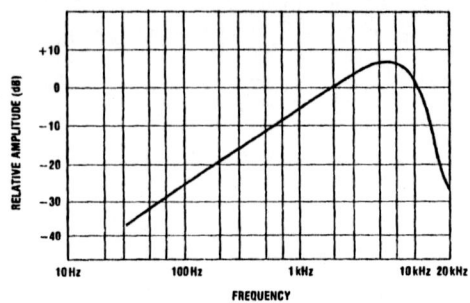

Figure 2.12.24 CCIR/ARM Noise Weighting Filter Characteristic

To make comparative S/N measurements more meaningful, several filters have been used to weight the contribution of the noise spectra over the frequency band of interest (N.A.B. A-Weighting Curve, D.I.N. 45405 for example) so that the S/N numbers correlate better to the subjective impression gained in listening tests. Recently the CCIR adopted a weighting filter (Recommendation 468-1) with the characteristic shape shown in Figure 2.12.24 which is based on the obtrusiveness as well as the level of different kinds of noise. While this filter is normally used with a quasi-peak reading meter to derive consistent readings with all types of noise (including clicks, pops and whistles as well as broad spectrum noise), for typical audio equipment such as tape decks and amplifiers, an average responding meter has been found to give equally consistent results.[7]

The filter characteristic of Figure 2.12.24 is known as the CCIR/ARM filter and is currently used by Dolby Laboratories for measurements on their Dolby® B-Type noise reduction units. Note that the 0 dB reference frequency is 2 kHz instead of the more conventional 1 kHz — so that S/N ratios obtained by this method are numerically close to the S/N ratios obtained by earlier methods and which are considered commercially acceptable for the quality of equipment being measured. Without the reference frequency shift the S/N ratios obtained with the CCIR/ARM filter would be several dB below the expected number.

REFERENCES

1. Meyer, R. G., "Notes on Noise," EECS Department, University of California, Berkeley, 1973.
2. Fitchen, F.C., *Low Noise Electronic Design,* John Wiley & Sons, New York, 1973.
3. Cherry, E. M. and Hooper, D. E., *Amplifying Devices and Low Pass Amplifier Design,* John Wiley & Sons, New York, 1968.
4. Sherwin, J., *Noise Specs Confusing?*, Application Note AN-104, National Semiconductor, 1975.
5. Roedel, R., "Reduction of Popcorn Noise in Integrated Circuits," *IEEE Trans. Electron Devices (Corresp.),* vol. ED-22, October 1975, pp. 962-964.
6. ISO/R226-1961 (E)
7. Dolby, R., Robinson, D., and Gundry, K. *A Practical Noise Measurement Method,* AES Preprint 1353 (F-3).

2.4 AUDIO RECTIFICATION
Or, "How Come My Phono Detects AM?"

Audio rectification refers to the phenomenon of RF signals being picked up, rectified, and amplified by audio circuits — notably by high-gain preamplifiers. Of all types of interference possible to plague a hi-fi system, audio rectification remains the most slippery and troublesome. A common occurrence of audio rectification is to turn on a phonograph and discover you are listening to your local AM radio station instead. There exist four main sources of interference, each with a unique character: If it is clearly audible through the speaker then AM radio stations are probably the source; if the interference is audible but garbled then suspect SSB and amateur radio equipment; a decrease in volume can be produced by FM pickup; and if buzzing occurs, then RADAR or TV is being received. Whatever the source, the approaches to eliminating it are similar.

Commonly, the rectification occurs at the first non-linear, high gain, wide bandwidth transistor encountered by the incoming signal. The signal may travel in unshielded or improperly grounded input cables; it may be picked up through the air by long, poorly routed wires; or it may enter on the AC power lines. It is rectified by the first stage transistor acting as a detector diode, subsequently amplified by the remaining circuitry, and finally delivered to the speaker. Bad solder joints can detect the RF just well as transistors and must be avoided (or suspected).

The following list should be consulted when seeking to eliminate audio rectification from existing equipment. For new designs, keep input leads short and shielded, with the shield grounded only at one point; make good clean solder connections; avoid loops created by multiple ground points; and make ground connections close to the IC or transistor that they associate with.

Audio Rectification Elimination Tips (Figure 2.4.1).

- Reduce input impedance.
- Place capacitor to ground close to input pin or base (~ 10-300pF).
- Use ceramic capacitors.
- Put ferrite bead on input lead close to the device input.
- Use RF choke in series with input (~ 10µH).
- Use RF choke (or ferrite bead) *and* capacitor to ground.
- Pray.

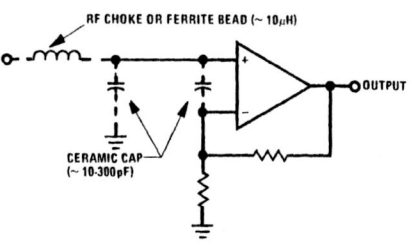

FIGURE 2.4.1 Audio Rectification Elimination Tips

A particularly successful technique is uniquely possible with the LM381 since both base and emitter points of the input transistor are available. A ceramic capacitor is mounted very close to the IC from pin 1 to pin 3, shorting base to emitter at RF frequencies (see Figure 2.4.2).

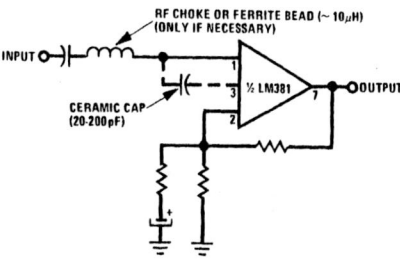

FIGURE 2.4.2 LM381 Audio Rectification Correction

2.5 DUAL PREAMPLIFIER SELECTION

National Semiconductor's line of integrated circuits designed specifically to be used as audio preamplifiers consists of the LM381, LM382, LM387, and the LM1303. All are dual amplifiers in recognition of their major use in two channel applications. In addition there exists the LM389 which has three discrete NPN transistors that can be configured into a low noise monaural preamplifier for minimum parts count mono systems (Section 4.11). Table 2.5.1 shows the major electrical characteristics of each of the dual preamps offered. A detailed description of each amplifier follows, where the individual traits and operating requirements are presented.

TABLE 2.5.1 Dual Preamplifier Characteristics

PARAMETER	LM381N (14 Pin DIP)			LM382N (14 Pin DIP)			LM387N (8 Pin DIP)			LM1303N (14 Pin DIP)			UNITS
	MIN	TYP	MAX	MIN	TYP	MAX	MIN	TYP	MAX	MIN	TYP	MAX	
Supply Voltage	9		40	9		40	9		30/40[6]	±4.5		±15	V
Quiescent Supply Current		10			10	16		10			15		mA
Input Resistance (open loop)													
Positive Input		100k			100k		50k	100k			25k		Ω
Negative Input		200k			200k			200k			25k		Ω
Open Loop Gain		104			100			104		76	80		dB
Output Voltage Swing $R_L = 10k\Omega$		$V_s - 2$			$V_s - 2$			$V_s - 2$		11.3	15.6		V_{p-p}
Output Current													
Source		8[2]			8[2]			8[2]		0.6	0.8		mA
Sink		2			2			2		0.6	0.8		mA
Output Resistance (open loop)		150			150			150			4k		Ω
Slew Rate ($A_v = 40$ dB)		4.7			4.7			4.7			5.0[7]		V/μs
Power Bandwidth													
20V_{p-p} ($V_s = 24$V)		75			75			75					kHz
11.3V_{p-p} ($V_s = \pm13$V)											100		kHz
Unity Gain Bandwidth		15			15			15			20		MHz
Input Voltage													
Positive Input			300			300			300				mV$_{RMS}$
Either Input											±5		V
Supply Rejection Ratio (Input Referred, 1 kHz)		120			120			110					dB
Channel Separation (f = 1 kHz)		60		40	60		40	60		60	70		dB
Total Harmonic Distortion (f = 1 kHz)[3]		0.1			0.1	0.3		0.1	0.5		0.1		%
Total Equivalent Input Noise ($R_s = 600\Omega$, 10-10k Hz)		0.5[4] 0.5[4,5]	1.0[4] 0.7[4,5]		0.8	1.2		0.8 0.65[6]	1.2 0.9[6]				μV_{RMS} μV_{RMS}
Total NAB[8] Output Noise ($R_s = 600\Omega$, 10-10k Hz)		190 140[5]						230 180[6]					μV_{RMS} μV_{RMS}

1. Specifications apply for $T_A = 25°C$ with $V_s = +14$V for LM381/382/387 and $V_s = \pm13$V for LM1303, unless otherwise noted.
2. DC current; symmetrical AC current = 2mA$_{p-p}$.
3. LM381 & LM387: Gain = 60dB; LM382: Gain = 60dB; LM1303: Gain = 40dB.
4. Single ended input biasing.
5. LM381AN.
6. LM387AN.
7. Frequency Compensation: C = 0.0047μF, Pins 3 to 4.
8. NAB reference level: 37dBV Gain at 1kHz. Tape Playback Circuit.

2.6 LM381 LOW NOISE DUAL PREAMPLIFIER

2.6.1 Introduction

The LM381 is a dual preamplifier expressly designed to meet the requirements of amplifying low level signals in low noise applications. Total equivalent input noise is typically $0.5\mu V_{RMS}$ (R_s = 600Ω, 10-10,000 Hz).

Each of the two amplifiers is completely independent, with an internal power supply decoupler-regulator, providing 120 dB supply rejection and 60 dB channel separation. Other outstanding features include high gain (112 dB), large output voltage swing (V_{CC} – 2V) p-p, and wide power bandwidth (75 kHz, $20 V_{p-p}$). The LM381 operates from a single supply across the wide range of 9 to 40 V. The amplifier is internally compensated and short-circuit protected.

Attempts have been made to fill this function with selected operational amplifiers. However, due to the many special requirements of this application, these recharacterizations have not adequately met the need.

With the low output level of magnetic tape heads and phonograph cartridges, amplifier noise becomes critical in achieving an acceptable signal-to-noise ratio. This is a major deficiency of the op amp in this application. Other inadequacies of the op amp are insufficient power supply rejection, limited small-signal and power bandwidths, and excessive external components.

2.6.2 Circuit Description

To achieve low noise performance, special consideration must be taken in the design of the input stage. First, the input should be capable of being operated single ended, since both transistors contribute noise in a differential stage degrading input noise by the factor $\sqrt{2}$. (See Section 2.3.) Secondly, both the load and biasing elements must be resistive, since active components would each contribute as much noise as the input device.

The basic input stage, Figure 2.6.1, can operate as a differential or single ended amplifier. For optimum noise performance Q_2 is turned OFF and feedback is brought to the emitter of Q_1.

In applications where noise is less critical, Q_1 and Q_2 can be used in the differential configuration. This has the advantage of higher impedance at the feedback summing point, allowing the use of larger resistors and smaller capacitors in the tone control and equalization networks.

The voltage gain of the single ended input stage is given by:

$$A_{V(AC)} = \frac{R_L}{r_e} = \frac{200k}{1.25k} = 160 \qquad (2.6.1)$$

where: $r_e = \frac{KT}{q I_E} \approx 1.25 \times 10^3$ at 25°C, $I_E \approx 20\mu A$

The voltage gain of the differential input stage is:

$$A_V = \frac{1}{2} \frac{R_L}{r_e} = \frac{1}{2} \frac{R_L q I_E}{KT} \approx 80 \qquad (2.6.2)$$

The schematic diagram of the LM381, Figure 2.6.2, is divided into separate groups by function — first and second voltage gain stages, third current gain stage, and the bias regulator.

The second stage is a common-emitter amplifier (Q_5) with a current source load (Q_6). The Darlington emitter-follower

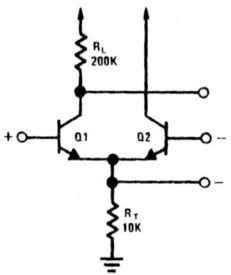

FIGURE 2.6.1 Input Stage

Q_3, Q_4 provides level shifting and current gain to the common-emitter stage (Q_5) and the output current sink (Q_7). The voltage gain of the second stage is approximately 2,000, making the total gain of the amplifier typically 160,000 in the differential input configuration.

The preamplifier is internally compensated with the pole-splitting capacitor, C_1. This compensates to unity gain at 15 MHz. The compensation is adequate to preserve stability to a closed loop gain of 10. Compensation for unity gain closure may be provided with the addition of an external capacitor in parallel with C_1 between pins 5 and 6, 10 and 11.

Three basic compensation schemes are possible for this amplifier: first stage pole, second stage pole and pole-splitting. First stage compensation will cause an increase in high frequency noise because the first stage gain is reduced, allowing the second stage to contribute noise. Second stage compensation causes poor slew rate (power bandwidth) because the capacitor must swing the full output voltage. Pole-splitting overcomes both these deficiencies and has the advantage that a small monolithic compensation capacitor can be used.

The output stage is a Darlington emitter-follower (Q_8, Q_9) with an active current sink (Q_7). Transistor Q_{10} provides short-circuit protection by limiting the output to 12 mA.

The biasing reference is a zener diode (Z_2) driven from a constant current source (Q_{11}). Supply decoupling is the ratio of the current source impedance to the zener impedance. To achieve the high current source impedance necessary for 120 dB supply rejection, a cascode configuration is used (Q_{11} and Q_{12}). The reference voltage is used to power the first stages of the amplifier through emitter-followers Q_{14} and Q_{15}. Resistor R_1 and zener Z_1 provide the starting mechanism for the regulator. After starting, zero volts appears across D_1, taking it out of conduction.

2.6.3 Biasing

Figure 2.6.3 shows an AC equivalent circuit of the LM381. The non-inverting input, Q_1, is referenced to a voltage source two V_{BE} above ground. The output quiescent point is established through negative DC feedback through the external divider R_4/R_5 (Figure 2.6.4).

For bias stability, the current through R_5 is made ten times the input current of Q_2 ($\approx 0.5\mu A$). Then, for the differential input, resistors R_5 and R_4 are:

$$R_5 = \frac{2 V_{BE}}{10 I_{Q2}} = \frac{1.3}{5 \times 10^{-6}} = 260 k\Omega \text{ maximum} \qquad (2.6.3)$$

$$R_4 = \left(\frac{V_{CC}}{2.6} - 1\right) R_5 \qquad (2.6.4)$$

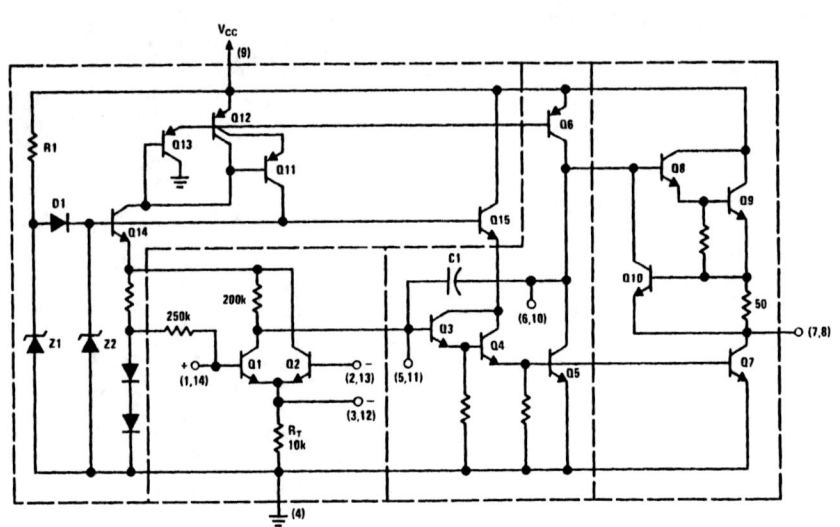

FIGURE 2.6.2 Schematic Diagram

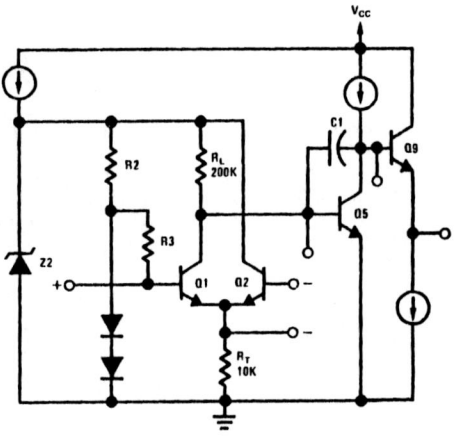

FIGURE 2.6.3 AC Equivalent Circuit

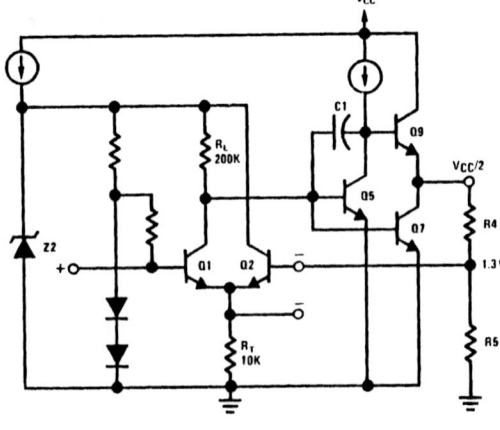

FIGURE 2.6.4 Differential Input Biasing

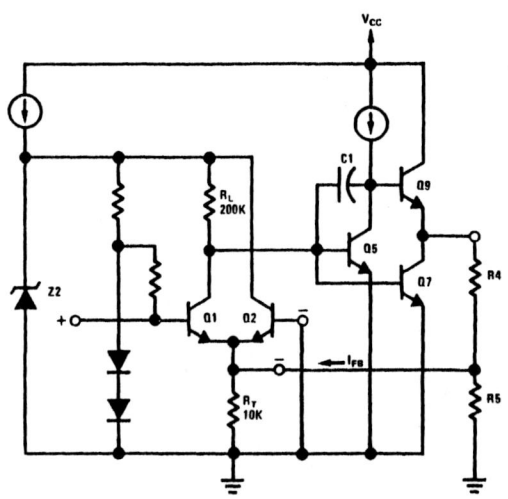

FIGURE 2.6.5 Single Ended Input Biasing

When using the single ended input, Q_2 is turned OFF and DC feedback is brought to the emitter of Q_1 (Figure 2.6.5). The impedance of the feedback summing point is now two orders of magnitude lower than the base of Q_2 ($\approx 10k\Omega$). Therefore, to preserve bias stability, the impedance of the feedback network must be decreased. In keeping with reasonable resistance values, the impedance of the feedback voltage source can be 1/5 the summing point impedance.

The feedback current is $< 100\mu A$ worst case. Therefore, for single ended input, resistors R_5 and R_4 are:

$$R_5 = \frac{V_{BE}}{5\, I_{FB}} = \frac{0.65}{5 \times 10^{-4}} = 1300\Omega \text{ maximum} \quad (2.6.5)$$

$$R_4 = \left(\frac{V_{CC}}{1.3} - 1\right) R_5 \quad (2.6.6)$$

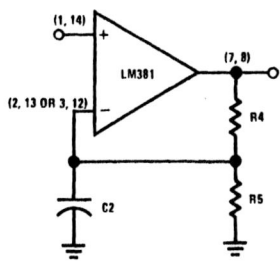

FIGURE 2.6.6 AC Open Loop

The circuits of Figures 2.6.4 and 2.6.5 have an AC and DC gain equal to the ratio R_4/R_5. To open the AC gain, capacitor C_2 is used to shunt R_5 (Figure 2.6.6). The AC gain now approaches open loop. The low frequency 3dB corner, f_o, is given by:

$$f_o = \frac{A_o}{2 \pi C_2 R_4} \quad (2.6.7)$$

where: A_o = open loop gain

2.6.4 Split Supply Operation

Although designed for single supply operation, the LM381 may be operated from split supplies just as well. (A trade-off exists when unregulated negative supplies are used since the inputs are biased to the negative rail without supply rejection techniques and hum may be introduced.) All that is necessary is to apply the negative supply (V_{EE}) to the ground pin and return the biasing resistor R_5 to V_{EE} instead of ground. Equations (2.6.3) and (2.6.5) still hold, while the only change in Equations (2.6.4) and (2.6.6) is to recognize that V_{CC} represents the total potential across the LM381 and equals the absolute sum of the split supplies used, e.g., V_{CC} = 30 volts for ±15 volt supplies. Figure 2.6.7 shows a typical split supply application; both differential and single ended input biasing are shown. (Note that while the output DC voltage will be approximately zero volts the positive input DC potential is about 1.3 volts above the negative supply, necessitating capacitive coupling into the input.)

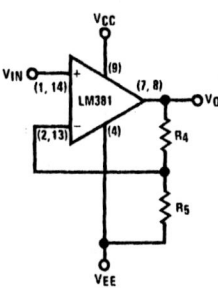

Differential Input Biasing

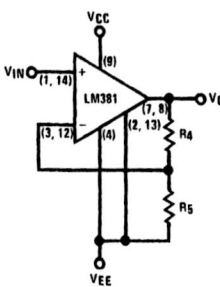

Single Ended Input Biasing

$V_{O\,DC} \approx 0$ VOLTS
$V_{IN\,DC} \approx V_{EE} + 1.2$ VOLTS

FIGURE 2.6.7 Split Supply Operation

2.6.5 Non-Inverting AC Amplifier

Perhaps the most common application of the LM381 is as a flat gain, non-inverting AC amplifier operating from a single supply. Such a configuration is shown in Figure 2.6.8. Resistors R_4 and R_5 provide the necessary biasing and establish the DC gain, A_{VDC}, per Equation (2.6.8).

$$A_{VDC} = 1 + \frac{R_4}{R_5} \qquad (2.6.8)$$

AC gain is set by resistor R_6 with low frequency roll-off at f_o being determined by capacitor C_2.

$$A_{VAC} = 1 + \frac{R_4}{R_6} \quad (R_6 \ll R_5) \qquad (2.6.9)$$

$$C_2 = \frac{1}{2\pi f_o R_6} \quad (C_c R_L \gg C_2 R_6) \qquad (2.6.10)$$

Since the LM381 is a high gain amplifier, proper power supply decoupling is required. For most applications a $0.1\mu F$ ceramic capacitor (C_s) with short leads and located close (within one inch) to the integrated circuit is sufficient. When used non-inverting, the maximum input voltage of $300\,mV_{RMS}$ ($850\,mV_{p-p}$) must be observed to maintain linear operation and avoid excessive distortion. Such is not the case when used inverting.

2.6.6 Inverting AC Amplifier

The inverting configuration (2.6.9) is very useful since it retains the excellent low noise characteristics without the limit on input voltage and has the additional advantage of being inherently unity gain stable. This is achieved by the voltage divider action of R_6 and R_5 on the input voltage. For normal values of R_4 and R_5 (with typical supply voltages) the gain of the amplifier itself, i.e., the voltage gain relative to pins 2 or 13 rather than the input, is always around ten — which is stable. The real importance is that while the addition of C_3 will guarantee unity gain stability (and roll-off high frequencies), it does so at the expense of slew rate.

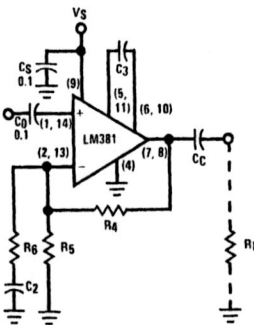

FIGURE 2.6.8 Non-inverting AC Amplifier

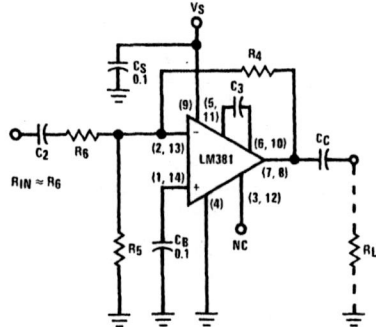

FIGURE 2.6.9 Inverting AC Amplifier

The small-signal bandwidth of the LM381 is nominally 20 MHz, making the preamp suitable for wide-band instrumentation applications. However, in narrow-band applications it is desirable to limit the amplifier bandwidth and thus eliminate high frequency noise. Capacitor C_3 accomplishes this by shunting the internal pole-splitting capacitor (C_1), limiting the bandwidth of the amplifier. Thus, the high frequency $-3dB$ corner is set by C_3 according to Equation (2.6.11).

$$C_3 = \frac{1}{2\pi f_3\, 2\,r_e\, A_{VAC}} - 4 \times 10^{-12} \qquad (2.6.11)$$

where: f_3 = high frequency $-3dB$ corner

r_e = first stage small-signal emitter resistance $\approx 1.3\,k\Omega$

A_{VAC} = mid-band gain in V/V

Capacitor C_o acts as an input AC coupling capacitor to block DC potentials in both directions and can equal $0.1\mu F$ (or larger). Output coupling capacitor C_c is determined by the load resistance and low frequency corner f per Equation (2.6.12).

$$C_c = \frac{1}{2\pi f R_L} \qquad (2.6.12)$$

Note: To avoid affecting f_o, $f \ll f_o$. For example, $f = 0.25 f_o$ will cause a 0.25 dB drop at f_o.

Using Figure 2.6.9 without C_3 at any gain retains the full slew rate of $4.7 V/\mu s$. The new gain equations follow:

$$A_{VDC} = -\frac{R_4}{R_5} \qquad (2.6.13)$$

$$A_{VAC} = -\frac{R_4}{R_6} \qquad (2.6.14)$$

Capacitor C_2 is still found from Equation (2.6.10), and C_c and C_s are as before. Capacitor C_B is added to provide AC decoupling of the positive input and can be made equal to $0.1\mu F$. Observe that pins 3 and 12 are not used, since the inverting configuration is not normally used with single ended input biasing techniques.

2.7 LM381A DUAL PREAMPLIFIER FOR ULTRA-LOW NOISE APPLICATIONS

2.7.1 Introduction

The LM381A is a dual preamplifier expressly designed to meet the requirements of amplifying low level signals in noise critical applications. Such applications include hydrophones, scientific and instrumentation recorders, low level wideband gain blocks, tape recorders, studio sound equipment, etc.

The LM381A can be externally biased for optimum noise performance in ultra-low noise applications. When this is done the LM381A provides a wideband, high gain amplifier with excellent noise performance.

The amplifier can be operated in either the differential or single ended input configuration. However, for optimum noise performance, the input must be operated single ended, since both transistors contribute noise in a differential stage, degrading input noise by the factor $\sqrt{2}$. (See Section 2.3) A second consideration is the design of the input bias circuitry. Both the load and biasing elements must be resistive, since active components would each contribute additional noise equal to that of the input device.

2.8 LM387/387A LOW NOISE MINIDIP DUAL PRE-AMPLIFIER

2.8.1 Introduction

The LM387 is a low cost, dual preamplifier supplied in the popular 8 lead minidip package. The internal circuitry is identical to the LM381 and has comparable performance. By omitting the external compensation and single ended biasing pins it has been possible to package this dual amplifier into the 8 pin minidip, making for very little board space requirement. Like the LM381, this preamplifier is 100% noise tested and guaranteed, when purchased through authorized distributors. Total equivalent input noise is typically $0.65 \mu V_{RMS}$ ($R_s = 600\Omega$, 100Hz-10kHz) and supply rejection ratio is typically 110dB (f = 1kHz). All other parameters are identical to the LM381. Biasing, compensation and split-supply operation are as previously explained.

2.8.2 Non-Inverting AC Amplifier

For low level signal applications requiring optimum noise performance the non-inverting configuration remains the most popular. The LM387 used as a non-inverting AC amplifier is configured similar to the LM381 and has the same design equations. Figure 2.8.1 shows the circuit with the equations duplicated for convenience.

2.8.3 Inverting AC Amplifier

For high level signals (greater than 300mV), the inverting configuration may be used to overcome the positive input overload limit. Voltage gains of less than 20dB are possible with the inverting configuration since the DC biasing resistor R_5 acts to voltage divide the incoming signal as previously described for the LM381. Design equations are the same as for the LM381 and are duplicated along with the inverting circuit in Figure 2.8.2.

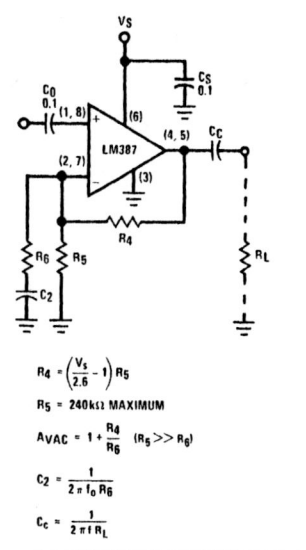

$R_4 = \left(\dfrac{V_s}{2.6} - 1\right) R_5$

$R_5 = 240 k\Omega$ MAXIMUM

$A_{VAC} = 1 + \dfrac{R_4}{R_6}$ ($R_5 \gg R_6$)

$C_2 = \dfrac{1}{2\pi f_0 R_6}$

$C_c = \dfrac{1}{2\pi f R_L}$

$f_0 =$ LOW FREQUENCY –3dB CORNER ($f \ll f_0$)

FIGURE 2.8.1 LM387 Non-inverting AC Amplifier

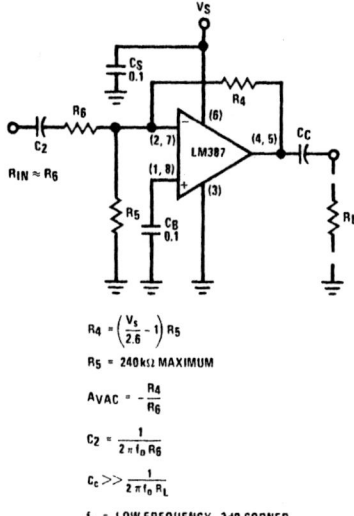

$R_4 = \left(\dfrac{V_s}{2.6} - 1\right) R_5$

$R_5 = 240 k\Omega$ MAXIMUM

$A_{VAC} = -\dfrac{R_4}{R_6}$

$C_2 = \dfrac{1}{2\pi f_0 R_6}$

$C_c \gg \dfrac{1}{2\pi f_0 R_L}$

$f_0 =$ LOW FREQUENCY –3dB CORNER

FIGURE 2.8.2 LM387 Inverting AC Amplifier

2.8.4 Unity Gain Inverting Amplifier

The requirement for unity gain stability is that the gain of the amplifier from pin 2 (or 7) to pin 4 (or 5) must be at least ten at all frequencies. This gain is the ratio of the feedback resistor R_4 divided by the total net impedance seen by the inverting input with respect to ground. The assumption is made that the driving, or source, impedance is small and may be neglected. In Figure 2.8.2 the net impedance looking back from the inverting input is $R_5 \| R_6$, at high frequencies. (At low frequencies where loop gain is large the impedance at the inverting input is very small and R_5 is effectively not present; at higher frequencies loop gain decreases, causing the inverting impedance to rise to the limit set by R_5. At these frequencies R_5 acts as a voltage divider for the input voltage guaranteeing amplifier gain of 10 when properly selected.) If the ratio of R_4 divided by $R_5 \| R_6$ is at least ten, then stability is assured. Since R_4 is typically ten times R_5 (for large supply voltages) and R_6 equals R_4 (for unity gain), then the circuit is stable without additional components. For low voltage applications where the ratio of R_4 to R_5 is less than ten, it becomes necessary to parallel R_5 with a series R-C network so the ratio at high frequencies satisfies the gain requirement. Figure 2.8.3 shows such an arrangement with the constraints on R_7 being given by Equations (2.8.1)-(2.8.3).

$$|A_V|(\text{pin 2 to 4}) = \frac{R_4}{R_5 \| R_6 \| R_7} \geq 10 \qquad (2.8.1)$$

$$RY = R_5 \| R_6 \qquad (2.8.2)$$

$$R_7 \leq \frac{RY \, R_4}{10 \, RY - R_4} \qquad (2.8.3)$$

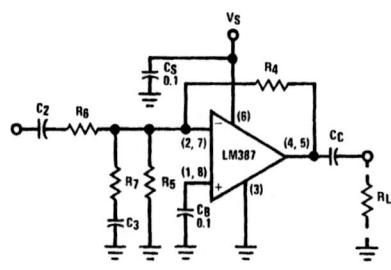

FIGURE 2.8.3 Unity Gain Amplifier for Low Supply Voltage

Example 2.8.1

Design a low noise unity gain inverting amplifier to operate from $V_S = 12\,V$, with low frequency capabilities to 20 Hz, input impedance equal to 20 kΩ, and a load impedance of 100 kΩ.

Solution:

1. $R_{in} = R_6 = 20\,k\Omega$.
2. For unity gain $R_4 = R_6$, $R_4 = 20k$.
3. From Figure 2.8.2:

$$R_4 = \left(\frac{V_S}{2.6} - 1\right) R_5 = \left(\frac{12}{2.6} - 1\right) R_5$$

$$R_4 = 3.62 \, R_5$$

Therefore:

$$R_5 = \frac{R_4}{3.62} = \frac{20k}{3.62} = 5,525\,\Omega$$

Use $R_5 = 5.6k$.

4. From Equation (2.8.2):

$$RY = R_5 \| R_6 = \frac{5.6k \times 20k}{5.6k + 20k} = 4375$$

5. From Equation (2.8.3):

$$R_7 \leq \frac{RY \, R_4}{10 \, RY - R_4} = \frac{4375 \times 20 \times 10^3}{10 \times 4375 - (20 \times 10^3)} = 3684$$

Use $R_7 = 3.6k$.

6. For $f_o = 20\,Hz$,

$$C_2 = \frac{1}{2\pi f_o R_6} = \frac{1}{2\pi \times 20 \times 20k} = 3.98 \times 10^{-7}$$

Use $C_2 = 0.5\,\mu F$.

For $f_{-3dB} = 20\,Hz$, the low frequency corner given by C_C and R_L must be at least a factor of 4 lower, i.e., $f < 5\,Hz$.

$$C_C = \frac{1}{2\pi f R_L} = \frac{1}{2\pi \times 5 \times 100k} = 3.18 \times 10^{-7}$$

Use $C_C = 0.33\,\mu F$.

7. The selection of C_3 is somewhat arbitrary, as its effect is only necessary at high frequencies. A convenient frequency for calculation purposes is 20 kHz.

$$C_3 = \frac{1}{2\pi(20\,kHz) R_7} = \frac{1}{2\pi \times 20k \times 3.6k} = 2.21 \times 10^{-9}$$

Use $C_C = 0.0022\,\mu F$

2.8.5 Application to Feedback Tone Controls

One of the most common audio circuits requiring unity gain stability is active tone controls. Complete design details are given in Section 2.14. An example of modified Baxandall tone controls using an LM387 appears as Figure 2.14.17 and should be consulted as an application of the stabilizing methods discussed in Section 2.8.4.

2.9 LM382 LOW NOISE DUAL PREAMPLIFIER WITH RESISTOR MATRIX

2.9.1 Introduction

The LM382 is a dual preamplifier patterned after the LM381 low noise circuitry but with the addition of an internal resistor matrix. The resistor matrix allows the user to select a variety of closed loop gain options and frequency response characteristics such as flat-band, NAB (tape), or RIAA (phonograph) equalization. The LM382 possesses all of the features of the LM381 with two exceptions: no single ended input biasing option and no external pins for adding additional compensation capacitance. The internal resistors provide for biasing of the negative input automatically, so no external resistors are necessary and use of the LM382 creates the lowest parts count possible for standard designs. Originally developed for the automotive tape player market with a nominal supply voltage of +12V, the output is self queuing to about +6V (regardless of applied voltage — but this can be defeated, as will be discussed later). A diagram of the LM382 showing the resistor matrix appears as Figure 2.9.1.

2-18

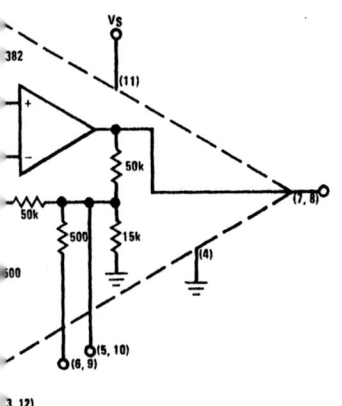

FIGURE 2.9.1 LM382 Resistor Matrix

Non-inverting AC Amplifier

A flat-response configuration of the LM382 shows that with just two or three capacitors a high-gain, low noise preamplifier is created.

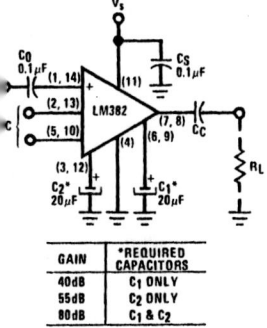

GAIN	*REQUIRED CAPACITORS
40dB	C_1 ONLY
55dB	C_2 ONLY
80dB	C_1 & C_2

FIGURE 2.9.2 LM382 as Fixed Gain-Flat Response Non-inverting Amplifier

To show the gains of Figure 2.9.2 are calculated it is easiest to redraw each case with the capacitors in and include only the relevant portion of the network per Figure 2.9.1. The redrawn 40dB gain (C_1 only) appears as Figure 2.9.3.

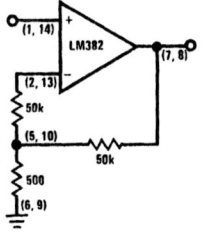

FIGURE 2.9.3 Equivalent Circuit for 40 dB Gain (C_1 Only)

Since bias currents are small and may be ignored in gain calculations, the 50k input resistor does not affect gain. Therefore, the gain is given by:

$$A_{V1} = 1 + \frac{50k}{500} = 101 \approx 40dB$$

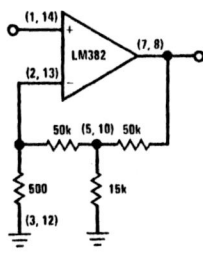

FIGURE 2.9.4 Equivalent Circuit for 55 dB Gain (C_2 Only)

With C_2 only, the redrawn equivalent circuit looks like Figure 2.9.4. Since the feedback network is wye-connected, it is easiest to perform a wye-delta transformation (see Appendix A3) in order to find an effective feedback resistor so the gain may be calculated. A complete transformation produces three equivalent resistors, two of which may be ignored. These are the ones that connect from the ends of each 50kΩ resistor to ground; one acts as a load on the amplifier and doesn't enter into the gain calculations, and the other parallels 500Ω and is large enough to have no effect. The remaining transformed resistor connects directly from the output to the input and is the equivalent feedback resistor, R_f. Its value is found from:

$$R_f \text{ (equivalent)} = 50k + 50k + \frac{(50k)^2}{15k} = 267k$$

The gain is now simply

$$A_{V2} = 1 + \frac{267k}{500} = 535 \approx 55dB$$

Adding both C_1 and C_2 gives the equivalent circuit of Figure 2.9.5.

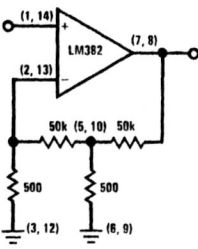

FIGURE 2.9.5 Equivalent Circuit for 80 dB Gain (C_1 and C_2)

Treating Figure 2.9.5 similarly to Figure 2.9.4, an equivalent feedback resistor is calculated:

$$R_f \text{ (equivalent)} = 50k + 50k + \frac{(50k)^2}{500} = 5.1 \text{ Meg}$$

Therefore, the gain is:

$$A_{V12} = 1 + \frac{5.1 \text{ Meg}}{500} = 10201 \approx 80 \text{ dB}$$

2.9.3 Adjustable Gain for Non-Inverting Case

As can be learned from the preceding paragraphs, there are many combinations of ways to configure the resistor matrix. By adding a resistor in series with the capacitors it is possible to vary the gain. Care must be taken in attempting low gains (< 20dB), as the LM382 is not unity gain stable and should not be operated below gains of 20dB. (Under certain specialized applications unity gain is possible, as will be demonstrated later.) A general circuit allowing adjustable gain and requiring only one capacitor appears as Figure 2.9.6.

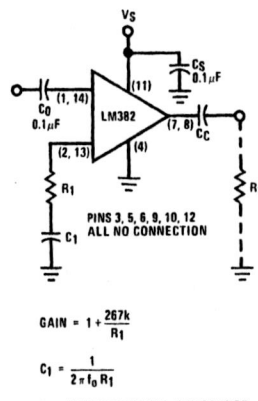

FIGURE 2.9.6 Adjustable Gain Non-inverting Amplifier

Referring to Figure 2.9.1, it is seen that the R_1-C_1 combination is used instead of the internal 500Ω resistor and that the remaining pins are left unconnected. The equivalent resistance of the 50k-50k-15k wye feedback network was found previously to equal 267kΩ, so the gain is now given by Equation (2.9.1).

$$\text{Gain} = 1 + \frac{267k}{R_1} \qquad (2.9.1)$$

And C_1 is found from Equation (2.9.2):

$$C_1 = \frac{1}{2\pi f_o R_1} \qquad (2.9.2)$$

where: f_o = low frequency –3dB corner.

2.9.4 Internal Bias Override

As mentioned in the introduction, it is possible to override the internal bias resistor which causes the output quiescent point to sit at +6V regardless of applied voltage. This is done by adding a resistor at pin 5 (or 10) which parallels the internal 15kΩ resistor and defeats its effect (Figure 2.9.7).

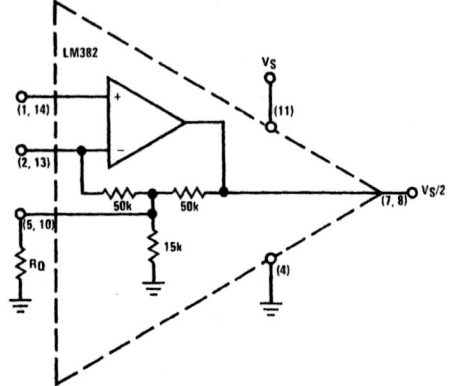

FIGURE 2.9.7 Internal Bias Override Resistor

Since the positive input is biased internally to a potential of +1.3V (see circuit description for LM381), it is necessary that the DC potential at the negative input equal +1.3V also. Because bias current is small (0.5μA), the voltage drop across the 50k resistor may be ignored, which says there is +1.3V across RQ. The current developed by this potential across RQ is drawn from the output stage, through the 50k resistor, through RQ and to ground. The subsequent voltage drop across the 50k resistor is additive to the +1.3V and determines the output DC level. Stated mathematically,

$$\frac{V_S}{2} = \left(\frac{50k}{RX}\right) 1.3V + 1.3V \qquad (2.9.3)$$

where: RX = RQ‖15k

From Equation (2.9.3) the relationships of RX and RQ may be expressed.

$$RX = \frac{50k}{\frac{V_S}{2.6} - 1} \qquad (2.9.4)$$

$$RQ = \frac{RX(15k)}{15k - RX} \qquad (2.9.5)$$

Example 2.9.1

Select RQ such that the output of a LM382 will center at $12 V_{DC}$ when operated from a supply of $V_S = 24 V_{DC}$.

Solution

1. Calculate RX from Equation (2.9.4).

$$RX = \frac{50 \times 10^3}{\frac{24}{2.6} - 1} = 6075 \Omega$$

2. Calculate RQ from Equation (2.9.5).

$$RQ = \frac{(6075)(15 \times 10^3)}{(15 \times 10^3) - 6075} = 10210 \Omega$$

Use RQ = 10kΩ.

Since RQ parallels the 15k resistor, then the AC gains due to the addition of capacitor C_1 or C_2 (or both) as given in Figure 2.9.2) are changed. The new gain equations become a function of RQ and are given as Equations (2.9.6)-(2.9.8) and refer to Figure 2.9.8.

C_1 Only: Gain $\approx 1 + \dfrac{50k}{RQ\|500}$ (2.9.6)

C_2 Only: Gain $= 201 + \dfrac{5 \times 10^6}{RX}$ (2.9.7)

C_1 & C_2:   Gain $\approx 201 + \dfrac{5 \times 10^6}{RQ\|500}$ (2.9.8)

where: RX and RQ are given by Equations (2.9.4) and (2.9.5).

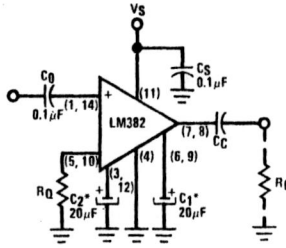

* – IF REQUIRED
PINS 2 & 13 NO CONNECTION

FIGURE 2.9.8 Fixed Gain Amplifier with Internal Bias Override

Continuing the previous example to find the effect of RQ on the gain yields:

3. C_1 Only: Gain $= 1 + \dfrac{50k}{10k\|500} = 53.6\,dB$

4. C_2 Only: Gain $= 201 + \dfrac{5 \times 10^6}{6075} = 60.2\,dB$

5. C_1 & C_2:   Gain $= 201 + \dfrac{5 \times 10^6}{10k\|500} = 80.6\,dB$

2.9.5 Inverting AC Amplifier

Examination of the resistor matrix (Figure 2.9.1) reveals that an inverting AC amplifier can be created with just one resistor (Figure 2.9.9).

The gain is found by calculating the equivalent feedback resistance as before, and appears in Figure 2.9.9. Higher gains are possible (while retaining large input resistance = R_1) by adding capacitor C_1 as shown in Figure 2.9.10. The internal bias override technique discussed for the non-inverting configuration may be applied to the inverting case as well. The required value of RQ is calculated from Equations (2.9.4) and (2.9.5) and affects the gain relation shown in Figures 2.9.9 and 2.9.10. The new gain equations are:

Without C_1:   Gain $= \left(-\dfrac{1}{R_1}\right)\left(10^5 + \dfrac{2.5 \times 10^9}{RQ\|15k}\right)$ (2.9.9)

With C_1:   Gain $= \left(-\dfrac{1}{R_1}\right)\left(10^5 + \dfrac{2.5 \times 10^9}{RQ\|500}\right)$ (2.9.10)

and the circuit is shown in Figure 2.9.11.

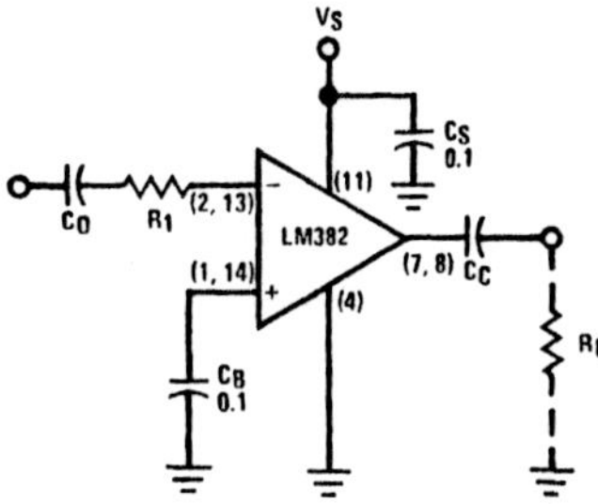

GAIN $= -\dfrac{267k}{R_1}$ ($\geq$ 20dB FOR STABILITY)

$C_0 = \dfrac{1}{2\pi f_0 R_1}$

f_0 = LOW FREQUENCY -3dB CORNER
INPUT IMPEDANCE = R_1
PINS 3, 5, 6, 9, 10, 12 NOT USED

FIGURE 2.9.9 LM382 as Inverting AC Amplifier

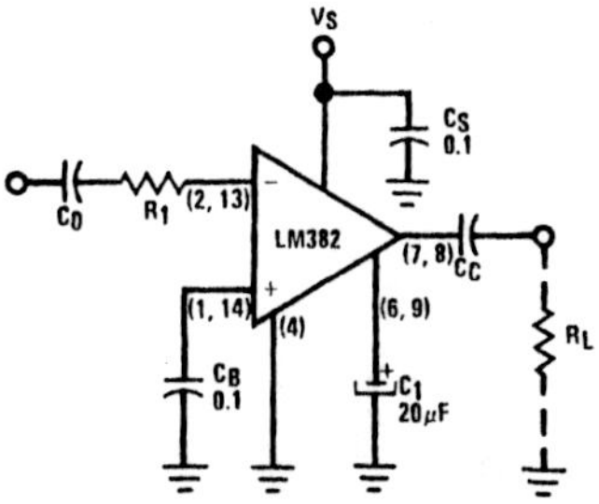

GAIN $= -\dfrac{5.1 \times 10^6}{R_1}$

$C_0 = \dfrac{1}{2\pi f_0 R_1}$

f_0 = LOW FREQUENCY -3dB CORNER ($C_C R_L \gg C_0 R_1$)
INPUT IMPEDANCE = R_1
PINS 3, 5, 10, 12 NOT USED

FIGURE 2.9.10 High Gain Inverting AC Amplifier

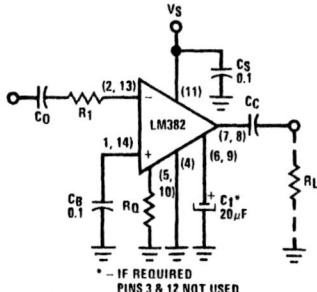

* – IF REQUIRED
PINS 3 & 12 NOT USED

FIGURE 2.9.11 Inverting Amplifier with Internal Bias Override

Example 2.9.2

Design an inverting amplifier to operate from a supply of $V_S = 24\,V_{DC}$, with output quiescent point equal to $12\,V_{DC}$, gain equal to 40 dB, input impedance greater than $10\,k\Omega$, low frequency performance flat to 20 Hz, and a load impedance equal to $100\,k\Omega$.

1. From the previous example $R_Q = 10\,k\Omega$.
2. Add C_1 for high gain and input impedance.
3. Calculate R_1 from Equation (2.9.10).

$$R_1 = \left(\frac{1}{\text{Gain}}\right)\left(10^5 + \frac{2.5 \times 10^9}{R_Q \| 500}\right)$$

$$R_1 = \left(\frac{1}{10^2}\right)\left(10^5 + \frac{2.5 \times 10^9}{10k \| 500}\right) \quad (\text{Note: } 40\,dB = 10^2\,V/V)$$

$$R_1 = 5.35 \times 10^4$$

Use $R_1 = 56\,k\Omega$.

4. Calculate C_O from equation shown in Figure 2.9.9.

$$C_O = \frac{1}{2\pi f_O R_1} = \frac{1}{(2\pi)(20)(56k)} = 1.42 \times 10^{-7}$$

Use $C_O = 0.15\,\mu F$.

5. Calculate C_C from Equation (2.6.12).

$$C_C = \frac{1}{2\pi f R_1} = \frac{1}{(2\pi)(5)(10^5)} = 3.18 \times 10^{-7}$$

Use $C_C = 0.33\,\mu F$.

The complete amplifier is shown in Figure 2.9.12.

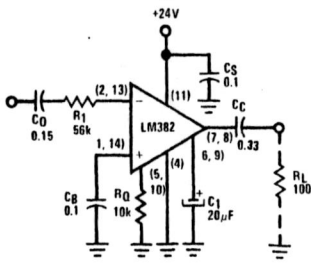

FIGURE 2.9.12 Inverting Amplifier with Gain = 40 dB and $V_S = +24\,V$

2.9.6 Unity Gain Inverting Amplifier

Referring back to Figure 2.9.1, it can be seen that by shorting pin 2 (or 13) to 5 (or 10) the feedback network reduces to a single $50\,k\Omega$ resistor connected from the output to the inverting input, plus the $15\,k\Omega$ biasing resistor from the inverting input to ground. To create unity gain then, a resistor equal to $50\,k\Omega$ is connected to the minus input. Simple enough; however, the amplifier is not stable. Since the 15k resistor acts as a voltage divider to the input, the gain of the amplifier (pin 7 to pin 2) is only 50k divided by 15k, or $3.33\,V/V$. Minimum required gain for stability is $10\,V/V$, so it becomes necessary to shunt the 15k resistor with a new resistor such that the parallel combination equals $5\,k\Omega$. This may be done AC or DC,

depending upon supply voltage. If done DC (tied from pin 2 (or 13) directly to ground), then it becomes RQ (from Figure 2.9.7) and affects the output DC level. Placing a capacitor in series with this resistor makes it effective only for AC voltages and does not change the output level. The required resistor equals $9.1\,k\Omega$, which is close enough to the required RQ for $V_S = 24\,V$. Two examples of unity gain amplifiers appear as Figure 2.9.13 and should satisfy the majority of applications.

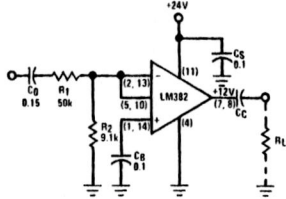

(a) Supply Voltage = 24 Volts

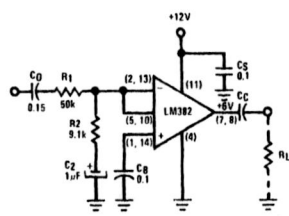

(b) Supply Voltage = 12 Volts

FIGURE 2.9.13 Unity Gain Inverting Amplifier

2.9.7 Remarks

The above application hints are not meant to be all-inclusive, but rather are offered as an aid to LM382 users to familiarize them with its many possibilities. Once understood, the internal resistor matrix allows for many possible configurations, only a few of which have been described in this section.

2.10 LM1303 STEREO PREAMPLIFIER

2.10.1 Introduction

The LM1303 is a dual preamplifier designed to be operated from split supplies ranging from $\pm 4.5\,V$ up to $\pm 15\,V$. It has "op amp" type inputs allowing large input signals with low distortion performance. The wideband noise performance is superior to traditional operational amplifiers, being typically $0.9\,\mu V_{RMS}$ (10 kHz bandwidth). Compensation is done externally and offers the user a variety of choices, since three compensation points are brought out for each amplifier. The LM1303 is pin-for-pin compatible with "739" type dual preamplifiers and in most applications serves as a direct replacement.

2.10.2 Non-Inverting AC Amplifier

The LM1303 used as a non-inverting amplifier (Figure 2.10.1) with split supplies allows for economical direct-coupled designs if the DC levels between stages are maintained at zero volts. Gain and C_1 equations are shown in the figure. Resistor R_3 is made equal to R_1 and provides DC bias currents to the positive input. Compensation capacitor C_2 is equal to $0.022\mu F$ and guarantees unity gain stability with a slew rate of approximately $1 V/\mu s$. Higher slew rates are possible when higher gains are used by reducing C_2 proportionally to the increase in gain, e.g., with a gain of ten, C_2 can equal $0.0022\mu F$, increasing the slew rate to around $10 V/\mu s$. Some layouts may dictate the addition of C_3 for added stability. It should be picked according to equation (2.10.1) where f_H is the high frequency $-3 dB$ corner.

$$C_3 = \frac{1}{2\pi f_H R_1} \quad (2.10.1)$$

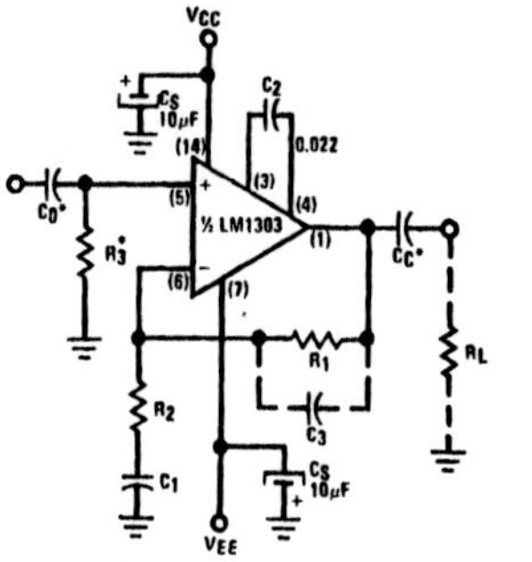

$$A_{VAC} = 1 + \frac{R_1}{R_2}$$

$$C_1 = \frac{1}{2\pi f_0 R_2} \quad \begin{array}{l} C_c R_L \gg C_1 R_2 \\ C_0 R_3 \gg C_1 R_2 \end{array}$$

f_0 = LOW FREQUENCY $-3 dB$ CORNER

* – MAY BE OMITTED FOR DIRECT-COUPLED DESIGNS.

FIGURE 2.10.1 LM1303 Non-inverting AC Amplifier

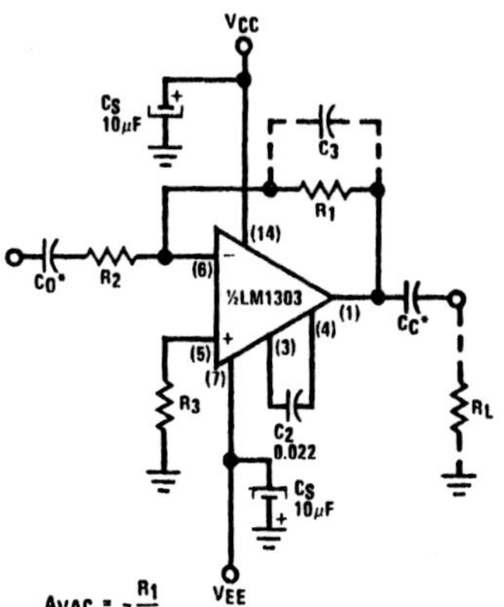

$$A_{VAC} = -\frac{R_1}{R_2}$$

$$C_0 = \frac{1}{2\pi f_0 R_2} \quad (C_c R_L \gg C_0 R_2)$$

f_0 = LOW FREQUENCY $-3 dB$ CORNER

* – MAY BE OMITTED FOR DIRECT-COUPLED DESIGNS.

FIGURE 2.10.2 LM1303 Inverting AC Amplifier

2.10.3 Inverting AC Amplifier

For applications requiring inverting operation, Figure 2.10.2 should be used. Capacitors C_2 and C_3 have the same considerations as the non-inverting case. Resistor R_3 is made equal to R_1 again, minimizing offsets and providing bias current. The same slew rate-gain stability trade-offs are possible as before.

2.11 PHONO PREAMPLIFIERS AND RIAA EQUALIZATION

2.11.1 Introduction

Phono preamplifiers differ from other preamplifiers only in their frequency response, which is tailored in a special manner to compensate, or equalize, for the recorded characteristic. If a fixed amplitude input signal is used to record a phonograph disc, while the frequency of the signal is varied from 20 Hz to 20 kHz, the playback response curve of Figure 2.11.1 will result. Figure 2.11.1 shows a plot of phono cartridge output amplitude versus frequency, indicating a severe alteration to the applied fixed amplitude signal. *Playback equalization* corrects for this alteration and re-creates the applied flat amplitude frequency response. To understand why Figure 2.11.1 appears as it does, an explanation of the recording process is necessary.

2.11.2 Recording Process and RIAA

The grooves in a stereo phonograph disc are cut by a chisel shaped cutting stylus driven by two vibrating systems arranged at right angles to each other (Figure 2.11.2). The cutting stylus vibrates mechanically from side to side in accordance with the signal impressed on the cutter. This is termed a "lateral cut" as opposed to the older method of "vertical cut." The resultant movement of the groove back and forth about its center is known as groove modulation. The amplitude of this modulation cannot exceed a fixed amount or "cutover" occurs. (Cutover, or overmodulation, describes the breaking through the wall of one groove into the wall of the previous groove.) The ratio of the maximum groove signal amplitude possible before cutover, to the effective groove noise amplitude caused by the surface of the disc material, determines the dynamic range of a record (typically 58 dB). The latter requirement results from the grainy characteristic of the disc surface acting as a noise generator. (The cutting stylus is heated in recording to impart a smooth side wall to minimize the noise.) Of interest in phono preamp design is that the record noise performance tends to be ten times worse than that of the preamp, with typical wideband levels equal to $10\mu V$.

Amplitude and frequency characterize an audio signal. Both must be recorded and recovered accurately for high quality music reproduction. Audio amplitude information translates to groove modulation amplitude, while the frequency of the audio signal appears as the rate of change of the groove modulations. Sounds simple enough, but Figure 2.11.1 should, therefore, be a horizontal straight line centered on 0 dB, since it represents a fixed amplitude input signal. The trouble results from the characteristics of the cutting head. Without the negative feedback coils (Figure 2.11.2) the velocity frequency response has a resonant peak at 700 Hz due to its construction. Adding the feedback coils produces a velocity output independent of frequency; therefore, the cutting head is known as a constant velocity device (Figure 2.11.2a).

Figure 2.11.1 appears as it does because the cutting amplifier is pre-equalized to provide the recording character-

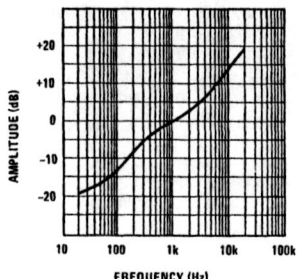

FIGURE 2.11.1 Typical Phono Playback Characteristic for a Fixed Amplitude Recorded Signal

istic shown. Two reasons account for the shape: first, low frequency attenuation prevents cutover; second, high frequency boosting improves signal-to-noise ratio. The unanswered question is why is all this necessary?

The not-so-simple answer begins with the driving coils of the cutting head. Being primarily inductive, their impedance characteristic is frequency dependent. If a fixed amplitude input signal translates to a fixed voltage used to drive the coils (called "constant velocity") then the resulting current, i.e., magnetic field, hence amplitude of vibration, becomes frequency dependent (Figure 2.11.2a); if a fixed amplitude input signal translates to a fixed current, i.e., fixed amplitude of vibration, used to drive the coils (called "constant amplitude) then the resulting voltage, i.e., cutting velocity, becomes frequency dependent (Figure 2.11.2b). With respect to frequency, for a given input amplitude the cutting head has only one degree of freedom: vibrating *rate* (constant velocity = voltage drive) or vibrating *distance* (constant amplitude = current drive).

The terms constant velocity and constant amplitude create confusion until it is understood that they have meaning only for a *fixed amplitude input signal*, and are used strictly to describe the resultant behavior of the cutting head as a *function of frequency*. It is to be understood that changing the *input* level results in an *amplitude* change for constant amplitude recording and a *velocity* change for constant velocity recording *independent* of frequency. For example,

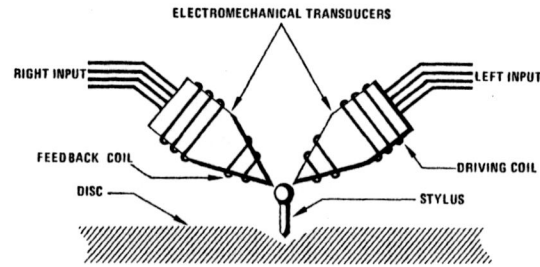

FIGURE 2.11.2 Stereo Cutting Head

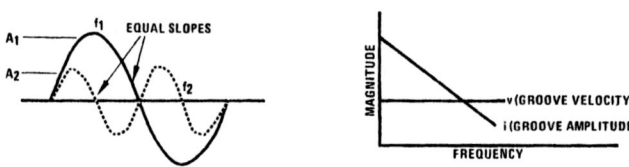

FIGURE 2.11.2A Constant Velocity Recording

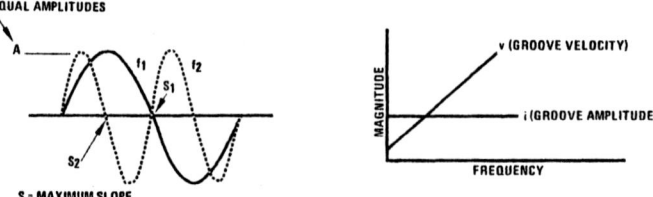

FIGURE 2.11.2B Constant Amplitude Recording

if an input level of 10 mV results in 0.1 mil amplitude change for constant amplitude recording and a velocity of 5 cm/s for constant velocity recording, then a change of input level to 20 mV would result in 0.2 mil and 10 cm/sec respectively — independent of frequency.

Each of these techniques when used to drive the vibrating mechanism suffers from dynamic range problems. Figures 2.11.2a and 2.11.2b diagram each case for two frequencies an octave apart. The discussion that follows assumes a fixed amplitude input signal and considers only the effect of frequency change on the cutting mechanism.

Constant velocity recording (Figure 2.11.2a) displays two readily observable characteristics. The amplitude varies inversely with frequency and the maximum slope is constant with frequency. The second characteristic is ideal since magnetic pickups (the most common type) are constant velocity devices. They consist of an active generator such as a magnetic element moving in a coil (or vice versa) with the output being proportional to the speed of movement through the magnetic field, i.e., proportional to groove velocity. However, the variable amplitude creates serious problems at both frequency extremes. For the ten octaves existing between 20 Hz and 20 kHz, the variation in amplitude is 1024 to 1! If 1 kHz is taken as a reference point to establish nominal cutter amplitude modulation, then at low frequencies the amplitudes are so great that cutover occurs. At high frequencies the amplitude becomes so small that acceptable signal-to-noise ratios are not possible — indeed, if any displacement exists at all. So much for constant velocity.

Looking at Figure 2.11.2b, two new observations are seen with regard to constant amplitude. Amplitude is constant with frequency (which corrects most of the ills of constant velocity), but the maximum slope varies directly with frequency, i.e., groove velocity is directly proportional to frequency. So now velocity varies 1024 to 1 over the audio band — swell! Recall that magnetic cartridges are constant velocity devices, not constant amplitude, so the output will rise at the rate of +6 dB/octave. (6 dB increase equals twice the amplitude.) To equalize such a system would require 60 dB of headroom in the preamp — not too practical. The solution is to try to get the best of both systems, which results in a modified constant amplitude curve where the midband region is allowed to operate constant velocity.

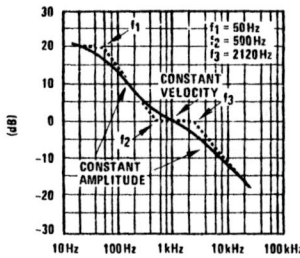

FIGURE 2.11.3 RIAA Playback Equalization

The required RIAA (Record Industry Association of America) playback equalization curve (Figure 2.11.3) shows the idealized case dotted and the actual realization drawn solid. Three frequencies are noted as standard design reference points and are sometimes referred to as time constants. This is a carryover from the practice of specifying corner frequencies by the equivalent RC circuit (t = RC) that realized the response. Conversion is done simply with the expression $t = 1/2\pi f$ and results in time constants of $3180\mu s$ for f_1, $318\mu s$ for f_2, and $75\mu s$ for f_3. Frequency f_2 is referred to as the *turnover* frequency since this is the point where the system changes from constant amplitude to constant velocity. (Likewise, f_3 is another turnover frequency.) Table 2.11.1 is included as a convenience in checking phono preamp RIAA response.

TABLE 2.11.1 RIAA Standard Response

Hz	dB	Hz	dB
20	+19.3	800	+0.7
30	+18.6	1k	0.0*
40	+17.8	1.5k	−1.4
50	+17.0	2k	−2.6
60	+16.1	3k	−4.8
80	+14.5	4k	−6.6
100	+13.1	5k	−8.2
150	+10.3	6k	−9.6
200	+8.2	8k	−11.9
300	+5.5	10k	*−13.7
400	+3.8	15k	−17.2
500	+2.6	20k	−19.6

* Reference frequency.

2.11.3 Ceramic and Crystal Cartridges

Before getting into the details of designing RIAA feedback networks for magnetic phono cartridges, a few words about crystal and ceramic cartridges are appropriate. In contradistinction to the constant velocity magnetic pickups, ceramic pickups are constant amplitude devices and therefore do not require equalization, since their output is inherently flat. Referring to Figure 2.11.3 indicates that the last sentence is not entirely true. Since the region between f_2 and f_3 is constant velocity, the output of a ceramic device will drop 12 dB between 500 Hz and 2000 Hz. While this appears to be a serious problem, in reality it is not. This is true due to the inherently poor frequency response of ceramic and restriction of its use to lo-fi and mid-fi market places. Since the output levels are so large (100 mV-2 V), a preamp is not necessary for ceramic pickups; the output is fed directly to the power amplifier via passive tone (if used) and volume controls.

2.11.4 LM387 or LM381 Phono Preamp

Magnetic cartridges have very low output levels and require low noise devices to amplify their signals without appreciably degrading the system noise performance. Nevertheless, note that usually the noise of the cartridge and loading resistor is comparable to the active device and should be included in the calculations (see Appendix A5).

Typical cartridge output levels are given in Table 2.11.2.

Output voltage is specified for a given modulation velocity. The magnetic pickup is a velocity device, therefore output is proportional to velocity. For example, a cartridge producing 5 mV at 5 cm/s will produce 1 mV at 1 cm/s and is specified as having a sensitivity of 1 mV/cm/s.

In order to transform cartridge sensitivity into useful preamp design information, we need to know typical and maximum modulation velocity limits of stereo records.

TABLE 2.11.2

Manufacturer	Model	Output at 5cm/sec
Empire Scientific	999	5mV
	888	8mV
Shure	V-15	3.5mV
	M91	5mV
Pickering	V-15 AT3	5mV

The RIAA recording characteristic establishes a maximum recording velocity of 25cm/s in the range of 800 to 2500Hz. Typically, good quality records are recorded at a velocity of 3 to 5cm/s.

Figure 2.11.3 shows the RIAA playback equalization. To obtain this, the desired transfer function of the preamplifier is given by:

$$\frac{V_{OUT}}{V_{IN}} = \frac{A(s + 2\pi \cdot 500)}{(s + 2\pi \cdot 50)(s + 2\pi \cdot 2120)} \quad (2.11.1)$$

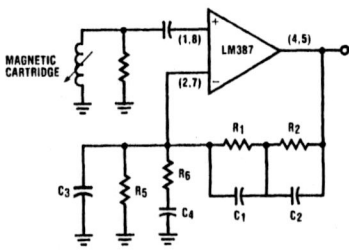

FIGURE 2.11.4 RIAA Phono Preamp

From Figure 2.11.4:

$$\frac{V_{OUT}}{V_{IN}} = \frac{K\left[s + \frac{R_1 + R_2}{(C_1 + C_2)R_1 R_2}\right]}{R_6\left(s + \frac{1}{C_1 R_1}\right)\left(s + \frac{1}{C_2 R_2}\right)} + 1 \quad (2.11.2)$$

Equating coefficients of (2.11.1) and (2.11.2),

$$R_1 C_1 = \frac{1}{2\pi \cdot 50} = 3180\,\mu s \quad (2.11.3)$$

$$R_2 C_2 = \frac{1}{2\pi \cdot 2120} = 75\,\mu s \quad (2.11.4)$$

$$\frac{R_1 R_2 (C_1 + C_2)}{R_1 + R_2} = \frac{1}{2\pi \cdot 500} = 318\,\mu s \quad (2.11.5)$$

Substituting (2.11.3) and (2.11.4) in (2.11.5):

$$R_1 = 11.78 R_2 \quad (2.11.6)$$

0 dB reference gain $= \dfrac{z + R_6}{R_6}$ (2.11.7)

where: $z = \left(R_1 \parallel \dfrac{1}{2\pi f C_1}\right) + \left(R_2 \parallel \dfrac{1}{2\pi f C_2}\right)$

Resistor R_5 together with R_1 and R_2 sets the DC bias (Section 2.6) and C_3 stabilizes the amplifier by rolling off the feedback at higher frequencies since the LM387 is not compensated for unity gain.

Example 2.11.1

Design a phonograph preamp operating from a 24V supply, with a cartridge of 0.5mV/cm/s sensitivity, to drive a power amplifier with an input overload limit of $1.25\,V_{RMS}$.

Solution

1. The maximum cartridge output of 25cm/s is $(0.5\,mV/cm/s) \times (25\,cm/sec) = 12.5\,mV$. The required midband gain is:

$$\frac{1.25\,V_{RMS}}{12.5\,mV_{RMS}} = 100$$

2. Before selecting R_6 to give a gain of 40dB at 1kHz, we must determine the complex impedance of the $R_1 R_2$, $C_1 C_2$ network at 1kHz. Ideally this should be such that R_6 is relatively low to minimize any noise contributions from the feedback network.

3. If we assume the amplifier output must be able to drive the feedback equalization network to the rated output at 20kHz, the slew rate required is:

$$S.R. = 2\pi E p f, \text{ where } E_p = 1.25 \times \sqrt{2}$$

$$= 2\pi \times 1.77 \times 20 \times 10^3$$
$$= 0.22\,V/\mu s$$

Using $1\,V/\mu s$ as a safety margin and noting that the output sink current of the LM387 is 2mA, the capacitance of the feedback network should be:

$$\leq \frac{2 \times 10^{-3}}{1 \times 10^{-6}}$$

$$\leq 0.002\,\mu F$$

Since C_2 will dominate the series arrangement of C_1 and C_2, put:

$C_2 = 0.0027\,\mu F$

4. From Equation (2.11.4):

$$R_2 = \frac{75 \times 10^{-6}}{0.0027 \times 10^{-6}} = 28\,k\Omega$$

Put $R_2 = 30\,k\Omega$

5. Equation (2.11.6):

$R_1 = 11.78 R_2$

$= 11.78 \times 30 \times 10^3 = 353\,k\Omega$

Put $R_1 = 360\,k\Omega$

6. Equation (2.11.3):

$$C_1 = \frac{3180 \times 10^{-6}}{360 \times 10^3} = 0.0088\,\mu F$$

Put $C_1 = 0.01\,\mu F$

7. At 1 kHz the feedback network impedance (z) = 37.6k ∠49°. Equation (2.11.7):

$$0\,dB \text{ reference gain} = 100 = \frac{37.6 \times 10^3}{R_6} + 1$$

$$\therefore R_6 = \frac{37.6 \times 10^3}{99} = 379\,\Omega$$

Put $R_6 = 390\,\Omega$

8. From Equation (2.6.4):

$$\left(\frac{V_{CC}}{2.6} - 1\right) R_5 = R_1 + R_2$$

$$\therefore R_5 = \frac{390 \times 10^3}{8.23} = 47\,k\Omega$$

Note: This value of R_5 will center the output at the mid supply point. However, for symmetrical clipping it is worth noting that the LM387 can swing to within 0.3V of ground and 1.7V of V_{CC}. To put the output midway between these points (11.2 V_{DC} with V_{CC} = 24V), put $R_5 = 56\,k\Omega$.

9. From Equation (2.6.10):

$$C_4 = \frac{1}{2\pi f_o R_6}$$

$$= \frac{1}{2\pi \cdot 10 \cdot 390}$$

$$= 40.8 \times 10^{-6}$$

Put $C_4 = 47\,\mu F$

The completed design is shown in Figure 2.11.5 where a 47 kΩ input resistor has been included to provide the RIAA standard cartridge load.

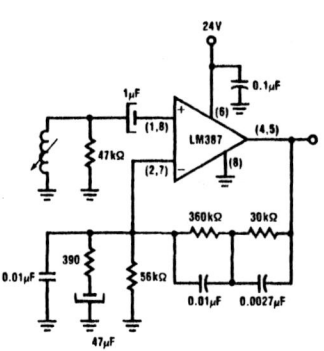

FIGURE 2.11.5 LM387 Phono Preamp (RIAA)

The LM381 integrated circuit may be substituted for the LM387 in Figure 2.11.5 by making the appropriate pin number changes.

2.11.5 LM382 Phono Preamp

By making use of the internal resistor matrix, a minimum parts count low noise phono preamp is possible using the LM382 (Figure 2.11.6). The circuit has been optimized for a supply voltage equal to 12-14 V. The midband 0 dB reference gain equals 46 dB (200 V/V) and cannot easily be altered. For designs requiring either gain or supply voltage changes, the required extra parts make selection of a LM381 or LM387 more appropriate.

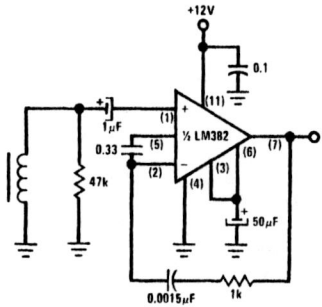

FIGURE 2.11.6 LM382 Phono Preamp. (RIAA)

2.11.6 LM1303 Phono Preamp

The LM1303 allows a convenient low noise phono preamp design when operating from split supplies. The circuit appears as Figure 2.11.7. For trimming purposes and/or gain changes the relevant formulas follow:

$$0\,dB \text{ Ref Gain} = 1 + \frac{R_2}{R_3} \quad (2.11.5)$$

$$f_1 = \frac{1}{2\pi R_1 C_1} \quad (2.11.6)$$

$$f_2 \approx \frac{1}{2\pi R_2 C_1} \quad (2.11.7)$$

$$f_3 = \frac{1}{2\pi R_2 C_2} \quad (2.11.8)$$

As shown in Figure 2.11.7, the 0 dB reference gain (1 kHz) equals about 34 dB and the feedback values have been altered slightly to minimize pole-zero interactions.

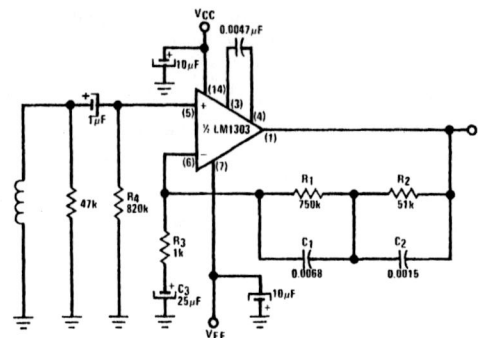

FIGURE 2.11.7 LM1303 Phono Preamp. (RIAA)

2.11.7 Inverse RIAA Response Generator

A useful test box to have handy while designing and building phono preamps is one which will yield the opposite of the playback characteristic, i.e., an inverse RIAA (or record) characteristic. The circuit (Figure 2.11.9) is achieved by adding a passive filter to the output of an LM387, used as a flat-response adjustable gain block. Gain is adjustable over a range of 24 dB to 60 dB and is set in accordance with the 0 dB reference gain (1 kHz) of the phono preamp under test. For example, assume the preamp being tested has +34 dB gain at 1 kHz. Connect a 1 kHz generator to the input of Figure 2.11.9. The passive filter has a loss of −40 dB at 1 kHz, which is corrected by the LM387 gain, so if a 1 kHz test output level of 1V is desired from a generator input level of 10 mV, then the gain of the LM387 is set at +46 dB (+46 dB − 40 dB + 34 dB = 40 dB (×100); 10 mV × 100 = 1V). Break frequencies of the filter are determined by Equations (2.11.9)-(2.11.11).

$$f_1 = 50\,Hz = \frac{1}{2\pi R_9 C_4} \qquad (2.11.9)$$

$$f_2 = 500\,Hz \approx \frac{1}{2\pi R_{10} C_4} \qquad (2.11.10)$$

$$f_3 = 2120\,Hz = \frac{1}{2\pi R_{10} C_5} \qquad (2.11.11)$$

The R_7-C_3 network is necessary to reduce the amount of feedback for AC and is effective for all frequencies beyond 20 Hz. With the values shown the inverse RIAA curve falls within 0.75 dB of Table 2.11.1.

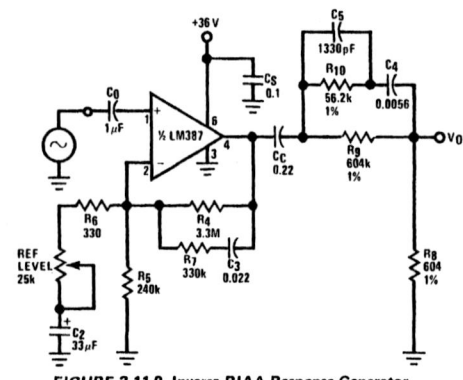

FIGURE 2.11.9 Inverse RIAA Response Generator

2.12 TAPE PREAMPLIFIERS

2.12.1 Introduction

A simplified diagram of a tape recording system is shown in Figure 2.12.1. The tape itself consists of a plastic backing coated with a ferromagnetic material. Both the record and erase heads are essentially inductors with circular metal cores having a narrow gap at the point of contact with the tape. The tape coating then forms a low reluctance path to complete the magnetic circuit. As the tape moves across the record head gap, the magnetic field at the trailing edge of the gap leaves the tape coating permanently magnetized with a remanent flux level (Φ_R) proportional to the signal current in the record head windings.

The bias and erase currents (I_B and I_E) are constant amplitude and frequency waveforms (between 50 kHz and 200 kHz) generated by the bias oscillator. In the erase head, the amplitude of the waveform (from 30 Volts to 150 Volts typically) will determine the degree to which previously recorded signals are "erased" from the tape — in a good machine this will be from 60 dB to 75 dB below the normal recording level. This same waveform, reduced in amplitude to between 5 to 25 times the maximum recording signal level, is used in the record head to determine the "operating point" of the magnetic recording process. Distortion, maximum output level and sensitivity are strong functions of the bias level. To gain some insight into the need for bias, let us take a closer look at the recording process.

Figure 2.12.2(a) shows the permanent magnetization Br (or remanent flux) of a short section of magnetic tape, obtained by applying a magnetizing field H produced by a dc current in the record head winding. This curve is clearly non-linear and if an ac signal current was used in the head winding a highly distorted recording would be made. One solution would be to apply a steady dc bias to the record head along with the ac signal so that the tape was always magnetized in a linear region of the curve (between points A and B for example). This method, called *dc bias*, uses only one part of the curve and reduces the distortion but has a very poor S/N ratio. An improvement may be obtained by pre-magnetizing the tape to saturation and using a dc bias on the record head to bring the magnetization back to zero. Even so, S/N ratios above 30 dB are not easy to achieve.

For high S/N ratios and low distortion, another method called *ac bias* is used.

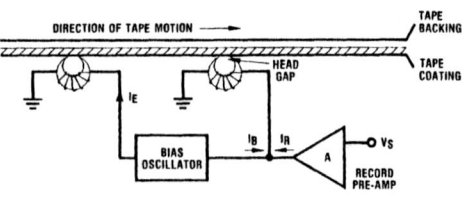

FIGURE 2.12.1 Simplified Recording System

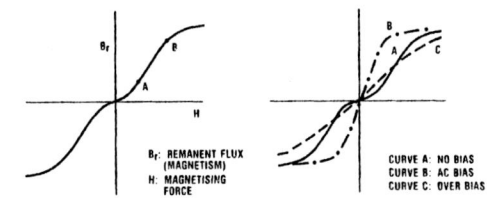

(a) Br-H Curve for Recording Tape (b) Br-H Curve with AC Bias

FIGURE 2.12.2

Figure 2.12.2(b) shows the remanent flux characteristic when a high level ac magnetic field is applied along with the signal. The sensitivity of the tape (curve B) has increased and the magnetization is a linear function of the signal over a much wider range. Note however, that if the bias signal is increased even more (curve C), the tape sensitivity falls off and the non-linearity increases again. The choice of "best" bias current level will depend on a number of factors including the characteristics of the tape and the record/playback heads. Also the ac bias waveform must be free of even order harmonics as these would add an effective dc component to the bias causing distortion for large signal swings and degrading the S/N ratio.

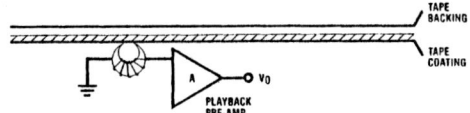

FIGURE 2.12.3 Simplified Playback System

Figure 2.12.3 shows a simplified diagram of a tape playback system. In a two head system (the majority of cassette recorders) the playback head also functions as the record head with appropriate switching. A three head system (record/playback/erase) allows monitoring of the actual recorded signal and the playback head gap can be optimized to improve its frequency response.

Magnetic tape is recorded "constant current" — i.e., constant recording current with frequency, implying a constant recorded magnetic flux level for a given signal amplitude at all frequencies. Since the heads can be regarded as primarily inductive, the impedance of a playback head rises at a 6dB/octave rate with respect to increasing frequency. Therefore the signal voltage from the playback head to the playback preamplifier does not have a flat frequency response, but instead shows a steadily increasing level with increasing signal frequency (Figure 2.12.4).

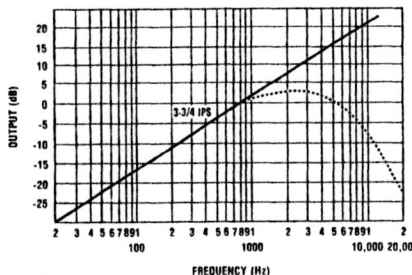

FIGURE 2.12.4 Playback Head Voltage Output vs. Frequency

For practical heads at high frequencies there is an abrupt change in response resulting in a severe decrease in amplitude with a continuing increase in frequency (dashed line on Figure 2.12.4). There are several reasons for this phenomenon — all different and unrelated, but each contributing to the loss of high frequency response. The first area of degradation is due to the effects of the decreasing recorded wavelengths of the higher frequencies.

$$\text{wavelength} = \lambda = \frac{\text{Tape speed (IPS)}}{\text{Frequency (Hz)}} \quad (2.12.1)$$

Two factors are important in minimizing recorded wavelength problems: recording tape speed (Figure 2.12.5) and *playback head* gap width (Figure 2.12.6).

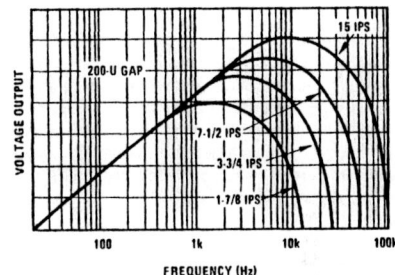

FIGURE 2.12.5 Effect of Tape Speed on Response

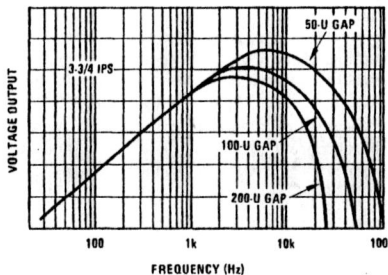

FIGURE 2.12.6 Effect of Head Gap on Response

The first of these is accounted for by the fact that for a given number of flux lines per unit cross-sectional area of the tape (corresponding to a given magnetizing force), higher tape speeds increase the total available flux in the head. For the playback head, when the gap length equals the recorded wavelength (100-U = 100 micro-inches = 0.0001 inches), no output signal is possible since both edges of the gap are at equal magnetic potentials. The gap loss for any given playback head gap and recorded wavelength can be calculated from Equation (2.12.2).

$$\text{Gap Loss (dB)} = 20 \log_{10} \frac{\sin \pi R}{\pi R} \quad (2.12.2)$$

where: $R = \dfrac{\text{Gap Width}}{\text{Wavelength}}$

Table 2.12.1 gives the calculated gap losses for typical gap widths at 1-7/8 I.P.S. and 3-3/4 I.P.S. tape speeds.

Tape Speed (IPS)	Gap Width Micro-inches	Gap Loss with Signal Freqency (dB)				
		1kHz	2kHz	4kHz	8kHz	16kHz
1-7/8	50 – U	−0.01	−0.04	−0.16	−0.66	−2.78
	100 – U	−0.04	−0.16	−0.66	−2.78	−15.61
3-3/4	100 – U	−0.01	−0.09	−0.16	−0.66	−2.78
	160 – U	−0.03	−0.10	−0.42	−1.73	−8.14

TABLE 2.12.1 Playback Head Gap Loss.

Other areas of serious high frequency loss are related to the thickness and formulation of the tape coating material. The thickness of the tape coating contributes to high frequency loss since only the surface layers of the coating contribute measurably to the recording of shorter wavelengths. As the signal frequency increases this effect becomes more pronounced and can be approximated as a −6dB/octave roll-off with a corner frequency equivalent to a time constant T given by:

$$T = \frac{\text{Magnetic Coating Thickness}}{\text{Tape Speed}} \quad (2.12.3)$$

The particular coating formulation used affects the high frequency response because as the magnetic flux variations increase in intensity, a point is reached at which the tape saturates and higher flux levels cannot produce a corresponding higher permanent magnetization of the tape. This effect is particularly significant at higher freqencies and can be explained by regarding the tape coating material as a large number of individual bar magnets in line with each other. At higher frequencies more of these bar magnets are recorded per inch of tape: thus each one grows shorter. As the effective length of the bar magnets decrease, more and more magnetic cancellation occurs due to the close proximity of north and south poles — hence self demagnetization and weaker recorded signals.

Finally the ac bias current used to avoid tape distortion will also contribute to high frequency loss — the technical term is bias erasure and can be significant.

2.12.2 Frequency Equalization

If the tape/head system were "ideal", the application of an unequalized signal current I_R to the recording head would result in a recorded flux Φ_R on the tape exactly proportional to the input signal current and independent of frequency — Figure 2.12.7(a). Similarly the recorded flux on the tape would produce a playback current I_p also independent of frequency. The voltage on the playback head terminals will be proportional to the rate of change of flux (Figure 2.12.7(b)). To compensate for this 6dB/octave rise in amplitude with increasing frequency, the playback preamplifier is equalized for the response shown in Figure 2.12.7(c).

Because of the miscellaneous losses in a real tape recording system further frequency equalization is necessary in both the record and playback preamplifiers. Additionally, despite variations in tape formulations and thicknesses, there are internationally recognized frequency equalization standards (similar to the R.I.A.A. equalization standard for phonograph discs — see Section 2.11.2). For open-reel and cartridge tape formats the N.A.B. standard reproducing characteristic is shown in Figure 2.12.8.

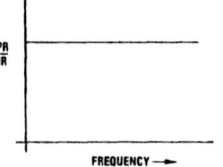

a) TAPE FLUX FOR CONSTANT RECORDING CURRENT

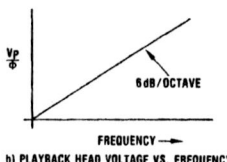

b) PLAYBACK HEAD VOLTAGE VS. FREQUENCY

c) PLAYBACK PRE-AMP RESPONSE

FIGURE 2.12.7 "Ideal" Record/Play System

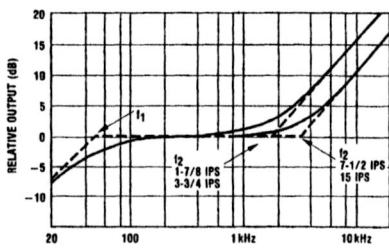

FIGURE 2.12.8 N.A.B. Standard Reproducing Characteristic

Regardless of tape speed the lower frequency (f_1) corner is 50 Hz, below which the amplifier output (for constant flux in the head) should fall off at a −6dB/octave rate. For tape speeds 1-7/8 I.P.S. and 3-3/4 I.P.S., the upper corner frequency (f_2) is 1.77 kHz, above which the output amplitude increases at a +6dB/octave rate. For 7-1/2 I.P.S. and 15 I.P.S. tape speeds, the upper corner frequency (f_2) is 3.18 kHz. Cassette format tapes are equalized to a slightly different standard discussed in a later section.

Ignoring tape/head losses for the moment, if we take the N.A.B. standard response and add the integrating function necessary to compensate the dΦ/dt characteristic of the playback head, we arrive at the overall playback preamplifier frequency response shown in Figure 2.12.9 (the 0dB reference amplitude is defined at the upper corner frequency f_2).

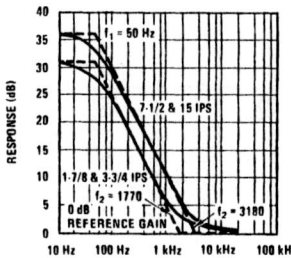

FIGURE 2.12.9 N.A.B. Playback Equalization Including Integration

In the same way the N.A.B. record characteristic will be the complement of the N.A.B. playback characteristic of Figure 2.12.8. This is shown in Figure 2.12.10 where the record current is boosted at +6dB/octave below 50Hz and cut at 6dB/octave above either 1.77kHz or 3.18kHz, depending on the tape speed.

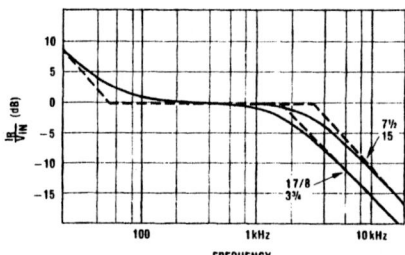

FIGURE 2.12.10 N.A.B. Record Equalization

Both the record and the playback preamplifier responses can be modified to accommodate the losses previously described, but in order to ensure compatibility of tapes recorded to N.A.B. standards on one machine with playback on a different machine, the necessary equalization for losses is obtained in a specific way. In the record preamplifier compensation is made for record head losses and the high frequency losses caused by a particular tape formulation and coating thickness, as well as providing the inverse of the N.A.B. playback characteristic and a current drive to the record head. The playback preamplifier is used to provide the standard playback characteristic, to compensate for playback head gap losses and to integrate the playback head voltage.

2.12.3 LM381 OR LM387 Tape Record Preamp

The frequency response of the record preamp is the complement of the N.A.B. playback equalization including record head and tape loss compensation. A practical design method is to equalize the playback preamplifier first for a flat response with a standard equalized reference tape of the type to be used. Then the record preamplifier is equalized using the same tape formulation and appropriate bias level adjustments for an overall flat response between record and playback. However, to illustrate where loss compensation occurs we will first design a record preamplifier for a 3-3/4 I.P.S. tape speed.

Figure 2.12.11 shows a typical response curve for an eight track record/play head at 3-3/4 I.P.S., obtained with an input signal level −12dB below tape saturation, peak biased at 1kHz and with no record and playback equalization. For a head gap width of 100−U, Table 2.12.1 shows that the playback head gap loss at 16Hz is still under −3dB. Using Equation (2.12.2) for a 440 microinch thick tape coating at 3-3/4 I.P.S.;

$$T = \frac{440 \times 10^{-6}}{3.75} = 117 \mu \text{Seconds}$$

or f − 3dB = 1.36kHz.

From this, we would expect the tape thickness loss to predominate, causing a −6dB/octave fall in frequency response above 1.4kHz. This is confirmed by Figure 2.12.11 where the overall response rises at 6dB/octave to 1.36kHz (f_2)

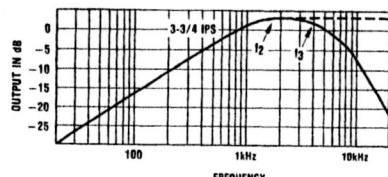

FIGURE 2.12.11 Record/Playback Head Frequency Response

and is then flattened out until head and tape high frequency losses cause the output to fall off (f_3). Note that the −3dB corner frequency of 1.36kHz corresponds almost exactly to the corner frequency (f_2) required by the N.A.B. standard for 3-3/4 I.P.S. tape speed, which calls for a −6dB/octave response (Figure 2.12.10). As a result, out preamplifier will not require any equalization for the N.A.B. standard but will require compensation above 4kHz for the head and other tape losses. This can be obtained with a 6dB/octave boost from 4kHz up to the upper desired frequency limit for the system. The response required can be met with the circuit shown in Figure 2.12.12.

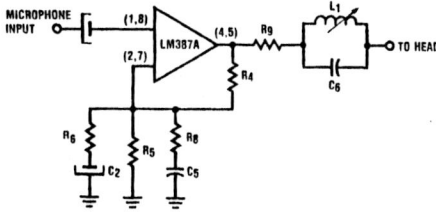

Resistors R_4 and R_5 set the dc bias and resistor R_6 and capacitor C_2 set the mid band gain as before (see Section 2.6). Capacitor C_5 sets the +3dB corner frequency f_3 at which the preamplifier compensates for the head losses.

$$f_3 = \frac{1}{2\pi C_5 R_6} \quad (2.12.4)$$

The preamp gain increases at +6dB/octave above f_3 until the desired high freqency cut-off is reached (f_4)

$$R_8 = \frac{1}{2\pi f_4 C_5} \quad (2.12.5)$$

Resistor R_9 is chosen to provide the proper head recording current

$$R_9 = \frac{V_O}{I_{R(MAX)}} \quad (2.12.6)$$

L_1 and C_6 form a parallel resonant trap at the bias frequency to present a high impedance to the record bias waveform and prevent intermodulation distortion.

Example 2.12.1

A recorder having a 24V power supply uses recording heads with the response characteristic of Figure 2.12.11 requiring 38μA ac drive current. A microphone of 10mV peak output is used. Single ended input is required for best noise performance.

Solution.

1. From Equation (2.6.5) let R_5 = 1.2kΩ.

2. Equation (2.6.6)

$$R_4 = \left(\frac{V_{CC}}{1.3} - 1\right) R_5$$

$$R_4 = \left(\frac{24}{1.3} - 1\right) 1200$$

$$R_4 = 2.09 \times 10^4 \cong 22\text{k}\Omega$$

3. The maximum output of the LM381 is (V_{CC} − 2V) p-p. With a 24V power supply, the maximum output is 22V (p-p) or 7.8V_{RMS}. Therefore an output swing of 6V_{RMS} is reasonable. For the specified head with a proper recording bias level a record current I_R of 38μA gives a recorded level −12dB below saturation. The preamp output should be able to deliver about four times this current without distortion. Therefore $I_{R(MAX)}$ = 0.152mA.

From Equation (2.12.6)

$$R_9 = \frac{V_O}{I_{R(MAX)}}$$

$$R_9 = \frac{6}{0.152 \times 10^{-3}} = 39.5 \times 10^3 \cong 39\text{k}\Omega$$

4. Let the high frequency cut-off be at 16kHz. Since the recording head response begins to fall off at 4kHz, the preamp gain should increase at 6dB/octave for the two octaves between 4kHz and 16kHz. If we allow 6V_{RMS} output voltage swing, then the peak gain = $\frac{6}{10 \times 10^{-3}}$ = 600 or 55.6dB

The midband gain is 12dB below this or 43.6dB (151V/V)

5. From Equation (2.6.9) the midband gain is

$$\frac{R_4 + R_6}{R_6} = 151$$

$$R_6 = \frac{R_4}{150} = \frac{22 \times 10^3}{150} = 146.7$$

R_6 = 150Ω

6. Equation (2.6.10)

$$C_2 = \frac{1}{2\pi f_o R_6} = \frac{1}{6.28 \times 30 \times 150}$$

$C_2 \cong 33\mu F$

7. Equation (2.12.4)

$$C_5 = \frac{1}{2\pi f_3 R_6} = \frac{1}{6.28 \times 4 \times 10^3 \times 150}$$

$C_5 \cong 0.27\mu F$

8. Equation (2.12.5)

$$R_8 = \frac{1}{2\pi f_4 C_5} = \frac{1}{6.28 \times 16 \times 10^3 \times 2.7 \times 10^{-7}}$$

R_8 = 36.8 ≅ 39Ω

The completed circuit is shown in Figure 2.12.13 with the addition of the bias trap $L_1 C_6$. R_{10} and C_9 couple the bias waveform from the bias oscillator to the head with R_{10} being used to adjust the actual bias level. For the specified head, peak bias current is 0.48mA with a head bias impedance of 27kΩ. Therefore the bias voltage on the head will be around 13V_{RMS}. To allow adjustment by R_{10}, the bias oscillator should be able to deliver about 38V_{RMS} or around 100V(p-p). Over biasing will reduce distortion but also cause a drop in high frequency response. Under biasing will allow the high frequencies to be increased but at the expense of higher distortion (the bias level used in cassette recorders is much more critical as discussed later).

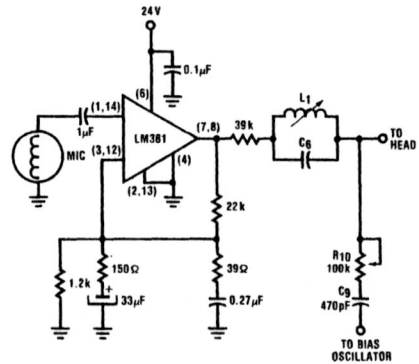

FIGURE 2.12.13 Typical Tape Recording Amplifier

2.12.4 LM387 OR LM381 Tape Playback Preamps

The N.A.B. playback response of Figure 2.12.8 can be obtained with the circuit of Figure 2.12.14. Resistors R_4 and R_5 set the dc bias according to Section 2.6. The reference gain of the preamp, at the upper corner frequency f_2, is set by the ratio

$$0\text{dB reference gain} = \frac{R_7 + R_6}{R_6} \quad (2.12.7)$$

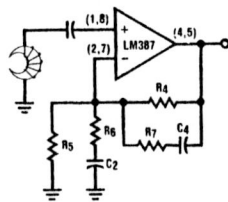

FIGURE 2.12.14 N.A.B. Tape Preamp

The corner frequency f_2 is determined when the impedance of C_4 equals the value of resistor R_7

$$f_2 = \frac{1}{2\pi C_4 R_7} \quad (2.12.8)$$

Corner frequency f_1 is determined when $X_{C_4} = R_4$:

i.e. $f_1 = \frac{1}{2\pi C_4 R_4} \quad (2.12.9)$

The low frequency -3dB roll off point, f_0, is set where $X_{C_2} = R_6$

$$f_0 = \frac{1}{2\pi C_2 R_6} \quad (2.12.10)$$

Example 2.12.2

Using the R/P head specified in Example 2.12.1, design a N.A.B. equalized preamp using the LM387. At 1kHz, 3-3/4 I.P.S. the head sensitivity is $800\mu V$ and the required preamp output is $0.5V_{RMS}$.

Solution

1. From Equation (2.6.3) let $R_5 = 240\text{k}\Omega$.

2. Equation (2.6.4):

$$R_4 = \left(\frac{V_{CC}}{2.6} - 1\right) R_5$$

$$R_4 = \left(\frac{24}{2.6} - 1\right) 2.4 \times 10^5$$

$$R_4 = 1.98 \times 10^6 \approx 2.2\text{M}\Omega$$

3. For a corner frequency, f_1, equal to 50Hz, Equation (2.12.9) is used.

$$C_4 = \frac{1}{2\pi f_1 R_4} = \frac{1}{6.28 \times 50 \times 2.2 \times 10^6}$$

$$= 1.45 \times 10^{-9} \approx 1500\text{pF}$$

4. From Figure 2.12.8, the corner frequency $f_2 = 1770\text{Hz}$ at 3-3/4 IPS. Resistor R_7 is found from Equation (2.12.8).

$$R_7 = \frac{1}{2\pi f_2 C_4}$$

$$R_7 = \frac{1}{6.28 \times 1770 \times 1.5 \times 10^{-9}} = 59.9 \times 10^3$$

$$R_7 \approx 62\text{k}\Omega$$

5. The required voltage gain at 1kHz is:

$$A_V = \frac{0.5 V_{RMS}}{800\mu V_{RMS}} = 6.25 \times 10^2 \text{V/V} = 56\text{dB}$$

6. From Figure 2.12.9 we see the reference frequency gain, above f_2, is 5dB down from the 1kHz value or 51dB (355V/V).

From Equation (2.12.7):

$$0\text{dB Ref Gain} = \frac{R_7 + R_6}{R_6} = 355$$

$$R_6 = \frac{R_7}{355 - 1} = \frac{62\text{k}}{354} = 175$$

$$R_6 \approx 180\Omega$$

7. For low frequency corner $f_0 = 40\text{Hz}$, Equation (2.12.10):

$$C_2 = \frac{1}{2\pi f_0 R_6} = \frac{1}{6.28 \times 40 \times 180} = 2.21 \times 10^{-5}$$

$$C_2 \approx 20\mu F$$

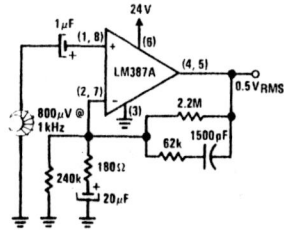

FIGURE 2.12.15 Typical Tape Playback Amplifier

An example of a LM387A tape playback preamp designed for 12 volt operation is shown in Figure 2.12.16 along with its frequency response.

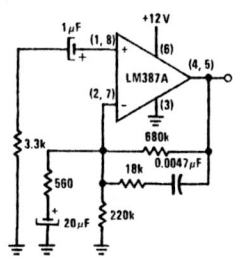

(a) NAB Tape Circuit

FIGURE 2.12.16 (a) LM387 Tape Preamp

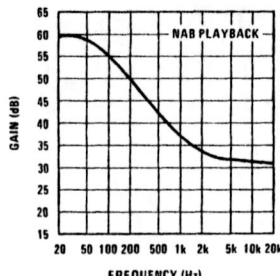

(b) Frequency Response of NAB Circuit
FIGURE 2.12.16 (b) LM387 Tape Preamp

2.12.5 Fast Turn-On NAB Tape Playback Preamp

The circuit shown in Figure 2.12.15 requires approximately 2.5 seconds to turn on for the gain and supply voltage chosen in the example. Turn-on time can closely be approximated by:

$$t_{ON} \approx -R_4 C_2 \ln\left(1 - \frac{1.2}{V_{CC}}\right) \quad (2.12.11)$$

As seen by Equation (2.12.11), increasing the supply voltage decreases turn-on time. Decreasing the amplifier gain also decreases turn-on time by reducing the $R_4 C_2$ product.

Where the turn-on time of the circuit of Figure 2.12.14 is too long, the time may be shortened by using the circuit of Figure 2.12.17. The addition of resistor R_D forms a voltage divider with R_6'. This divider is chosen so that zero DC voltage appears across C_2. The parallel resistance of R_6' and R_D is made equal to the value of R_6 found by Equation (2.12.7). In most cases the shunting effect of R_D is negligible and $R_6' \approx R_6$.

For differential input, R_D is given by:

$$R_D = \frac{(V_{CC} - 1.2) R_6'}{1.2} \quad (2.12.12)$$

For single ended input:

$$R_D = \frac{(V_{CC} - 0.6) R_6'}{0.6} \quad (2.12.13)$$

In cases where power supply ripple is excessive, the circuit of Figure 2.12.17 cannot be used since the ripple is coupled into the input of the preamplifier through the divider.

The circuit of Figure 2.12.18 provides fast turn-on while preserving the 120dB power supply rejection.

The DC operating point is still established by R_4/R_5. However, Equations (2.6.3) and (2.6.5) are modified by a factor of 10 to preserve DC bias stability.

For differential input, Equation (2.6.3) is modified as:

$$R_5 = \frac{2 V_{BE}}{100 \, I_{Q2}} = \frac{1.2}{50 \times 10^{-6}} \quad (2.6.3a)$$

$$= 24 \, k\Omega \text{ maximum}$$

For single ended input:

$$R_5 = \frac{V_{BE}}{50 \, I_{FB}} = \frac{0.6}{50 \times 10^{-4}} \quad (2.6.5a)$$

$$= 120 \, \Omega \text{ maximum}$$

Equations (2.12.7), (2.12.8), and (2.12.10) describe the high frequency gain and corner frequencies f_2 and f_0 as before.

Frequency f_1 now occurs where X_{C4} equals the composite impedance of the R_4, R_6, C_2 network as given by Equation (2.12.14).

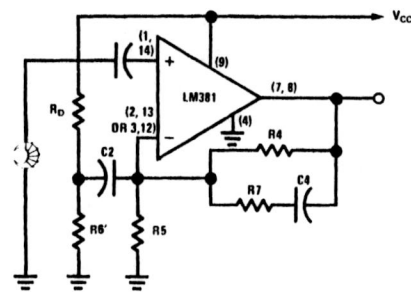

FIGURE 2.12.17 Fast Turn-On NAB Tape Preamp

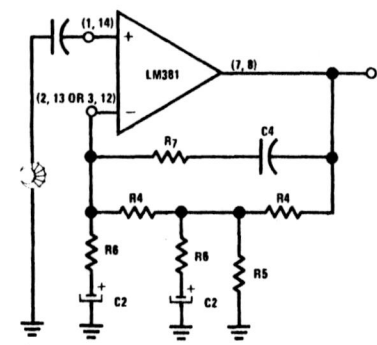

FIGURE 2.12.18 Two-Pole Fast Turn-On NAB Tape Preamp

$$C_4 = \frac{1}{2 \pi f_1 R_6 \left[\left(\frac{R_4 + R_6}{R_6}\right)^2 - 1\right]} \quad (2.12.14)$$

The turn-on time becomes:

$$t_{ON} \approx -2\sqrt{R_4 C_2} \ln\left(1 - \frac{1.2}{V_{CC}}\right) \quad (2.12.15)$$

Example 2.12.3

Design an NAB equalized preamp with the fast turn-on circuit of Figure 2.18.18 for the same requirements as given in Example 2.12.2.

Solution

1. From Equation (2.6.3a) let $R_5 = 24 \, k\Omega$.

2. Equation (2.6.4):

$$R_4 = \left(\frac{V_{CC}}{2.6} - 1\right) R_5$$

$$= \left(\frac{24}{2.6} - 1\right) 24 \times 10^3 = 1.98 \times 10^5$$

$$R_4 \approx 220 \text{k}\Omega$$

3. From Example 2.12.2, the reference frequency gain, above f_2, is 51 dB or 355 V/V.

Equation (2.12.7):

$$\frac{R_7 + R_6}{R_6} = 355$$

4. The corner frequency f_2 is 1770 Hz for 3-3/4 IPS.

Equation (2.12.8):

$$C_4 = \frac{1}{2\pi f_2 R_7}$$

5. The corner frequency f_1 is 50 Hz and is given by Equation (2.12.14).

$$C_4 = \frac{1}{2\pi f_1 R_6 \left[\left(\frac{R_4 + R_6}{R_6}\right)^2 - 1\right]}$$

6. Solving Equations (2.12.7), (2.12.8), and (2.12.14) simultaneously gives:

$$R_6 = \frac{R_4 \left[f_1 + \sqrt{f_1^2 + f_1 f_2 (\text{Ref Gain})}\right]}{f_2 (\text{Ref Gain})} \quad (2.12.15)$$

$$R_6 = \frac{2.2 \times 10^5 (50 + \sqrt{2500 + 50 \times 1770 \times 355})}{1770 \times 355}$$

$$= 1.98 \times 10^3 \approx 2 \text{k}\Omega$$

7. From Equation (2.12.7):

$$R_7 = 354 R_6 = 708 \times 10^3$$

$$R_7 \approx 680 \text{k}\Omega$$

8. Equation (2.12.8):

$$C_4 = \frac{1}{2\pi f_2 R_7} = \frac{1}{6.28 \times 1770 \times 680 \times 10^3}$$

$$C_4 = 1.32 \times 10^{-10} \approx 120 \text{pF}$$

9. Equation (2.12.10):

$$C_2 = \frac{1}{2\pi f_0 R_6} = \frac{1}{6.28 \times 40 \times 2 \times 10^3}$$

$$C_2 = 1.99 \times 10^{-6} \approx 2 \mu F$$

This circuit is shown in Figure 2.12.19 and requires only 0.1 seconds to turn on. Note, however, that the non-inverting input has to charge the head coupling capacitance to 1.2V through an internal 250kΩ resistor and will increase the turn-on time slightly.

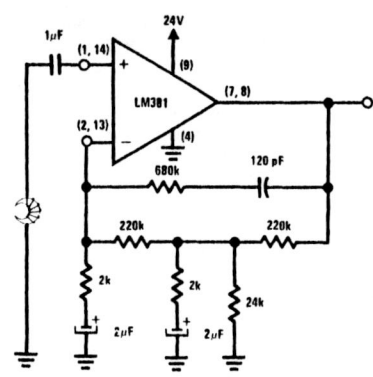

FIGURE 2.12.19

2.12.6 LM382 Tape Playback Preamp

With just one capacitor in addition to the gain setting capacitors, it is possible to design a complete low noise, NAB equalized tape playback preamp (Figure 2.12.20). The circuit is optimized for automotive use, i.e., V_s = 10-15V. The wideband 0dB reference gain is equal to 46dB (200V/V) and is not easily altered. For designs requiring either gain or supply voltage changes the required extra parts make selection of a LM387 a more appropriate choice.

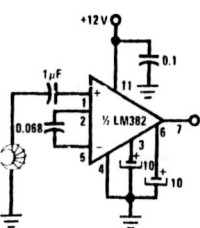

FIGURE 2.12.20 LM382 Tape Preamp (NAB, 1-7/8 & 3-3/4 IPS)

2.12.7 LM1303 Tape Playback Preamp

For split supply applications, the LM1303 may be used as a tape preamp as shown in Figure 2.12.21. Design equations are given below for trimming or alteration purposes. (Frequency points refer to Figure 2.12.9.)

$$0\text{dB Ref Gain} = 1 + \frac{R_2}{R_3} \quad (2.12.16)$$

$$f_1 = \frac{1}{2\pi R_1 C_1} \quad (2.16.17)$$

$$f_2 = \frac{1}{2\pi R_2 C_1} \quad (2.12.18)$$

response (e.g., in the play mode the microphone preamp is disabled so that its feedback network will not affect the playback preamp frequency response). Similarly, a common monitor amplifier drives two output stages for either the record amplifier function or for output signal amplification when in the play mode. Because only one output stage can be active at any time, both feedback networks can be connected to the monitor amplifier inverting input.

In the record mode — monitor amplifier output pin 10 active — the microphone amplifier output is connected to the ALC circuit as well as the monitor amplifier input. Above a predetermined threshold signal level the ALC circuit operates to attenuate the microphone signal and maintain a relatively constant level — a useful function in speech recorders. The rectifier and peak detector of the ALC circuit is also used to develop a recording level meter drive.

2.12.10 LM1818 Microphone and Playback Preamplifiers

Both the microphone and playback amplifiers are similar in design (Figure 2.12.28). The non-inverting inputs, pins 16 and 17, are biased at $1.2V_{DC}$ through $50k\Omega$ resistors. Normally these resistors would also source current ($24\mu A$) at turn-on for the input capacitor from the tape head. This capacitor is usually selected to be large in value to give a low impedance at low frequencies which will minimize the amplifier input noise current degrading the system noise figure. To prevent long turn-on times, an internal circuit will source $200\mu A$ to pin 17 at turn-on, enabling capacitors around $10\mu F$ to be used. At the same time, the R/P logic clamps the head to prevent this charge current from magnetizing the head at turn-on. The amplifiers have collector currents of $50\mu A$ — optimized for low noise with typical tape head source impedances — and are internally compensated for closed loop gains greater than 5. In the "Record" mode, Q_8 is saturated and Q_4 is "off" so that the microphone signal is amplified to the output stage Q_9. Pin 2 is held low, clamping the R/P head and sinking the bias current flowing in the head during the record mode.

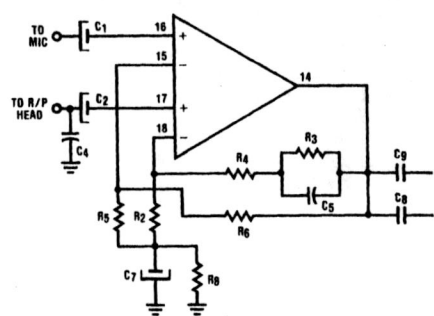

FIGURE 2.12.29 Microphone and Playback Preamplifier Feedback Networks

In the circuit of Figure 2.12.29, the microphone and playback amplifiers show the same low frequency roll-off capacitor C_7. Both inverting inputs are referenced at $1.2V - 0.7V = 0.5V_{DC}$. The output quiescent point, pin 14, is established by negative feedback through the external divider $(R_6 + R_5)/R_8$. For bias stability the current through R_8 is made ten times the current from the inverting inputs (pin 15 or pin 18). This current is the input stage collector current of $50\mu A$.

$$\therefore R_8 = \frac{0.5}{10 \times 50 \times 10^{-6}} = 1k\Omega \text{ maximum} \qquad (2.12.19)$$

For low values of R_2 and R_5, and large values of R_3 and R_4

$$(R_5 + R_6) = (2V_{DC} - 1)R_8 \qquad (2.12.20)$$

where V_{DC} = pin 14 voltage.

For the playback preamplifier, the midband gain above corner frequency f_2 (Figure 2.12.24).

$$A_{VAC} = 1 + \frac{R_4}{R_2} \qquad (2.12.21)$$

The low frequency corner f_1 (Figure 2.12.24) is determined where the impedance of C_5 equals R_3

$$f_1 = \frac{1}{2\pi C_5 R_3} \qquad (2.12.22)$$

The upper corner frequency f_2 is determined by

$$f_2 = \frac{1}{2\pi R_4 C_5} \qquad (2.12.23)$$

For a tape recorder to accomodate either ferric or CrO_2 tapes, Equation (2.12.23) has two solutions, depending on whether f_2 is 1.33kHz or 2.27kHz. Since C_5 also sets the lower corner frequency of f_1 which is common to either type of tape, R_4 should be switched. This will decrease the midband gain above f_2 [Equation (2.12.21)] for CrO_2 tapes by about 4.7dB.

In the "Record" mode the microphone amplifier is on and the dc output level at pin 14 is given by

$$V_{DC} = \frac{R_5 + R_6 + R_8}{2(R_5 + R_8)} \qquad (2.12.24)$$

The AC voltage gain is given by

$$A_{VAC}(MIC) = 1 + \frac{R_6}{R_5} \qquad (2.12.25)$$

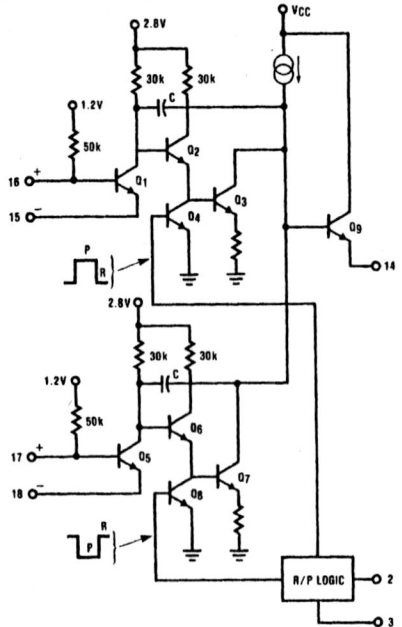

FIGURE 2.12.28 Microphone and Playback Amplifiers

The output amplifiers (Figure 2.12.30) are used to provide a low impedance drive to an audio power amplifier stage when in the "Play" mode, or to provide the necessary equalization for the tape head in the "Record" mode. Again the R/P logic (shown as switches in Figure 2.12.30) decides which stage will be active and therefore which feedback network is operational. The input from the microphone or playback preamplifiers is coupled to the non-inverting input of the differential pair Q_{10}, Q_{11}. The base of Q_{10} is biased externally via R_{10} from the supply voltage divider R_{11} and R_{12}. This divider is normally designed to place the amplifier quiescent outputs at half supply voltage to maximize the output signal swing capability. Both output stages Q_{16} and Q_{18} are Class A amplifiers with active current source loads Q_{15} and Q_{17}. The position of the R/P logic switches determines which current source is active and delivering 700µA.

The gain for the audio output stage (pin 9 output) is set by the ratio of R_{14} and R_{16}.

$$\text{Playback gain } V_{AC} = 1 + \frac{R_{16}}{R_{14}} \quad (2.12.26)$$

The low frequency 3dB corner f_0 is given by

$$f_0 = \frac{1}{2\pi C_{12} R_{14}} \quad (2.12.27)$$

When the R/P logic is in the record mode, pin 10 output is active and the midband gain

$$A_{V(REC)} = 1 + \frac{R_{15}}{R_{14}} \quad (2.12.28)$$

The resistor R_{22} provides the proper head recording current

$$R_9 = \frac{V_{O(MAX)}}{I_{R(MAX)}} \quad (2.12.29)$$

For inexpensive, monaural cassette recorders, C_{17} is used to shunt R_{22} to compensate for high frequency losses (Figure 2.12.22)

$$f_3 = \frac{1}{2\pi C_{17} R_{22}} \quad (2.12.30)$$

C_{26} is used to filter the bias waveform at the output from the record amplifier (in place of the more expensive bias trap of Figure 2.12.13).

During the record mode the output from the microphone preamplifier is used to supply a signal to the ALC circuit and meter drive circuit. In certain recording situations — speech for example where the speaker can vary in distance from the microphone — it is convenient to have a circuit (ALC), which continuously and automatically adjusts the overall gain of the recording (microphone) amplifier in order to maintain a proper recording signal level that is neither buried in noise nor high enough to cause tape saturation.

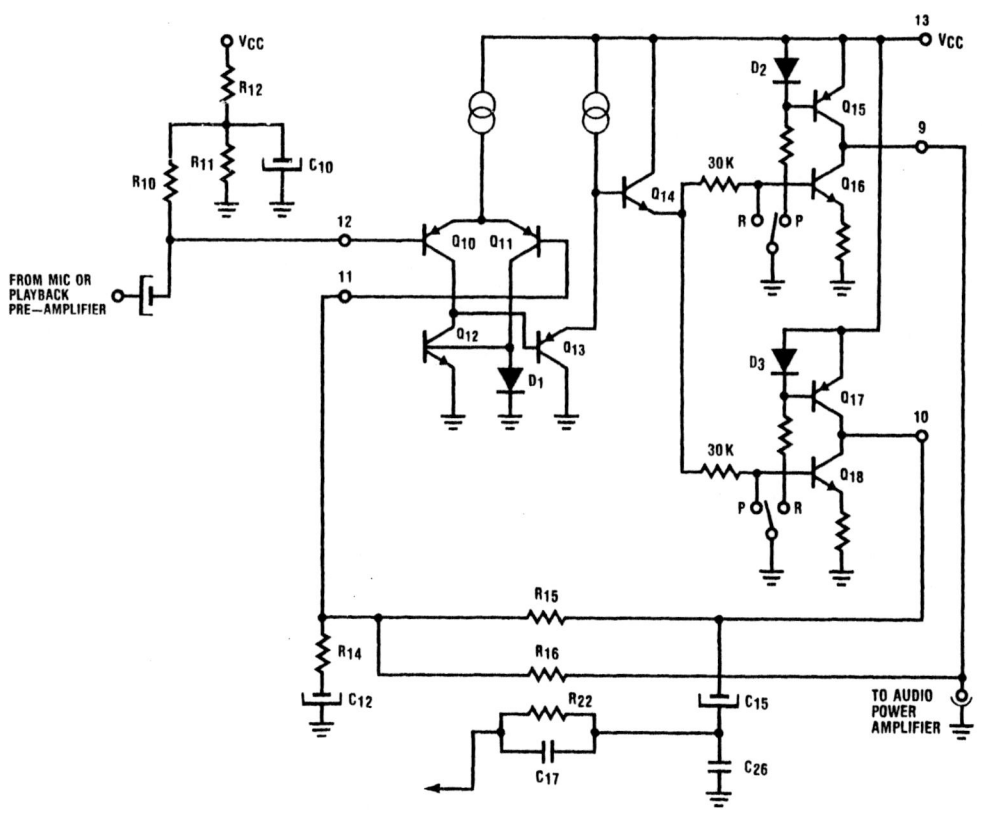

FIGURE 2.12.30 LM1818 Monitor and Output Amplifier
With Feedback and Biasing Components

Referring to Figure 2.12.31, the input signal is rectified by the action of Q_{19}, Q_{20} and Q_{21}. Q_{20} and Q_{21} are taking most of the current from the $50\mu A$ current source so that Q_{19} is "on" just enough to provide base current. When a signal is applied at pin 4, the negative swings will cut-off Q_{20} and Q_{21} allowing Q_{19} to conduct. The current that then flows in the $2k\Omega$ resistor at pin 4 is "mirrored" by Q_{22} and will develop a voltage across the $20k\Omega$ resistor in series with diodes D_4 through D_6. In the absence of signal, a current source biases the filter capacitor C_{13} through diode D_8 to two diode voltage drops above ground. When the signal current in the $20k\Omega$ resistor causes Q_{23} base to reach four diode voltage drops above ground Q_{26} can turn on. Q_{26} operates in a saturated mode and can sink or source current for small positive collector-emitter voltages. By connecting pin 5 to the microphone Q_{26} behaves as a variable resistor working against the $10k\Omega$ resistor (R_1) in series with the microphone to attenuate the microphone signal when it is above the ALC threshold. Typical ALC response curves are shown in Figure 2.12.32.

The same half wave rectifier is used to supply an input to the meter drive circuit, Q_{27} through Q_{21} in Figure 2.12.33. Since Q_{27} base is connected back to the collector of Q_{22}, with no signal present, the meter memory capacitor is held two diode voltage drops above ground – therefore the output (pin 8) starts at ground. For small signals, the memory capacitor C_{14} is charged by Q_{27} and discharged by R_{18} and a constant $50\mu A$ discharge current in Q_{30}. This allows fast, accurate response in the lower portion of the meter range. When larger signals cause the output voltage at pin 8 to get above 0.7V, Q_{30} is shut off. This will normally correspond to a recording signal level at 0"VU" and the increased discharge time (since there is no $50\mu A$ discharge current and only R_{18} is discharging C_{14}) allows more time for high recording levels to be identified. Should the signal level increase further to around 1.0V at pin 8 (+3"VU" for example), Q_{31} becomes active and rapidly discharges C_{14} to prevent damage to the meter movement.

The meter calibration is performed by setting the series resistor to produce 0"VU" on the meter scale when the voltage at pin 8 is $0.7V_{DC}$. This voltage level is obtained with a $70mV_{RMS}$ signal at pin 4.

Example 2.12.3

Using a combined R/P cassette tape head with a response similar to that shown in Figure 2.12.22, design a portable monaural record-playback system to operate from 6V supplies. The head sensitivity in playback is $0.3mV_{RMS}$ for a 1kHz signal recorded –12dB below tape saturation, obtained with a $60\mu A$ record current level. An output of 250mW into an 8Ω speaker is required, with a system bandwidth from 80Hz to 10kHz. A microphone with a 1.8mV output level will be used.

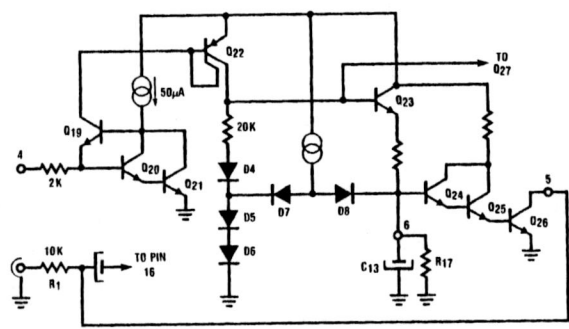

FIGURE 2.12.31 LM1818 Auto Level Circuit

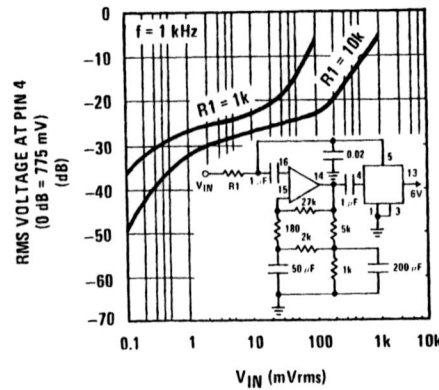

FIGURE 2.12.32 Automatic Level Control (ALC) Response Characteristics

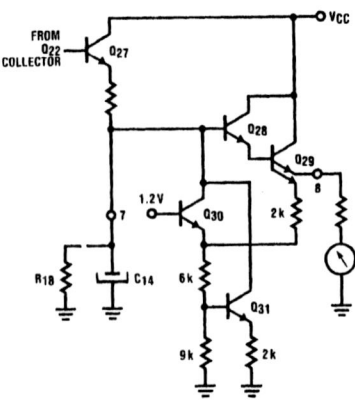

FIGURE 2.12.33 LM1818 Meter Drive Circuit

Solution

We will start by designing the playback system using the circuit configuration of Figure 2.12.29

1. From Equation (2.12.19)

   $R_8 = 1k\Omega$

2. To set the playback preamp output (Pin 14) at half the supply voltage

   $V_{DC} = \dfrac{6V}{2} = 3$ Volts

   Using Equation (2.12.20)

   $(R_5 + R_6) = (2V_{DC} - 1)R_8$
   $= 5k\Omega$

3. The value of R_2 should be low to minimize the feedback path noise contribution, but not so low that C_7 is excessively large to maintain the low frequency response. A suitable value is

   $R_2 = 180\Omega$

4. $C_7 = \dfrac{1}{2\pi \times 10 \times 180}$

   $= 88.4 \times 10^{-6}$

   Put $C_7 = 100\mu F$

5. When the level meter reads 0"VU", the signal level at pin 4 is 70mV$_{RMS}$. For a typical cassette recorder, 0"VU" is usually chosen to leave about 3dB to 6dB headroom before tape saturation. For the given head sensitivity at 1kHz, -3dB corresponds to 0.85mV$_{RMS}$.

   Playback preamplifier gain at 1kHz
   $= \dfrac{70}{0.85} = 82$V/V or 38.4dB

   The midband gain (above the equalization corner frequency f_2) for ferric tapes is 2.4dB below the 1kHz gain — Figure 2.12.24. Therefore the midband gain is $38.4 - 2.4 = 36$dB or 63V/V.
   Using Equation (2.12.21)

   $63 = 1 + \dfrac{R_4}{R_2}$

   $R_4 = 180 \times 62 \approx 12k\Omega$

6. Using Equation (2.12.23)

   $f_2 = \dfrac{1}{2\pi C_5 R_4}$

   $C_5 = \dfrac{1}{2\pi \times 1326 \times 12 \times 10^3}$

   $= 1 \times 10^{-8} = 0.01\mu F$

7. Equation (2.12.22)

   $f_1 = \dfrac{1}{2\pi C_5 R_3}$

   $R_3 = \dfrac{1}{2\pi .50 \times 0.01 \times 10^{-6}}$

   $= 318k\Omega$

   Put $R_3 = 330k\Omega$

8. To compensate for the playback head gap loss, C_4 is resonated with the playback head inductance to give a 4.5dB boost at 10kHz. The value of C_4 is best determined empirically using a calibrated test tape. Typically,

   $C_4 = 470$pF

9. In the record mode, we would like the microphone to produce the same level at pin 4, so that for a given flux level on the tape, the meter will indicate the same reading both in record and playback.

   Equation (2.12.25)

   $A_{VAC(MIC)} = 1 + \dfrac{R_6}{R_5}$

   $\dfrac{70}{1.8} = 1 + \dfrac{R_6}{R_5}$

   $R_6 = 39R_5$

   From step (2)

   $R_5 + R_6 = 5k\Omega$
   $\therefore R_5 = 120\Omega$
   $R_6 = 4.7k\Omega$

10. The gain required in the monitor amplifier during playback will depend on the sensitivity of the audio power amplifier driving the speaker. Using the LM386 amplifier in the configuration shown in Figure 2.12.34 (see also Section 4.7), an input level of 100mVrms is required to obtain 250mW output power in an 8Ω load. To ensure adequate playback volume at lower recording levels the monitor amplifier should provide this level to the volume control with head output signals — 12dB below 0"VU". Using Equation (2.12.26)

    $A_{V(MON)} = 6 = 1 + \dfrac{R_{16}}{R_{14}}$

    $R_{16} = 5R_{14}$

    Put $R_{14} = 50k$, $R_{16} = 250k\Omega$

11. For a quiescent output dc voltage of 3V

    $R_{11} = R_{12} = 10k\Omega$

    Put $R_{10} = 250k\Omega$ to balance input bias currents.

12. In the record mode, the monitor amplifier input from the microphone preamplifier is 70mV$_{RMS}$ for 0"VU" on the meter. Since the monitor amplifier output can typically swing 1.65V$_{RMS}$ on a 6V supply, Equation (2.12.28) gives the required gain as

    $A_{V(REC)} = 1 + \dfrac{R_{15}}{R_{14}}$

    $\dfrac{1.65}{70 \times 10^{-3}} = 1 + \dfrac{R_{15}}{R_{14}}$

    $R_{15} = 23R_{14}$

    $R_{15} = 1m\Omega$

13. A record current of 60μA produces a tape flux level −12dB below saturation. Therefore, for our specified 0"VU" level −3dB below saturation

$I_R = 170\mu A$

Equation (2.12.29)

$R_{22} = \dfrac{1.65}{170 \times 10^{-6}} = 10k\Omega$

14. The high frequency cut-off is 10kHz. The head response begins to fall off at 2.5kHz and is −16dB at 10kHz. Since 4.5dB of this loss is compensated for on playback (see step 8), the remaining −12.5dB loss is compensated for during recording by shunting R_{22} with C_{17}.

Equation (2.12.30)

$C_{17} = \dfrac{1}{2\pi.2.5 \times 10^3 \times 10 \times 10^3}$

$= 0.0068\mu F$

The complete schematic is shown in Figure 2.12.34 with the measured record/play response in Figure 2.12.35. The S/N ratio (CCIR/ARM weighted) is 59.5dB for the electronics, and 54dB with tape. Total current consumption (excluding tape drive and power amp) is 10.7mA in play, increasing to 45mA during record due to the bias oscillator.

No discussion of cassette recorder design would be complete without mention of the noise reduction systems that are a major contribution to the acceptability of the cassette format in higher quality applications. At present the most popular is the complementary noise reduction system developed by Dolby Laboratories — the consumer version of which is the Dolby® B-Type. The National LM1011/1011A is specifically designed to implement the functions of the Dolby B system, and as such is found in many commercially available cassette recorders. However, it should be emphasized that the use of the LM1011/1011A in Dolby systems is by license agreement with Dolby Labs* and LM1011's are available only to licensed manufacturers.

An alternative, non-complementary system suitable for cassette recorders is described in Section 5.8.

*License information available from Dolby Laboratories Licensing Corporation
731 Sansome St., San Fransisco, CA 94111

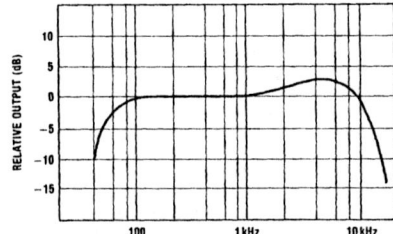

FIGURE 2.12.35 Record/Play Frequency Response

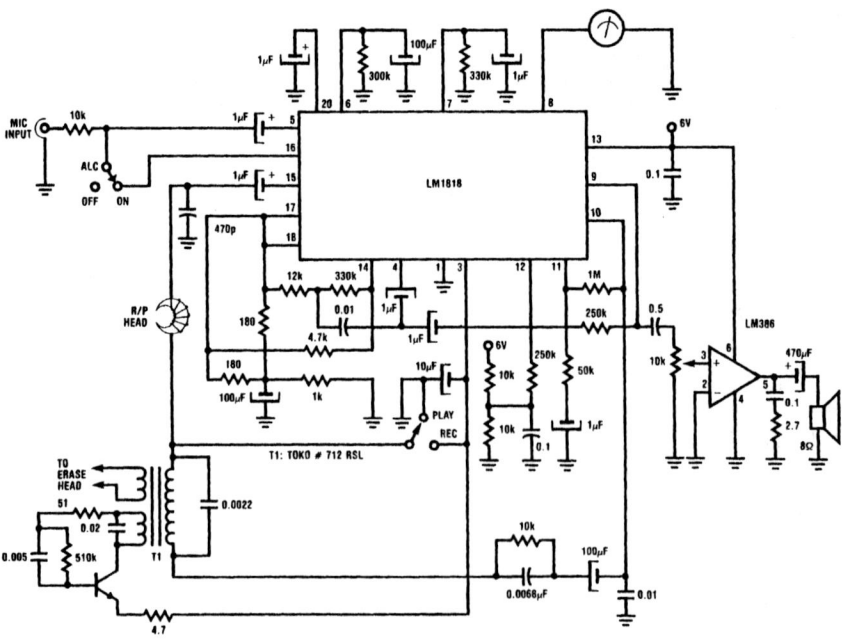

FIGURE 2.12.34 Monaural Cassette Recorder

2.13 MIC PREAMPS

2.13.1 Introduction

Microphones classify into two groups: high impedance (~20kΩ), high output (~20mV); and low impedance (~200Ω), low output (~2mV). The first category places no special requirements upon the preamp; amplification is done simply and effectively with the standard non-inverting or inverting amplifier configurations. The frequency response is reasonably flat and no equalization is necessary. Hum and noise requirements of the amplifier are minimal due to the large input levels. If everything is so easy, where is the hook? It surfaces with regard to hum and noise pickup of the microphone itself. Being a high impedance source, these mics are very susceptible to stray field pickup (e.g., 60Hz). Their use must be restricted to short distances (typically less than 10 feet of cable length), because of the potential high frequency roll-off caused by cable capacitance.

Low impedance microphones also have a flat frequency response, requiring no special equalization in the preamp section. Their low output levels do, however, impose rather stringent noise requirements upon the preamp. For a signal-to-noise ratio of 65dB with a 2mV input signal, the total equivalent input noise (EIN) of the preamp must be 1.12μV (10-10kHz). National's line of low noise dual preamps with their guaranteed EIN of ≤ 0.7μV (LM381A) and ≤ 0.9μV (LM387A) make excellent mic preamps, giving at least 67dB S/N (LM387A) performance (re: 2mV input level).

Low impedance mics take two forms: unbalanced two wire output, one of which is ground, and balanced three wire output, two signal and one ground. Balanced mics predominate usage since the three wire system facilitates minimizing hum and noise pickup by using differential input schemes. This takes the form of a transformer with a center-tapped primary (grounded), or use of a differential op amp. More about balanced mics in a moment, but first the simpler unbalanced preamps will be discussed.

2.13.2 Transformerless Unbalanced Designs

Low impedance unbalanced (or single-ended) mics may be amplified with the circuits appearing in Figure 2.13.1. The LM381A (Figure 2.13.1a) biased single-ended makes a simple, quiet preamp with noise performance −69dB below a 2mV input reference point. Resistors R_4 and R_5 provide negative input bias current and establish the DC output level at one-half supply. Gain is set by the ratio of R_4 to R_2, while C_2 establishes the low frequency −3dB corner. High frequency roll-off is done with C_3. Capacitor C_1 is made large to reduce the effects of 1/f noise currents at low

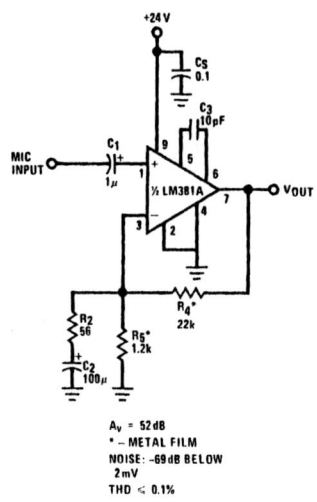

(a) LM381A S. E. Bias

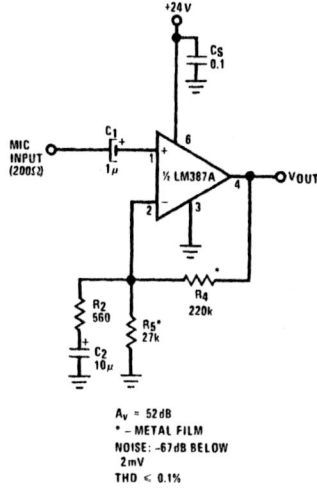

(b) LM387A

FIGURE 2.13.1 Transformerless Mic Preamps for Unbalanced Inputs

frequencies. (See Section 2.6 for details on biasing and gain adjust.)

The LM387A (Figure 2.13.1b) offers the advantage of fewer parts and a very compact layout, since it comes in the popular 8-pin minidip package. The noise degradation referenced to the LM381 is only +2dB, making it a desirable alternative for designs where space or cost are dominant factors. Biasing and gain resistors are similar to LM381A. (See Section 2.8 for details.)

2.13.3 Transformer-Input Balanced Designs

Balanced microphones are used where hum and noise must be kept at a minimum. This is achieved by using a three wire system — two for signal and a separate wire for ground. Proper grounding of microphones and their interconnecting cables is crucial since all noise and hum frequencies picked up along the way to the preamplifier will be amplified as signal. The rationale behind the twisted-pair concept is that all interference will be induced *equally* into each signal wire and will thus be applied to the preamp *common-mode*, while the actual transmitted signal appears differential. Balanced-input transformers with center-tapped primaries and single-ended secondaries (Figure 2.13.2) dominate balanced mic preamp designs. By grounding the center-tap all common-mode signals are shunted to ground, leaving the differential signal to be transformed across to the secondary winding, where it is converted into a single-ended output. Amplification of the secondary signal is done either with the LM381A (Figure 2.13.2a) or with the LM387A (Figure 2.13.2b). Looking back to Figure 2.13.1 shows the two circuits being the same with the exception of a change in gain to compensate for the added gain of the transformer. The net gain equals 52dB and produces 775mV output for a nominal 2mV input. Selection of the input transformer is fixed by two factors: mic impedance and amplifier optimum source impedance. For the cases shown the required impedance ratio is 200:10k, yielding a voltage gain (and turns ratio) of about seven ($\sqrt{10k/200}$).

Assuming an ideal noiseless transformer gives noise performance −80dB below a 2mV input level. Using a carefully designed transformer with electrostatic shielding, rejection of common-mode signals to 60dB can be expected (which is better than the cable manufacturer can match the twisting of the wires).

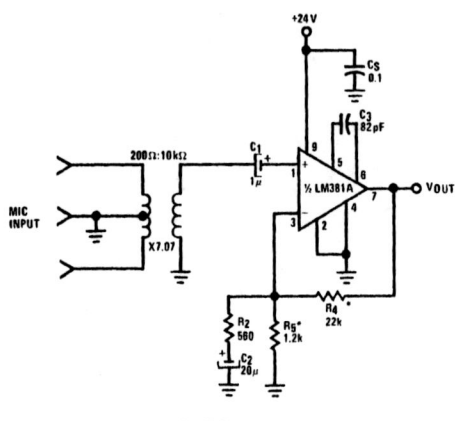

(a) LM381A S. E. Bias

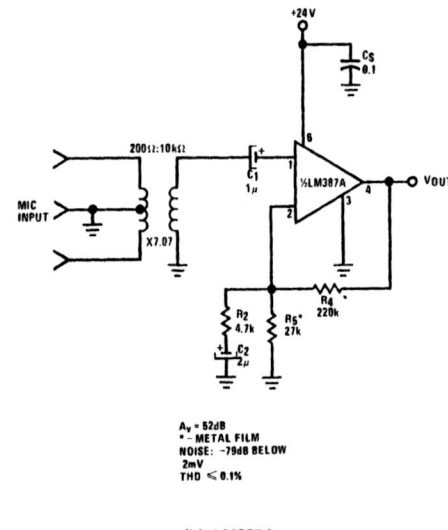

(b) LM387A

FIGURE 2.13.2 Transformer-Input Mic Preamps for Balanced Inputs

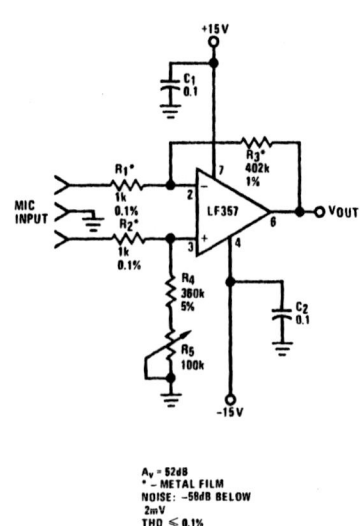

FIGURE 2.13.3 Transformerless Mic Preamp for Balanced Inputs

2.13.4 Transformerless Balanced Designs

Transformer input designs offer the advantage of nearly noise-free gain and do indeed yield the best noise performance for microphone applications; however, when the total performance of the preamplifier is examined, many deficiencies arise. Even the best transformers will introduce certain amounts of harmonic distortion; they are very susceptible to hum pickup; common-mode rejection is not optimum; and not a small problem is the expense of quality input transformers. For these reasons, transformerless designs are desirable. By utilizing the inherent ability of an operational amplifier to amplify differential signals while rejecting common-mode ones, it becomes possible to eliminate the input transformer.

Figure 2.13.3 shows the FET input op amp, LF357 (selected for its high slew rate and CMRR) configured as a difference amplifier. As shown, with $R_1 = R_2$ and $R_3 = R_4 + R_5$ the gain is set by the ratio of R_3 to R_1 (see Appendix A4) and equals 52dB. The LF357 is selected over the quieter LM387A due to its high common-mode rejection capability. The LM387A (or LM381A) requires special circuitry when used with balanced inputs since it was not designed to reject common-mode signals. (A design trade-off was made for lower noise.) See Figure 2.13.4.

Input resistors R_1 and R_2 are made large compared to the source impedance, yet kept as small as possible, to achieve an optimum balance between input loading effects and low noise. Making $R_1 + R_2$ equal to ten times the source impedance is a good compromise value. Matching impedances is *not* conducive to low noise design and should be avoided.[1] The common-mode rejection ratio (CMRR) of the LF357 is 100dB and can be viewed as the *"best case"* condition, i.e., with a perfect match in resistors, the CMRR will be 100dB. The effect of resistor mismatch on CMRR cannot be overemphasized. The amplifier's ability to reject common-mode assumes that *exactly* the same signal is simultaneously present at both the inverting and non-inverting inputs (pins 2 and 3). Any mismatch between resistors will show up as a *differential* signal present at the input terminals and will be amplified accordingly. By using 0.1% tolerance resistors, and adjusting R_5 for minimum output with a common-mode signal applied, a CMRR near 100dB is possible. Using 1% resistors will degrade CMRR to about 80dB. The LF356 may be substituted for the LF357 if desired with only a degradation in slew rate (12V/μs vs. 50V/μs) and gain bandwidth (5MHz vs. 20MHz).

Due to the thermal noise of the relatively large input resistors the noise performance of the Figure 2.13.3 circuit is poorer than the other circuits, but it offers superior hum rejection relative to Figure 2.13.1 and eliminates the costly transformer of Figure 2.13.2.

2.13.5 Low Noise Transformerless Balanced Designs

A low noise transformerless design can be obtained by using a LM387A in front of the LF356 (or LF357) as shown in Figure 2.13.4. This configuration is known as an instrumentation amplifier after its main usage in balanced bridge instrumentation applications. In this design each half of the LM387A is

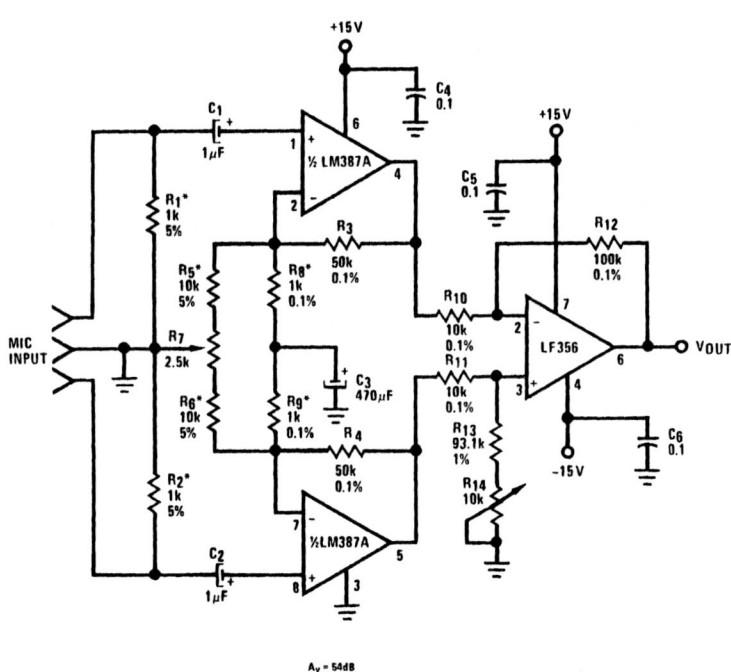

FIGURE 2.13.4 Low Noise Transformerless Balanced Mic Preamp

wired as a non-inverting amplifier with bias and gain setting resistors as before. Resistors R_1 and R_2 set the input impedance at 2kΩ (balanced). Potentiometer R_7 is used to set the output dc level at zero volts by matching the dc levels of pins 4 and 5 of the LM387A.

This allows direct coupling between the stages, thus eliminating the coupling capacitors and the associated matching problem for optimum CMRR. AC gain resistors, R_8 and R_9 are grounded by the common capacitor, C_3, eliminating another capacitor and assuring AC gain match. Close resistor tolerance is necessary around the LM387A in order to preserve common-mode signals appearing at the input. The function of the LM387A is to amplify the low level signal adding as little noise as possible, and leave common-mode rejection to the LF356.

By substituting a LM381A, a professional quality transformerless balanced mic preamp can be designed. The circuit is the same as Figure 2.13.4.

REFERENCES

1. Smith, D. A. and Wittman, P. H., "Design Considerations of Low-Noise Audio Input Circuitry for a Professional Microphone Mixer," *Jour. Aud. Eng. Soc.*, vol. 18, no. 2, April 1970, pp. 140-156.

2.14 TONE CONTROLS

2.14.1 Introduction

There are many reasons why a user of audio equipment may wish to alter the frequency response of the material being played. The purist will argue that he wants his amplifier "flat," i.e., no alteration of the source material's frequency response; hence, amplifiers with tone controls often have a FLAT position or a switch which bypasses the circuitry. The realist will argue that he wants the music to reach his ears "flat." This position recognizes that such parameters as room acoustics, speaker response, etc., affect the output of the amplifier and it becomes necessary to compensate for these effects if the listener is to "hear" the music "flat," i.e., as recorded. And there is simply the matter of personal taste (which is not simple): one person prefers "bassy" music; another prefers it "trebley."

2.14.2 Passive Design

Passive tone controls offer the advantages of lowest cost and minimum parts count while suffering from severe insertion loss which often creates the need for a tone recovery amplifier. The insertion loss is approximately equal to the amount of available boost, e.g., if the controls have +20dB of boost, then they will have about -20dB insertion loss. This is because passive tone controls work as AC voltage dividers and really only cut the signal.

2.14.3 Bass Control

The most popular bass control appears as Figure 2.14.1 along with its associated frequency response curve. The curve shown is the ideal case and can only be approximated. The corner frequencies f_1 and f_2 denote the half-power points and therefore represent the frequencies at which the relative magnitude of the signal has been reduced (or increased) by 3dB.

Passive tone controls require "audio taper" (logarithmic) potentiometers, i.e., at the 50% rotation point the slider splits the resistive element into two portions equal to 90% and 10% of the total value. This is represented in the figures by "0.9" and "0.1" about the wiper arm.

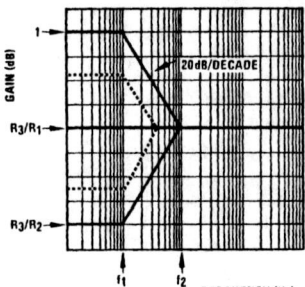

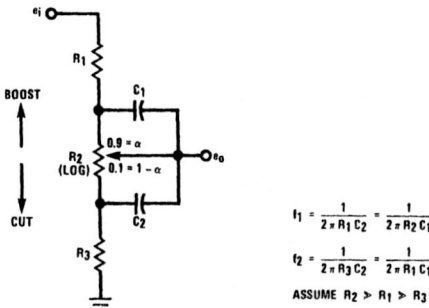

FIGURE 2.14.1 Bass Tone Control — General Circuit

$$f_1 = \frac{1}{2\pi R_1 C_2} = \frac{1}{2\pi R_2 C_1}$$

$$f_2 = \frac{1}{2\pi R_3 C_2} = \frac{1}{2\pi R_1 C_1}$$

ASSUME $R_2 \gg R_1 \gg R_3$

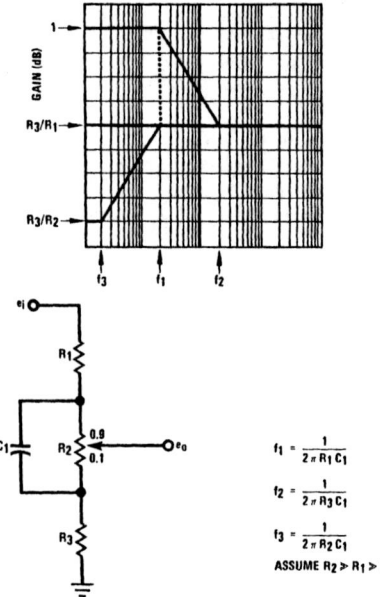

FIGURE 2.14.2 Minimum-Parts Bass Tone Control

$$f_1 = \frac{1}{2\pi R_1 C_1}$$

$$f_2 = \frac{1}{2\pi R_3 C_1}$$

$$f_3 = \frac{1}{2\pi R_2 C_1}$$

ASSUME $R_2 \gg R_1 \gg R_3$

For designs satisfying $R_2 \gg R_1 \gg R_3$, the amount of available boost or cut of the signal given by Figure 2.14.1 is set by the following component ratios:

$$\frac{R_1}{R_2} = \frac{R_3}{R_1} = \frac{C_1}{C_2} = \text{bass boost or cut amount} \quad (2.14.1)$$

The turnover frequency f_2 occurs when the reactance of C_1 equals R_1 and the reactance of C_2 equals R_3 (assuming $R_2 \gg R_1 \gg R_3$):

$$C_1 = \frac{1}{2\pi f_2 R_1} \quad (2.14.2)$$

$$C_2 = \frac{1}{2\pi f_2 R_3} \quad (2.14.3)$$

The frequency response will be accentuated or attenuated at the rate of ±20dB/decade = ±6dB/octave (single pole response) until f_1 is reached. This occurs when the limiting impedance is dominant, i.e., when the reactance of C_1 equals R_2 and the reactance of C_2 equals R_1:

$$f_1 = \frac{1}{2\pi R_1 C_2} = \frac{1}{2\pi R_2 C_1} \quad (2.14.4)$$

Note that Equations (2.14.1)-(2.14.4) are not independent but all relate to each other and that selection of boost/cut amount and corner frequency f_2 fixes the remaining parameters. Also of passing interest is the fact that f_2 is dependent upon the wiper position of R_2. The solid-line response of Figure 2.14.1 is only valid at the extreme ends of potentiometer R_2; at other positions the response changes as depicted by the dotted line response. The relevant time constants involved are $(1-\alpha)R_2 C_1$ and $\alpha R_2 C_2$, where α equals the fractional rotation of the wiper as shown in Figure 2.14.1. While this effect might appear to be undesirable, in practice it is quite acceptable and this design continues to dominate all others.

Figure 2.14.2 shows an alternate approach to bass tone control which offers the cost advantage of one less capacitor and the disadvantage of asymmetric boost and cut response. The degree of boost or cut is set by the same resistor ratios as in Figure 2.14.1.

$$\frac{R_2}{R_1} = \frac{R_1}{R_3} = \text{bass boost or cut amount} \quad (2.14.5)$$

assumes $R_2 \gg R_1 \gg R_3$

The boost turnover frequency f_2 occurs when the reactance of C_1 equals R_3:

$$C_1 = \frac{1}{2\pi f_2 R_3} \quad (2.14.6)$$

Maximum boost occurs at f_1, which also equals the cut turnover frequency. This occurs when the reactance of C_1 equals R_1, and maximum cut is achieved where $X_{C_1} = R_2$. Again, all relevant frequencies and the degree of boost or cut are related and interact. Since in practice most tone controls are used in their boost mode, Figure 2.14.2 is not as troublesome as it may first appear.

2.14.4 Treble Control

The treble control of Figure 2.14.3 represents the electrical analogue of Figure 2.14.1, i.e., resistors and capacitors inter-changed, and gives analogous performance. The amount of boost or cut is set by the following ratios:

$$\frac{R_3}{R_1} = \frac{C_1}{C_2} = \text{treble boost or cut amount} \quad (2.14.7)$$

assumes $R_2 \gg R_1 \gg R_3$

Treble turnover frequency f_1 occurs when the reactance of C_1 equals R_1 and the reactance of C_2 equals R_3:

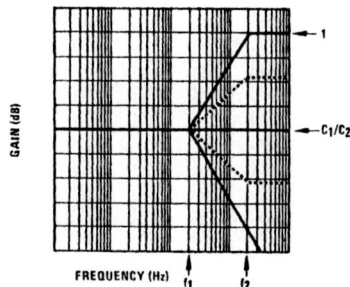

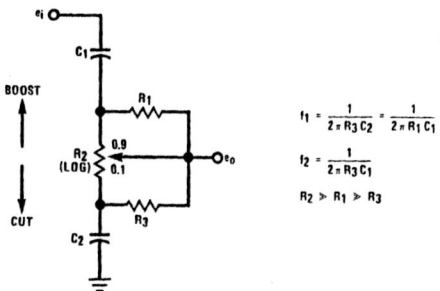

FIGURE 2.14.3 Treble Tone Control – General Circuit

$$C_1 = \frac{1}{2\pi f_1 R_1} \quad (2.14.8)$$

$$C_2 = \frac{1}{2\pi f_1 R_3} \quad (2.14.9)$$

The amount of available boost is reached at frequency f_2 and is determined when the reactance of C_1 equals R_3.

$$f_2 = \frac{1}{2\pi R_3 C_1} \quad (2.14.10)$$

In order for Equations (2.14.8) and (2.14.9) to remain valid, it is necessary for R_2 to be designed such that it is much larger than either R_1 or R_3. For designs that will not permit this condition, Equations (2.14.8) and (2.14.9) must be modified by replacing the R_1 and R_3 terms with $R_1 \| R_2$ and $R_3 \| R_2$ respectively. Unlike the bass control, f_1 is not dependent upon the wiper position of R_2, as indicated by the dotted lines shown in Figure 2.14.3. Note that in the full cut position attenuation tends toward zero without the shelf effect of the boost characteristic.

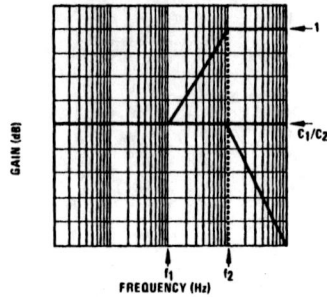

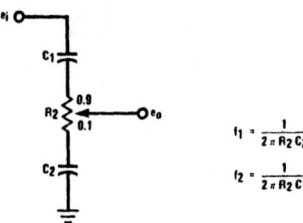

$$f_1 = \frac{1}{2\pi R_2 C_2}$$

$$f_2 = \frac{1}{2\pi R_2 C_1}$$

FIGURE 2.14.4 Minimum-Parts Treble Tone Control

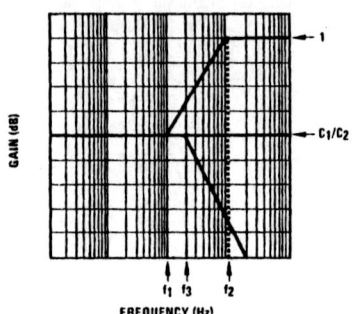

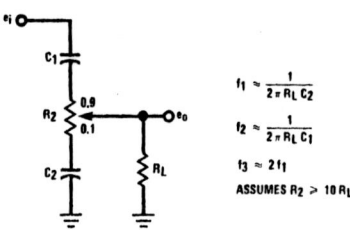

$$f_1 \approx \frac{1}{2\pi R_L C_2}$$

$$f_2 \approx \frac{1}{2\pi R_L C_1}$$

$$f_3 \approx 2f_1$$

ASSUMES $R_2 \geq 10 R_L$

FIGURE 2.14.5 Effect of Loading Treble Tone Control

It is possible to omit R_1 and R_3 for low cost systems. Figure 2.14.4 shows this design with the modified equations and frequency response curve. The obvious drawback appears to be that the turnover frequency for treble cut occurs a decade later (for ±20dB designs) than the boost point. As noted previously, most controls are used in their boost mode, which lessens this drawback, but probably more important is the effect of finite loads on the wiper of R_2.

Figure 2.14.5 shows the loading effect of R_L upon the frequency response of Figure 2.14.4. Examination of these two figures shows that the presence of low impedance (relative to R_2) on the slider changes the break points significantly. If R_L is 1/10 of R_2 then the break points shift a full decade higher. The equations given in Figure 2.14.5 hold for values of $R_2 \geq 10 R_L$. A distinct advantage of Figure 2.14.5 over Figure 2.14.4 is seen in the cut performance. R_L tends to pull the cut turnover frequency back toward the boost corner — a nice feature, and with two fewer resistors. Design becomes straightforward once R_L is known. C_1 and C_2 are calculated from Equations (2.14.11) and (2.14.12).

$$C_1 = \frac{1}{2\pi f_2 R_L} \quad (2.14.11)$$

$$C_2 = \frac{1}{2\pi f_1 R_L} \quad (2.14.12)$$

Here again, gain and turnover frequencies are related and fixed by each other.

Example 2.14.1

Design a passive, symmetrical bass and treble tone control circuit having 20dB boost and cut at 50Hz and 10kHz, relative to midband gain.

Solution

1. For symmetrical controls, combine Figures 2.14.1 and 2.14.3.

BASS (Figure 2.14.1):

2. From Equation (2.14.1):

$$\frac{R_1}{R_2} = \frac{R_3}{R_1} = \frac{C_1}{C_2} = \frac{1}{10} \quad (-20\text{dB})$$

$f_1 = 50\,\text{Hz}$ and $f_2 = 500\,\text{Hz}$

3. Let $R_2 = 100\text{k}$ (audio taper).

4. From Step 2:

$$R_1 = \frac{R_2}{10} = \frac{100\text{k}}{10} = 10\text{k}$$

$$R_3 = \frac{R_1}{10} = \frac{10\text{k}}{10} = 1\text{k}$$

5. From Equation (2.14.2) and Step 2:

$$C_1 = \frac{1}{2\pi f_2 R_1} = \frac{1}{(2\pi)(500)(10\text{k})} = 3.18 \times 10^{-8}$$

Use $C_1 = 0.033\,\mu\text{F}$

$C_2 = 10 C_1$

$C_2 = 0.33\,\mu\text{F}$

TREBLE (Figure 2.14.3):

6. From Equation (2.14.7):

$$\frac{R_3}{R_1} = \frac{C_1}{C_2} = \frac{1}{10} \quad (-20\text{dB})$$

$f_1 = 1\text{kHz}, f_2 = 10\text{kHz}$

7. Let $R_2 = 100\text{k}$ (audio taper).

8. Select $R_1 = 10\text{k}$ (satisfying $R_2 \gg R_1$ and minimizing component spread).

Then:

$$R_3 = \frac{R_1}{10} = \frac{10\text{k}}{10} = 1\text{k}$$

9. From Equation (2.14.8) and Step 6:

$$C_1 = \frac{1}{2\pi f_1 R_1} = \frac{1}{(2\pi)(1\text{k})(10\text{k})} = 1.59 \times 10^{-8}$$

Use $C_1 = 0.015\mu\text{F}$

$C_2 = 10 C_1$

$C_2 = 0.15\mu\text{F}$

The completed design appears as Figure 2.14.6, where R_I has been included to isolate the two control circuits, and C_0 is provided to block all DC voltages from the circuit — insuring the controls are not "scratchy," which results from DC charge currents in the capacitors and on the sliders. C_0 is selected to agree with system low frequency response:

$$C_0 = \frac{1}{(2\pi)(20\text{Hz})(10\text{k})} = 7.9 \times 10^{-7}$$

Use $C_0 = 1\mu\text{F}$.

2.14.5 Use of Passive Tone Controls with LM387 Preamp

A typical application of passive tone controls (Figure 2.14.7) involves a discrete transistor used following the circuit to further amplify the signal as compensation for the loss through the passive circuitry. While this is an acceptable practice, a more judicious placement of the same transistor results in a superior design without increasing parts count or cost.

Placing the transistor *ahead* of the LM387 phono or tape preamplifier (Figure 2.14.8) improves the S/N ratio by boosting the signal before equalizing. An improvement of at least 3dB can be expected (analogous to operating a LM381A with single-ended biasing). The transistor selected must be low-noise, but in quantity the difference in price becomes negligible. The only precaution necessary is to allow sufficient headroom in each stage to minimize transient clipping. However, due to the excellent open-loop gain and large output swing capability of the LM387, this is not difficult to achieve.

An alternative to the transistor is to use an LM381A selected low-noise preamp. Superior noise performance is possible. (See Section 2.7.) The large gain and output swing are adequate enough to allow sufficient single-stage gain to overcome the loss of the tone controls. Figure 2.14.9 shows an application of this concept where the LM381A is used differentially. Single-ended biasing may be used for even quieter noise voltage performance.

2.14.6 Loudness Control

A loudness control circuit compensates for the logarithmic nature of the human ear. Fletcher and Munson[1] published curves (Figure 2.14.10) demonstrating this effect. Without loudness correction, the listening experience is characterized by a pronounced loss of bass response accompanied by a slight loss of treble response as the volume level is decreased. Compensation consists of boosting the high and

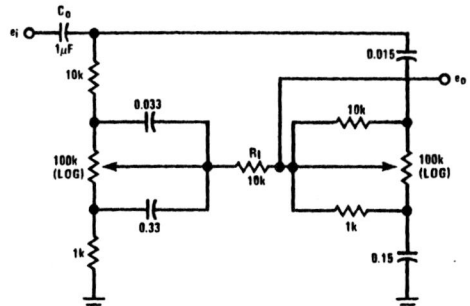

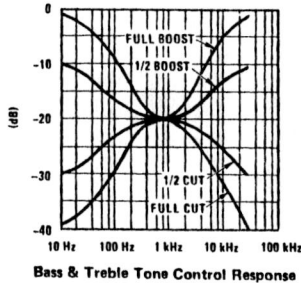

Bass & Treble Tone Control Response

FIGURE 2.14.6 Complete Passive Bass & Treble Tone Control

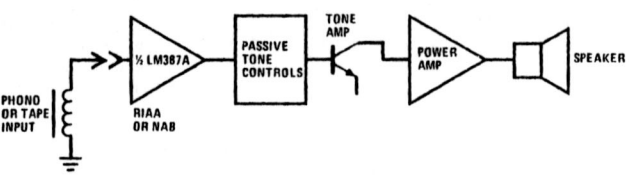

FIGURE 2.14.7 Typical Passive Tone Control Application

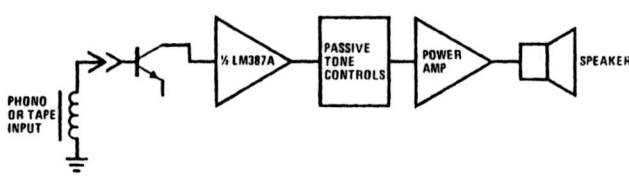

FIGURE 2.14.8 Improved Circuit Using Passive Tone Controls

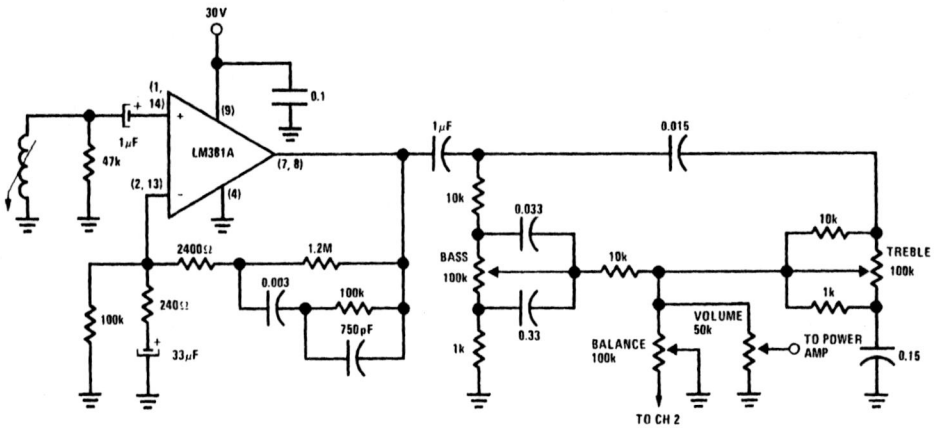

FIGURE 2.14.9 Single Channel of Complete Phono Preamp

low ends of the audio frequency band as an inverse function of volume control setting. One commonly used circuit appears as Figure 2.14.11 and uses a tapped volume pot (tap @ 10% resistance). The switchable R-C network paralleling the pot produces the frequency response shown in Figure 2.14.12 when the wiper is positioned at the tap point (i.e., mid-position for audio taper pot). As the wiper is moved further away from the tap point (louder) the paralleling circuit has less and less effect, resulting in a volume sensitive compensation scheme.

2.14.7 Active Design

Active tone control circuits offer many attractive advantages: they are inherently symmetrical about the axis in boost and cut operation; they have very low THD due to being incorporated into the negative feedback loop of the gain block, as opposed to the relatively high THD exhibited by a tone recovery transistor; and the component spread, i.e., range of values, is low.

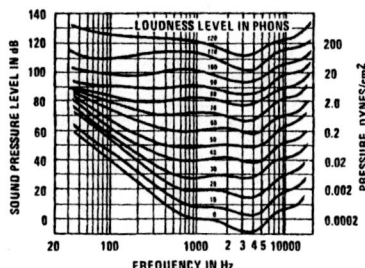

FIGURE 2.14.10 Fletcher-Munson Curves (USA). (Courtesy, Acoustical Society of America)

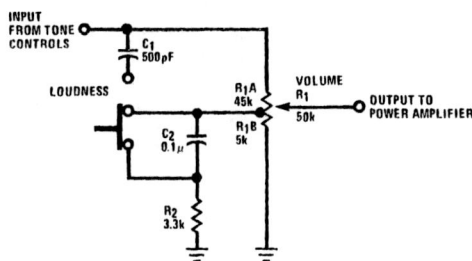

FIGURE 2.14.11 Loudness Control

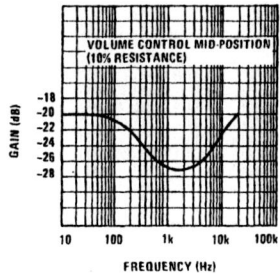

FIGURE 2.14.12 Loudness Control Frequency Response

The most common active tone control circuit is the so-called "Americanized" version of the Baxandall (1952)[2] negative feedback tone controls. A complete bass and treble active tone control circuit is given in Figure 2.14.13a. At very low frequencies the impedance of the capacitors is large enough that they may be considered open circuits, and the gain is controlled by the bass pot, being equal to Equations (2.14.13) and (2.14.14) at the extreme ends of travel.

$$|A_{VB}| = \frac{R_1 + R_2}{R_1} \quad \text{(max bass boost)} \qquad (2.14.13)$$

$$\left|\frac{1}{A_{VB}}\right| = \frac{R_1}{R_1 + R_2} \quad \text{(max bass cut)} \qquad (2.14.14)$$

At very high frequencies the impedance of the capacitors is small enough that they may be considered short circuits, and the gain is controlled by the treble pot, being equal to Equations (2.14.15) and (2.14.16) at the extreme ends of travel.

$$|A_{VT}| = \frac{R_3 + R_1 + 2R_5}{R_3} \quad \text{(max treble boost)} \qquad (2.14.15)$$

$$\left|\frac{1}{A_{VT}}\right| = \frac{R_3}{R_3 + R_1 + 2R_5} \quad \text{(max treble cut)} \qquad (2.14.16)$$

Equations (2.14.15) and (2.14.16) are best understood by recognizing that the bass circuit at high frequencies forms a wye-connected load across the treble circuit. By doing a wye-delta transformation (see Appendix A3), the effective loading resistor is found to be $(R_1 + 2R_5)$ which is in parallel with $(R_3 + R_4)$ and dominates the expression. (See Figure 2.14.13b.) This defines a constraint upon R_4 which is expressed as Equation (2.14.17).

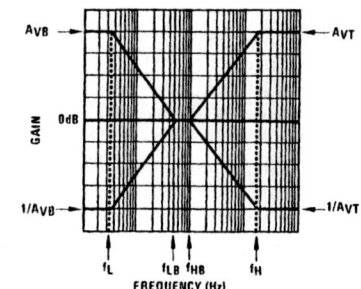

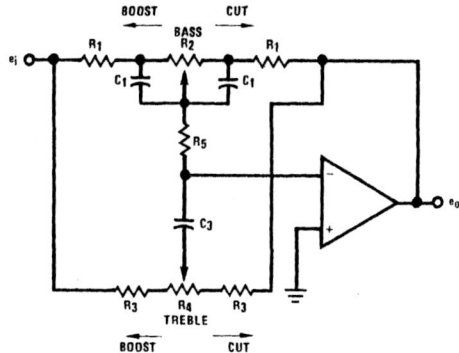

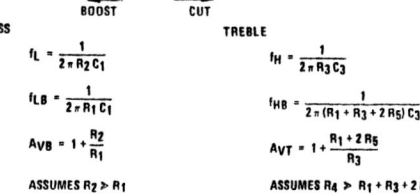

BASS

$f_L = \dfrac{1}{2\pi R_2 C_1}$

$f_{LB} = \dfrac{1}{2\pi R_1 C_1}$

$A_{VB} = 1 + \dfrac{R_2}{R_1}$

ASSUMES $R_2 \gg R_1$

TREBLE

$f_H = \dfrac{1}{2\pi R_3 C_3}$

$f_{HB} = \dfrac{1}{2\pi (R_1 + R_3 + 2R_5) C_3}$

$A_{VT} = 1 + \dfrac{R_1 + 2R_5}{R_3}$

ASSUMES $R_4 \gg R_1 + R_3 + 2R_5$

FIGURE 2.14.13a Bass and Treble Active Tone Control

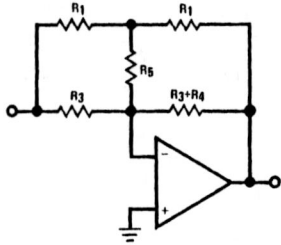

(a) High Frequency Max Treble Boost Equivalent Circuit

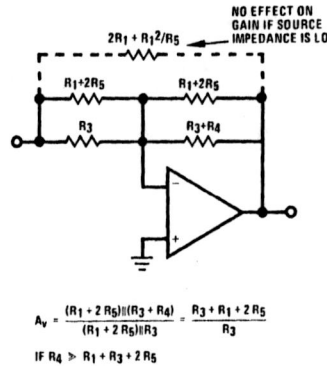

$$A_v = \frac{(R_1 + 2R_5)\|(R_3 + R_4)}{(R_1 + 2R_5)\|R_3} = \frac{R_3 + R_1 + 2R_5}{R_3}$$

IF $R_4 \gg R_1 + R_3 + 2R_5$

(b) High Frequency Circuit After Wye-Delta Transformation

FIGURE 2.14.13b Development of Max Treble Gain

$$R_4 \geq 10(R_3 + R_1 + 2R_5) \quad (2.14.17)$$

At low-to-middle frequencies the impedance of C_1 decreases at the rate of −6 dB/octave, and is in parallel with R_2, so the effective resistance reduces correspondingly, thereby reducing the gain. This process continues until the resistance of R_1 becomes dominant and the gain levels off at unity.

The action of the treble circuit is similar and stops when the resistance of R_3 becomes dominant. The design equations follow directly from the above.

$$C_1 = \frac{1}{2\pi f_{LB} R_1} \quad \text{assumes } R_2 \gg R_1 \quad (2.14.18)$$

$$R_2 = \frac{1}{2\pi f_L C_1} \quad (2.14.19)$$

$$C_3 = \frac{1}{2\pi f_H R_3} \quad (2.14.20)$$

$$R_5 = \frac{1}{2}\left(\frac{1}{2\pi f_{HB} C_3} - R_1 - R_3\right) \quad (2.14.21)$$

The relationship between f_L and f_{LB} and between f_H and f_{HB} is not as clear as it may first appear. As used here these frequencies represent the ±3 dB points relative to gain at midband and the extremes. To understand their relationship in the most common tone control design of ±20 dB at extremes, reference is made to Figure 2.14.14. Here it is seen what shape the frequency response will actually have. Note that the flat (or midband) gain is not unity but approximately ±2 dB. This is due to the close proximity of the poles and zeros of the transfer function. Another effect of this close proximity is that the slopes of the curves are not the expected ±6 dB/octave, but actually are closer to ±4 dB/octave. Knowing that f_L and f_{LB} are 14 dB apart in magnitude, and the slope of the response is 4 dB/octave, it is possible to relate the two. This relationship is given as Equation (2.14.22).

$$\frac{f_{LB}}{f_L} = \frac{f_H}{f_{HB}} \approx 10 \quad (2.14.22)$$

Example 2.14.2

Design a bass and treble active tone control circuit having ±20 dB gain with low frequency upper 3 dB corner at 30 Hz and high frequency upper 3 dB corner at 10 kHz.

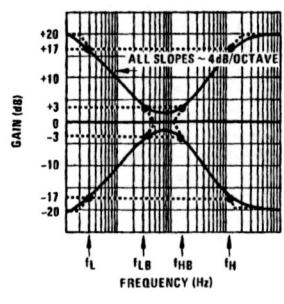

$$\frac{f_{LB}}{f_L} = \frac{f_H}{f_{HB}} \approx 10$$

FIGURE 2.14.14 Relationship Between Frequency Breakpoints of Active Tone Control Circuit

Solution

BASS DESIGN:

1. Select R_2 = 100k (linear). This is an arbitrary choice.

2. From Equation (2.14.13):

$$A_{VB} = 1 + \frac{R_2}{R_1} = 10 \, (+20 \text{dB})$$

$$R_1 = \frac{R_2}{10 - 1} = \frac{100k}{9} = 1.11 \times 10^4$$

$$R_1 = 11k$$

3. Given f_L = 30 Hz and from Equations (2.14.22) and (2.14.18):

$$f_{LB} = 10 f_L = 300 \text{ Hz}$$

$$C_1 = \frac{1}{2\pi f_{LB} R_1} = \frac{1}{(2\pi)(300)(11k)} = 4.82 \times 10^{-8}$$

$$C_1 = 0.05 \mu F$$

TREBLE DESIGN:

4. Let $R_5 = R_1$ = 11k. This also is an arbitrary choice.

2-52

5. From Equation (2.14.15):

$$A_{VT} = 1 + \frac{R_1 + 2R_5}{R_3} = 10 \ (+20\,dB)$$

$$R_3 = \frac{R_1 + 2R_5}{10 - 1} = \frac{11k + 2(11k)}{9} = 3.67 \times 10^3$$

$$R_3 = 3.6k$$

6. Given $f_H = 10\,kHz$ and from Equation (2.14.20):

$$C_3 = \frac{1}{2\pi f_H R_3} = \frac{1}{(2\pi)(10\,kHz)(3.6k)} = 4.42 \times 10^{-9}$$

$$C_3 = 0.005\,\mu F$$

7. From Equation (2.14.17):

$$R_4 \geq 10(R_3 + R_1 + 2R_5)$$
$$\geq 10(3.6k + 11k + 22k)$$
$$\geq 3.66 \times 10^5$$
$$R_4 = 500k$$

The completed design is shown in Figure 2.14.15, where the quad op amp LM349 has been chosen for the active element. The use of a quad makes for a single IC, stereo tone control circuit that is very compact and economical. The buffer amplifier is necessary to insure a low driving impedance for the tone control circuit and creates a high input impedance (100kΩ) for the source. The LM349 was chosen for its fast slew rate (2.5V/μs), allowing undistorted, full-swing performance out to >25kHz. Measured THD was typically 0.05% @ 775mV across the audio band. Resistors R_6 and R_7 were added to insure stability at unity gain since the LM349 is internally compensated for positive gains of five or greater. R_6 and R_7 act as input voltage dividers at high frequencies such that the actual input-to-output gain is never less than five (four is used inverting). Coupling capacitors C_4 and C_5 serve to block DC and establish low frequency roll-off of the system; they may be omitted for direct-coupled designs.

2.14.8 Alternate Active Bass Control

Figure 2.14.16 shows an alternate design for bass control, offering the advantage of one less capacitor while retaining identical performance to that shown in Figure 2.14.13. The development of Figure 2.14.16 follows immediately from Figure 2.14.13 once it is recognized that at the extreme wiper positions one of the C_1 capacitors is shorted out and the other bridges R_2.

The modifications necessary for application with the LM387 are shown in Figure 2.14.17 for a supply voltage of 24V. Resistors R_4 and R_5 are added to supply negative input bias as discussed in Section 2.8. The feedback coupling capacitor C_0 is necessary to block DC voltages from being fed back into the tone control circuitry and upsetting the DC bias, also to insure quiet pot operation since there are no DC level changes occurring across the capacitors, which

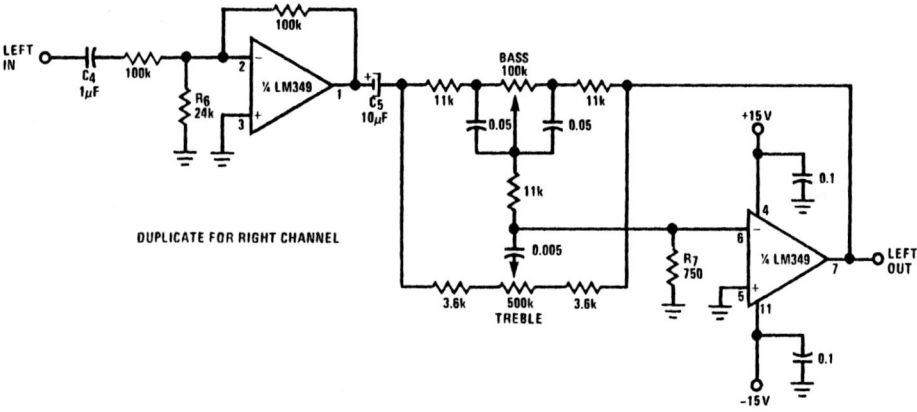

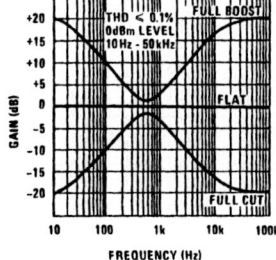

FIGURE 2.14.15 Typical Active Bass & Treble Tone Control with Buffer

would cause "scratchiness." The R_7-C_3 network creates the input attenuation at high frequencies for stability.

For other supply voltages R_4 is recalculated as before, leaving R_5 equal to 240 kΩ. It is not necessary to change R_7 since its value is dictated by the high frequency equivalent impedance seen by the inverting input (equals 33 kΩ).

2.14.9 Midrange Control

The addition of a midrange control which acts to boost or cut the midrange frequencies in a manner similar to the bass and treble controls offers greater flexibility in tone control.

The midrange control circuitry appears in Figure 2.14.18. It is seen that the control is a merging together of the bass and treble controls, incorporating the bass bridging capacitor and the treble slider capacitor to form a combined network. If the bass control is, in fact, a low pass filter, and the treble control a high pass filter, then the midrange is a combination of both, i.e., a bandpass filter.

While the additional circuitry appears simple enough, the resultant mathematics and design equations are not. In the bass and treble design of Figure 2.14.13 it is possible to include the loading effects of the bass control upon the treble circuit, make some convenient design rules, and obtain useful equations. (The treble control offers negligible load to the bass circuit.) This is possible, primarily because the frequencies of interest are far enough apart so as not to interfere with one another. Such is not the case with the midrange included. Any two of the controls appreciably loads the third. The equations that result from a detailed analysis of Figure 2.14.18 become so complex that they are useless for design. So, as is true with much of real-world engineering, design is accomplished by empirical (i.e., trial-and-error) methods. The circuit of Figure 2.14.18 gives the performance shown by the frequency plot, and should be optimum for most applications. For those who feel a change is necessary, the following guidelines should make it easier.

1. To increase (or decrease) midrange gain, decrease (increase) R_6. This will also shift the midrange center frequency higher (lower). (This change has minimal effect upon bass and treble controls.)

2. To move the midrange center frequency (while preserving gain, and with negligible change in bass and treble performance), change both C_4 and C_5. Maintain the relationship that $C_5 \approx 5C_4$. Increasing (decreasing) C_5 will decrease (increase) the center frequency. The amount of shift is approximately equal to the inverse ratio of the new capacitor to the old one. For example, if the original capacitor is C_5 and the original center frequency is f_o, and the new capacitor is C_5' with the new frequency being f_o', then

$$\frac{C_5'}{C_5} \approx \frac{f_o}{f_o'}$$

The remainder of Figure 2.14.18 is as previously described in Figure 2.14.15.

The temptation now arises to add a fourth section to the growing tone control circuitry. It should be avoided. Three paralleled sections appears to be the realistic limit to what can be expected with one gain block. Beyond three, it is best to separate the controls and use a separate op amp with each control and then sum the results. (See Section 2.17 on equalizers for details.)

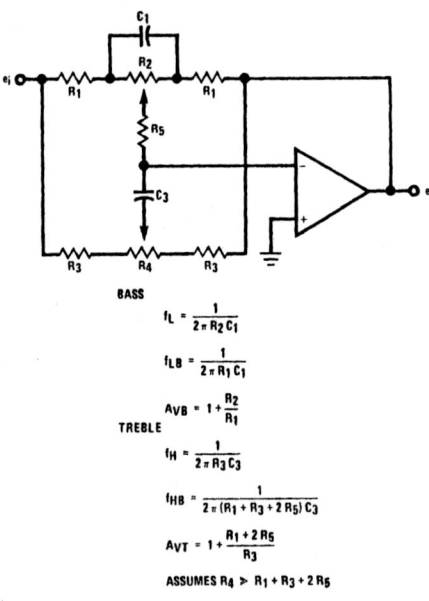

FIGURE 2.14.16 Alternate Bass Design Active Tone Control

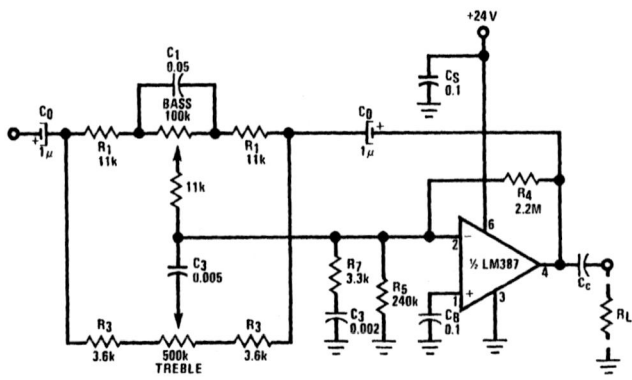

FIGURE 2.14.17 LM387 Feedback Tone Controls

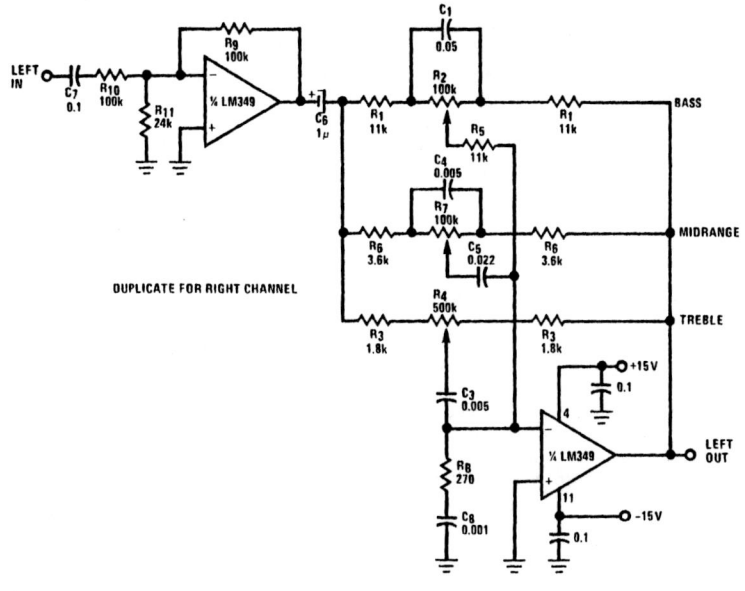

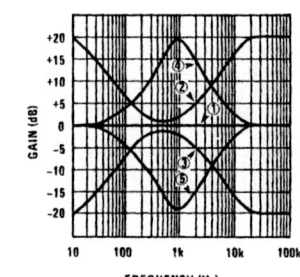

① ALL CONTROLS FLAT
② BASS & TREBLE BOOST, MID FLAT
③ BASS & TREBLE CUT, MID FLAT
④ MID BOOST, BASS & TREBLE FLAT
⑤ MID CUT, BASS & TREBLE FLAT

FIGURE 2.14.18 Three Band Active Tone Control (Bass, Midrange & Treble)

REFERENCES

1. Fletcher, H., and Munson, W. A., "Loudness, Its Definition, Measurement and Calculation," *J. Acoust. Soc. Am.*, vol. 5, p. 82, October 1933.
2. Baxandall, P. J., "Negative Feedback Tone Control — Independent Variation of Bass and Treble Without Switches," *Wireless World*, vol. 58, no. 10, October 1952, p. 402.

2.15 SCRATCH, RUMBLE AND SPEECH FILTERS

2.15.1 Introduction

Infinite-gain, multiple-feedback active filters using LM387 (or LM381) as the active element make simple low-cost audio filters. Two of the most popular filters found in audio equipment are SCRATCH (low pass), used to roll off excess high frequency noise appearing as hiss, ticks and pops from worn records, and RUMBLE (high pass), used to roll off low frequency noise associated with worn turntable and tape transport mechanisms. By combining low and high pass filter sections, a broadband bandpass filter is created such as that required to limit the audio bandwidth to include only speech frequencies (300 Hz-3 kHz).

2.15.2 Definition of ω_c and ω_o for 2-Pole Active Filters

When working with active filter equations, much confusion exists about the difference between the terms ω_o and ω_c. The center frequency, f_o, equals $\omega_o/2\pi$ and has meaning only for *bandpass* filters. The term ω_c and its associated frequency, f_c, is the cutoff frequency of a high or low pass filter defined as the point at which the magnitude of the response is -3dB from that of the passband (i.e., 0.707 times the passband value). Figure 2.15.1 illustrates the two cases for two-pole filters.

Equally confusing is the concept of "Q" in relation to high and low pass two-pole active filters. The design equations contain Q; therefore it must be determined before a filter can be realized — but what does it mean? For bandpass filters the meaning of Q is clear; it is the ratio of the center frequency, f_O, to the –3dB bandwidth. *For low and high pass filters, Q only has meaning with regard to the amount of peaking occurring at f_O and the relationship between the –3dB frequency, f_C, and f_O.*

The relationship that exists between ω_O and ω_C follows:

High Pass $\omega_C = \dfrac{\omega_O}{\beta}$ (2.15.1)

Low Pass $\omega_C = \beta \omega_O$ (2.15.2)

$\beta = \sqrt{\left(1 - \dfrac{1}{2Q^2}\right) + \sqrt{\left(1 - \dfrac{1}{2Q^2}\right)^2 + 1}}$ (2.15.3)

A table showing various values of ω_C for several different values of Q is provided for convenience (Table 2.15.1). Notice that $\omega_C = \omega_O$ only for the Butterworth case (Q = 0.707). Since Butterworth filters are characterized by a maximally flat response (no peaking like that diagrammed in Figure 2.15.1), they are used most often in audio systems.

Always use Equations (2.15.1)-(2.15.3) (or Table 2.15.1) when Q equals anything other than 0.707.

2.15.3 High Pass Design

An LM387 configured as a high-pass filter is shown in Figure 2.15.2. Design procedure is to select R_2 and R_3 per Section 2.8 to provide proper bias; then, knowing desired passband gain, A_O, the Q and the corner frequency f_C, the remaining components are calculated from the following:

Calculate ω_O from $\omega_C = 2\pi f_C$ and Q using Equations (2.15.1) and (2.15.3) (or Table 2.15.1).

Let $C_1 = C_3$

Then:

$C_1 = \dfrac{Q}{\omega_O R_2}(2A_O + 1)$ (2.15.4)

$C_2 = \dfrac{C_1}{A_O}$ (2.15.5)

$R_1 = \dfrac{1}{Q \omega_O C_1 (2A_O + 1)}$ (2.15.6)

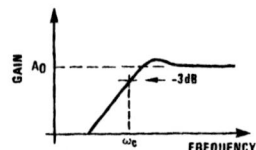

(a) High Pass

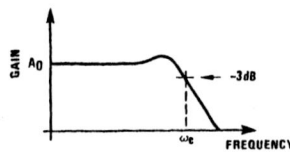

(b) Low Pass

FIGURE 2.15.1 Definition of ω_C for Low and High Pass Filters

TABLE 2.15.1 ω_C vs. Q

Q	ω_C Low-Pass	ω_C High-Pass
0.707*	1.000ω_O	1.000ω_O
1	1.272ω_O	0.786ω_O
2	1.498ω_O	0.668ω_O
3	1.523ω_O	0.657ω_O
4	1.537ω_O	0.651ω_O
5	1.543ω_O	0.648ω_O
10	1.551ω_O	0.645ω_O
100	1.554ω_O	0.644ω_O

* Butterworth

Substitution of f_C for f_O in Butterworth filter design equations is therefore permissible and experimental results will agree with calculations — but only for Butterworth.

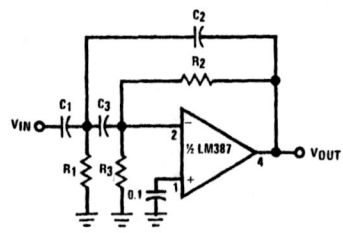

FIGURE 2.15.2 LM387 High Pass Active Filter

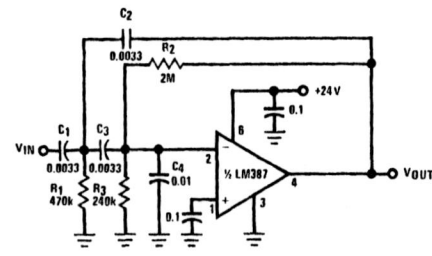

f_c = 50 Hz
SLOPE = –12dB/OCTAVE
A_O = –1
THD ≤ 0.1%

FIGURE 2.15.3 Rumble Filter Using LM387

Example 2.15.1

Design a two-pole active high pass filter for use as a rumble filter. Passband gain, A_O = 1, Q = 0.707 (Butterworth) and corner frequency, f_C = 50 Hz. Supply V_S = +24 V.

Solution
1. Select $R_3 = 240k$.
2. From Section 2.8,

$$R_2 = \left(\frac{V_s}{2.6} - 1\right) R_3 = \left(\frac{24}{2.6} - 1\right) 240k = 1.98 \times 10^6$$

Use $R_2 = 2M$

3. Since $Q = 0.707$, $\omega_o = \omega_c = 2\pi f_c$ (see Table 2.15.1).
4. Let $C_1 = C_3$.
5. From Equation (2.15.4):

$$C_1 = \frac{(0.707)(2 + 1)}{(2\pi)(50)(2 \times 10^6)} = 3.38 \times 10^{-9}$$

Use $C_1 = C_3 = 0.0033 \mu F$

6. From Equation (2.15.5):

$$C_2 = \frac{C_1}{(1)} = C_1 = 0.0033 \mu F$$

7. From Equation (2.15.6):

$$R_1 = \frac{1}{(0.707)(2\pi)(50)(0.0033 \times 10^{-6})(2 + 1)}$$

$$= 45.5 \times 10^4$$

Use $R_1 = 470 k\Omega$.

The final design appears as Figure 2.15.3. For checking and trimming purposes Equation (2.15.7) is useful:

$$f_c = \frac{1}{2\pi C_1 \sqrt{R_1 R_2}} \qquad (2.15.7)$$

Capacitor $C_4 = 0.01$ is included to guarantee high frequency stability for unity gain designs (required for $A_o \leqslant 10$).

2.15.4 Low Pass Design

The low pass configuration for a LM387 is shown in Figure 2.15.4. Design procedure is almost the reverse of the high pass case since biasing resistor R_4 will be selected last. Knowing A_o, Q and f_c, proceed by calculating a constant K per Equation (2.15.8).

$$K = \frac{1}{4Q^2(A_o + 1)} \qquad (2.15.8)$$

Arbitrarily select C_1 to be a convenient value.
Then: $C_2 = KC_1 \qquad (2.15.9)$

Calculate ω_o from $\omega_c = 2\pi f_c$ and Q using Equations (2.15.1) and (2.15.3) (or Table 2.15.1).
Then:

$$R_2 = \frac{1}{2Q\omega_o C_1 K} \qquad (2.15.10)$$

$$R_3 = \frac{R_2}{A_o + 1} \qquad (2.15.11)$$

$$R_1 = \frac{R_2}{A_o} \qquad (2.15.12)$$

$$R_4 = \frac{R_2 + R_3}{\left(\frac{V_s}{2.6} - 1\right)} \qquad (2.15.13)$$

Example 2.15.2

Design a two-pole active low-pass filter for use as a scratch filter. Passband gain, $A_o = 1$, $Q = 0.707$ (Butterworth) and corner frequency $f_c = 10 kHz$. Supply $V_s = +24 V$.

Solution

1. From Equation (2.15.8):

$$K = \frac{1}{(4)(0.707)^2(1 + 1)} = 0.25$$

2. Select $C_1 = 560 pF$ (arbitrary choice).
3. From Equation (2.15.9):

$$C_2 = KC_1 = (0.25)(560 pF) = 140 pF$$

Use $C_2 = 150 pF$

4. Since $Q = 0.707$, $\omega_o = \omega_c = 2\pi f_c$ (see Table 2.15.1).
5. From Equation (2.15.10):

$$R_2 = \frac{1}{(2)(0.707)(2\pi)(10 kHz)(560 pF)(0.25)} = 80.4k$$

Use $R_2 = 82k$

6. From Equation (2.15.11):

$$R_3 = \frac{82k}{2} = 41k$$

Use $R_3 = 39k$

7. From Equation (2.15.12):

$$R_1 = \frac{R_2}{1} = R_2 = 82k$$

8. From Equation (2.15.13):

$$R_4 = \frac{82k + 39k}{\left(\frac{24}{2.6} - 1\right)} = 14.7k$$

Use $R_4 = 15k$

The complete design (Figure 2.15.5) includes C_3 for stability and input blocking capacitor C_4. Checking and trimming can be done with the aid of Equation (2.15.14).

$$f_o = \frac{Q}{\pi C_1} \sqrt{\frac{A_o + 1}{R_2 R_3}} \qquad (2.15.14)$$

2.15.5 Speech Filter

A speech filter consisting of a highpass filter based on Section 2.15.3, in cascade with a low pass based on Section 2.15.4, is shown in Figure 2.15.6 with its frequency response as Figure 2.15.7. The corner frequencies are 300Hz and 3kHz with roll-off of $-40dB$/decade beyond the corners. Measured THD was 0.07% with a 0dBm signal of 1kHz. Total output noise with input shorted was $150\mu V$ and is

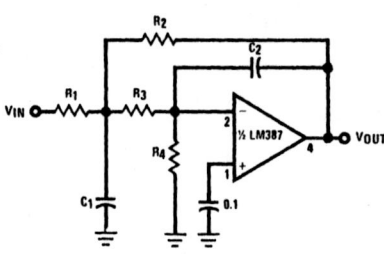

FIGURE 2.15.4 LM387 Low Pass Active Filter

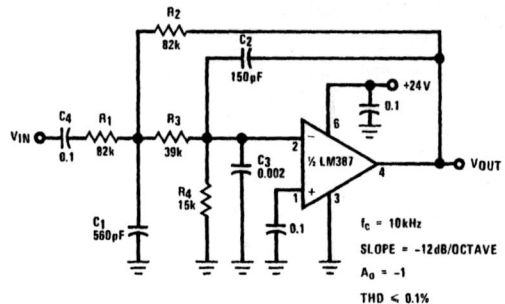

FIGURE 2.15.5 Scratch Filter Using LM387

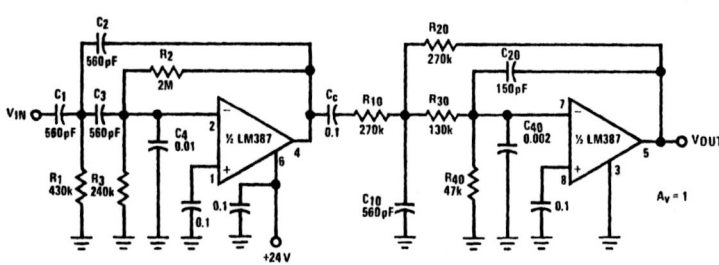

FIGURE 2.15.6 Speech Filter (300Hz-3kHz Bandpass)

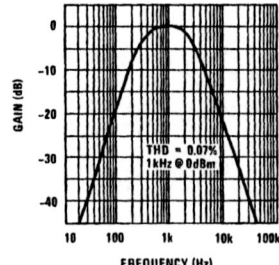

FIGURE 2.15.7 Speech Filter Frequency Response

due mostly to thermal noise of the resistors, yielding S/N of 74dBm. The whole filter is very compact since the LM387 dual preamp is packaged in the 8-pin minidip, making tight layout possible.

2.16 BANDPASS ACTIVE FILTERS

Narrow bandwidth bandpass active filters do not require cascading of low and high pass sections as described in Section 2.15.5. A single amplifier bandpass filter using the LM387 (Figure 2.16.1) is capable of $Q \leq 10$ for audio frequency low distortion applications. The wide gain bandwidth (20MHz) and large open loop gain (104dB) allow high frequency, low distortion performance not obtainable with conventional op amps.

Beginning with the desired f_o, A_o and Q, design is straightforward. Start by selecting R_3 and R_4 per Section 2.8, *except use* $24k\Omega$ *as an upper limit of* R_4 (instead of $240k\Omega$). This minimizes loading effects of the LM387 for high Q designs.

Let $C_1 = C_2$. Then:

$$R_1 = \frac{R_3}{2 A_o} \qquad (2.16.1)$$

$$C_1 = \frac{Q}{A_o \omega_o R_1} \qquad (2.16.2)$$

$$R_2 = \frac{Q}{(2Q^2 - A_0)\,\omega_0 C_1} \quad (2.16.3)$$

For checking and trimming, use the following:

$$A_0 = \frac{R_3}{2R_1} \quad (2.16.4)$$

$$f_0 = \frac{1}{2\pi C_1}\sqrt{\frac{R_1 + R_2}{R_1 R_2 R_3}} \quad (2.16.5)$$

$$Q = \frac{1}{2}\omega_0 R_3 C_1 \quad (2.16.6)$$

4. From Equation (2.16.1):

$$R_1 = \frac{R_3}{2A_0} = \frac{200k}{2} = 100k$$

$$R_1 = 100k$$

5. Let $C_1 = C_2$; then, from Equation (2.16.2):

$$C_1 = \frac{Q}{A_0 \omega_0 R_1} = \frac{10}{(1)(2\pi)(20k)(1 \times 10^5)} = 796\,pF$$

Use $C_1 = 820\,pF$

6. From Equation (2.16.3):

$$R_2 = \frac{Q}{(2Q^2 - A_0)\,\omega_0 C_1}$$

$$= \frac{10}{[(2)(10)^2 - 1](2\pi)(20k)(820\,pF)} = 488\,\Omega$$

Use $R_2 \approx 470\,\Omega$

The final design appears as Figure 2.16.2. Capacitor C_3 is used to AC ground the positive input and can be made equal to $0.1\,\mu F$ for all designs. Input shunting capacitor C_4 is included for stability since the design gain is less than 10.

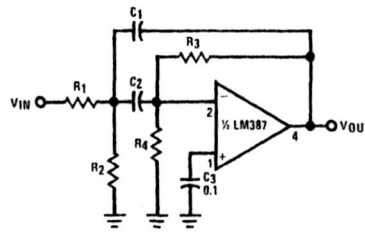

FIGURE 2.16.1 LM387 Bandpass Active Filter

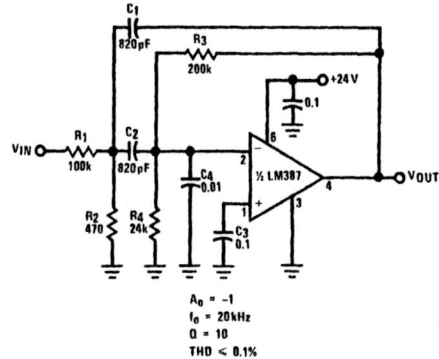

FIGURE 2.16.2 20kHz Bandpass Active Filter

2.17 OCTAVE EQUALIZERS

2.17.1 Ten Band Octave Equalizer

An octave equalizer offers the user several bands of tone control, separated an octave apart in frequency with independent adjustment of each band. It is designed to compensate for any unwanted amplitude-frequency or phase-frequency characteristics of an audio system.

A convenient ten band octave equalizer can be constructed based on the filter circuit shown in Figure 2.17.1 where the potentiometer R_2 can control the degree of boost or cut at the resonant frequency set by the series filter of $C_2 R_s$ and L, by varying the relative proportions of negative feedback and input signal to the amplifier section.

Example 2.16.1

Design a two-pole active bandpass filter with a center frequency $f_0 = 20\,kHz$, midband gain $A_0 = 1$, and a bandwidth of $2000\,Hz$. A single supply, $V_s = 24\,V$, is to be used.

Solution

1. $Q \triangleq \dfrac{f_0}{BW} = \dfrac{20\,kHz}{2000\,Hz} = 10, \quad \omega_0 = 2\pi f_0$

2. Let $R_4 = 24\,k\Omega$.

3. $R_3 = \left(\dfrac{V_s}{2.6} - 1\right) R_4 = \left(\dfrac{24}{2.6} - 1\right) 24k = 1.98 \times 10^5$

   Use $R_3 = 200k$

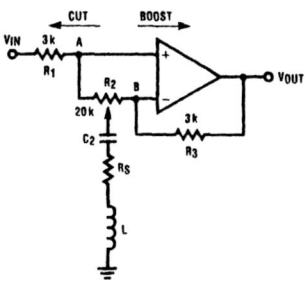

FIGURE 2.17.1 Typical Octave Equalizer Section

Assuming ideal elements, at the resonant frequency with R_2 slider set to the mid position, the amplifier is at unity gain. With the slider of R_2 moved such that C_2 is connected to the junction of R_1 and R_2, the $R_s L C_2$ network will attenuate the input such that

$$\frac{V_{OUT}}{V_{IN}} = \frac{R_s}{3k + R_s} \qquad (2.17.1)$$

If the slider is set to the other extreme, the gain at the resonant frequency is:

$$\frac{V_{OUT}}{V_{IN}} = \frac{3k + R_s}{R_s} \qquad (2.17.2)$$

In the final design, R_s is approximately 500Ω, giving a boost or attenuation factor of 7 ($\cong$ 17dB). However, other filter sections of the equalizer connected between A and B will reduce this factor to about 12dB.

To avoid trying to obtain ten inductors ranging in value from 3.9H to 7.95mH for the ten octave from 32Hz to 16kHz, a simulated inductor design will be used. Consider the equivalent circuit of an inductor with associated series and parallel resistance as shown in Figure 2.12.2. The input impedance of the network is given by:

$$Z_{IN} = \frac{sLR_p}{(sL + R_p)} + R_s$$

$$= \frac{(R_p + R_s)\left(sL + \frac{R_p R_s}{R_p + R_s}\right)}{(sL + R_p)}$$

$$\therefore Z_{IN} = \frac{(R_p + R_s)\left(s + \frac{R_p R_s}{L(R_p + R_s)}\right)}{(s + R_p/L)} \qquad (2.17.3)$$

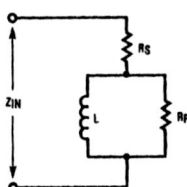

FIGURE 2.17.2 Ideal Inductor with Series and Parallel Resistances

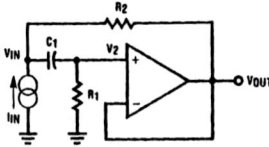

FIGURE 2.17.3 Simulated Inductor

This input impedance can be realised with the active circuit shown in Figure 2.17.3. Assuming an ideal amlifier with infinite gain and infinite input impedance,

$$V_2 = V_{OUT} = \frac{V_{IN} R_1}{(1/sC_1 + R_1)} \qquad (2.17.4)$$

The input current I_{IN} is given by,

$$I_{IN} = \frac{V_{IN} - V_2}{R_2} + \frac{V_{IN}}{(1/sC_1 + R_1)} \qquad (2.17.5)$$

Substituting (2.17.4) into this expression gives:

$$I_{IN} = V_{IN} \left\{ \frac{1}{R_2} + \frac{sC_1}{(1 + sC_1 R_1)} - \frac{R_1}{R_2\left(R_1 + \frac{1}{sC_1}\right)} \right\}$$

$$= V_{IN} \left\{ \frac{1 + sC_1 R_2}{R_2(sC_1 R_1 + 1)} \right\}$$

Since $Z_{IN} = \frac{V_{IN}}{I_{IN}}$

$$Z_{IN} = \frac{R_2 + sC_1 R_1 R_2}{1 + sC_1 R_2}$$

$$= \frac{R_1\left(\frac{1}{C_1 R_1} + s\right)}{\left(\frac{1}{C_1 R_2} + s\right)} \qquad (2.17.6)$$

Equating (2.17.3) and (2.17.6)

$$\frac{(R_p + R_s)\left[s + \frac{R_p R_s}{L(R_p + R_s)}\right]}{s + R_p/L} = \frac{R_1(s + \frac{1}{C_1 R_1})}{s + \frac{1}{C_1 R_2}}$$

$$\therefore R_1 = R_p + R_s \qquad (2.17.7)$$

$$\frac{R_p R_s}{L(R_p + R_s)} = \frac{1}{C_1 R_1}$$

$$\therefore C_1 = \frac{1}{R_p R_s} \qquad (2.17.8)$$

$$\frac{R_p}{L} = \frac{1}{C_1 R_2}$$

$$\therefore R_2 = \frac{L}{R_p} \times \frac{R_p R_s}{L} = R_s \qquad (2.17.9)$$

From the above equations it is apparent that R_1 should be large in order to reduce the effect of R_p on the filter operation, and to allow reasonably small capacitor values for each band (since capacitors will be non-polarized). R_1 should not be too large since it will carry the bias current for the non-inverting input of the amplifier.

The choice of Q for each of the filters depends on the permissible "ripple" in the boost or cut positions and the number of filters being used. For example, if we had only two filters separated by one octave, an ideal filter Q would be 1.414 so that the −3dB response frequencies will coincide, giving the same gain as that at the band centers. For the ten band equalizer a Q of 1.7 is better, since several filters will be affecting the gain at a given frequency. This will keep the maximum ripple at full boost or cut to less than ± 2dB.

EXAMPLE 2.17.1

Design a variable (± 12dB) octave equalizer section with a Q of 1.7 and a center frequency of 2kHz.

Solution

1. Select $R_1 = 68k$
2. From equations (2.17.1) and (2.17.2) $R_s = 470$
3. $L = \frac{QR_s}{2\pi f_o} = \frac{QR_2}{2\pi f_o}$

$$\therefore L = \frac{1.7 \times 470}{2\pi \times 2 \times 10^3} = 63.6 mH \qquad (2.17.10)$$

2-60

4. From equation (2.17.8)

$$C_1 = \frac{L}{R_p + R_s} = \frac{L}{(R_1 - R_2)R_2}$$

$$= \frac{63.6 \times 10^{-3}}{(68 \times 10^3 - 470)470}$$

$$\therefore C_1 = 2000\,\text{pF}$$

5. $$C_2 = \frac{1}{\omega_o^2 L}$$

$$= \frac{1}{(2\pi \times 2 \times 10^3)^2 63.5 \times 10^{-3}}$$

$$\therefore C_2 = 0.1\,\mu\text{F} \tag{2.17.11}$$

Table 2.17.1 summarizes the component values required for the other sections of the equalizer. The final design appears in Figure 2.17.4 and uses LM348 quad op-amps. Other unity gain stable amplifiers can be used. For example, LF356 will give lower distortion at the higher frequencies. Although linear taper potentiometers can be used, these will result in very rapid action near the full boost or full cut positions. S taper

TABLE 2.17.1

f_o(Hz)	C_1	C_2	R_1	R_2
32	0.12μF	4.7μF	75kΩ	560Ω
64	0.056μF	3.3μF	68kΩ	510Ω
125	0.033μF	1.5μF	62kΩ	510Ω
250	0.015μF	0.82μF	68kΩ	470Ω
500	8200pF	0.39μF	62kΩ	470Ω
1k	3900pF	0.22μF	68kΩ	470Ω
2k	2000pF	0.1μF	68kΩ	470Ω
4k	1100pF	0.056μF	62kΩ	470Ω
8k	510pF	0.022μF	68kΩ	510Ω
16k	330pF	0.012μF	51kΩ	510Ω

potentiometers (Allen Bradley #70A1G032 R2035) will give a better response. All the capacitors used for tuning the simulated inductors (C_2) should be non-polarized mylar or polystyrene.

Signal to noise ratio of the equalizer with the controls set "flat" is 73dB referred to a $1V_{RMS}$ input signal. THD is under 0.01% at 20kHz.

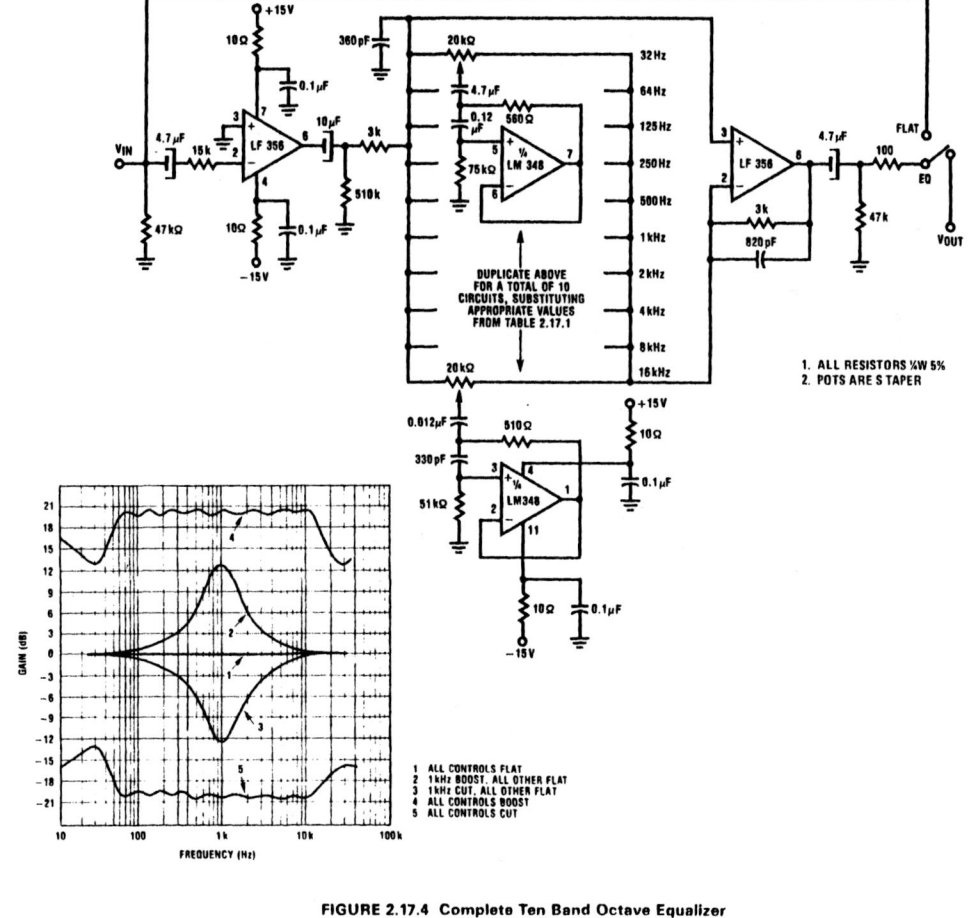

FIGURE 2.17.4 Complete Ten Band Octave Equalizer

2.17.2 Pink Noise Generator

Once an equalizer is incorporated into a music system the question quickly arises as to how best to use it. The most obvious way is as a "super tone control" unit, where control is now extended from the familiar two or three controls to ten controls (or even 30 if 1/3 octave equalizers are used). While this approach is most useful and the results are dramatic in their ability to "liven" up a room, there still remains, with many, the desire to have some controlled manner in which to equalize the listening area without resorting to the use of expensive (and complicated) spectrum or real-time analyzers.

The first step in generating a self-contained room equalizing instrument is to design a pink noise generator to be used as a controlled source of noise across the audio spectrum. With the advent of medium scale integration and MOS digital technology, it is quite easy to create a pink noise generator using only one IC and a few passive components.

The MM5837 digital noise source is an MOS/MSI pseudo-random sequence generator, designed to produce a broadband white noise signal for audio applications. Unlike traditional semiconductor junction noise sources, the MM5837 provides very uniform noise quality and output amplitude. Originally designed for electronic organ and synthesizer applications, it can be directly applied to room equalization. Figure 2.17.5 shows a block diagram of the internal circuitry of the MM5837.

The output of the MM5837 is broadband white noise. In order to generate pink noise it is necessary to understand the difference between the two. *White noise* is characterized by a +3dB rise in amplitude per octave of frequency change (equal energy per constant bandwidth). *Pink noise* has flat amplitude response per octave change of frequency (equal energy per octave). Pink noise allows correlation between successive octave equalizer stages by insuring the same voltage amplitude is used each time as a reference standard.

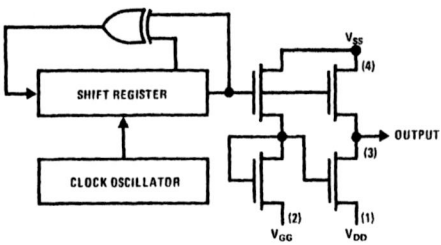

FIGURE 2.17.5 MM5837 Noise Source

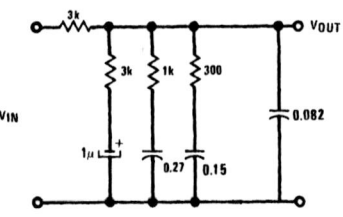

FIGURE 2.17.6 Passive –3dB/Octave Filter

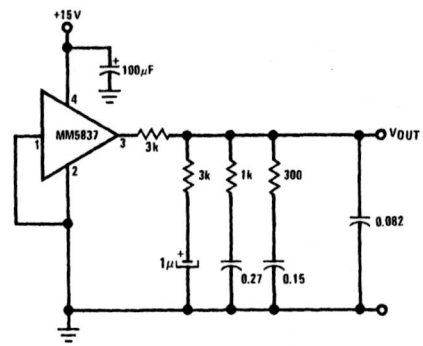

FIGURE 2.17.7 Pink Noise Generator

What is required to produce pink noise from a white noise source is simply a –3dB/octave filter. If capacitive reactance varies at a rate of –6dB/octave then how can a slope of *less* than –6dB/octave be achieved? The answer is by cascading several stages of lag compensation such that the zeros of one stage partially cancel the poles of the next stage, etc. Such a network is shown as Figure 2.17.6 and exhibits a –3dB/octave characteristic (±1/4dB) from 10Hz to 40kHz. The complete pink noise generator is given by Figure 2.17.7 and gives a flat spectral distribution over the audio band of 20Hz to 20kHz. The output at pin 3 is a 11.5V_{p-p} random pulse train which is attenuated by the filter. Actual output is about 1V_{p-p} AC pink noise riding on a 8.5V DC level.

2.17.3 Room Equalizing Instrument

For a room equalizing instrument, a different type of equalizer is required than that previously described under the Ten Band Octave Equalizer. The difference lies in the necessary condition that only one section must pass its bandwidth of frequencies at any time. The reason for this is that to use this instrument all but one band will be switched out and under this condition the pink noise will be passed through the remaining filter and it must pass only its octave of noise. The filtered noise is passed on to the power amplifier and reproduced into the room by the speaker. A microphone with flat audio band frequency response (but uncalibrated) is used to pick up the noise at some central listening point. The microphone input is amplified and used to drive a VU meter where some (arbitrary) level is established via the potentiometer of the filter section. This filter section is then switched out and the next one is switched in. Its potentiometer is adjusted such that the VU meter reads the same as before. Each filter section in turn is switched in, adjusted, and switched out, until all ten octaves have been set. The whole process takes about two minutes. When finished the room response will be equalized flat for each octave of frequencies. From here it becomes personal preference whether the high end is rolled off (a common practice) or the low end is boosted. It allows for greater experimentation since it is very easy to go back to a known (flat) position. It is also easy to correct for new alterations within the listening room (drape changes, new rugs, more furniture, different speaker placement, etc.). Since all adjustments are made relative to each other, the requirement for expensive, calibrated microphones is obviated. Almost any microphone with flat output over frequency will work.

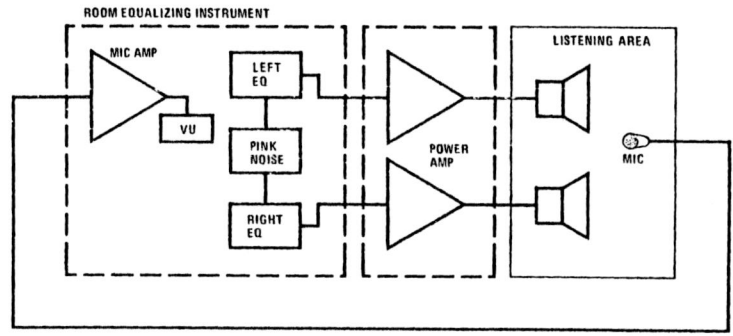

(a) Stereo Application

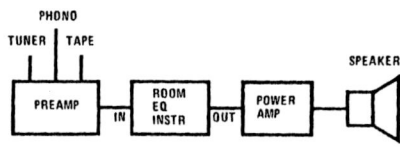

(b) Adding EQ to Component System   (c) Adding EQ to Receiver System

FIGURE 2.17.8 Typical Equalizing Instrument Application

For stereo applications, a two channel instrument is required as diagrammed in Figure 2.17.8a. Figures 2.17.8b and -c show typical placement of the equalizer unit within existing systems.

While any bandpass filter may be used for the filter sections, the multiple-feedback, infinite-gain configuration of Figure 2.17.9 is chosen for its low sensitivity factors. The design equations appear as follows:

$$R_1 = \frac{Q}{2\pi f_o A_o C_1} \quad (2.17.12)$$

$$R_2 = \frac{Q}{(2Q^2 - A_o) 2\pi f_o C_1} = \frac{A_o R_1}{2Q^2 - A_o} \quad (2.17.13)$$

$$R_3 = \frac{Q}{\pi f_o C_1} \quad (2.17.14)$$

$$A_o = \frac{R_3}{2R_1} \quad (2.17.15)$$

$$Q = \pi f_o C_1 R_3 \quad (2.17.16)$$

$$f_o = \frac{1}{2\pi C_1} \sqrt{\frac{R_1 + R_2}{R_1 R_2 R_3}} \quad (2.17.17)$$

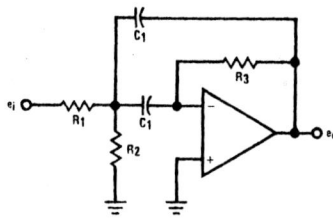

FIGURE 2.17.9 Bandpass Filter Section

Design

1. Select $A_o = 4$ (12 dB) and $Q = 2$.
2. Select R_1 for desired input resistance. (Note that net input impedance is $(R_1 + R_2)/10$, since there are 10 sections in parallel.)

   Let $R_1 = 120k$.
3. Calculate R_2 from Equations (2.17.13) and (2.17.12):

$$R_2 = \frac{Q}{(2Q^2 - A_o) 2\pi f_o C_1} = \frac{Q}{[2(2)^2 - 4] 2\pi f_o C_1}$$

$$= \frac{Q}{(4) 2\pi f_o C_1} = \frac{Q}{2\pi f_o A_o C_1} = R_1$$

$$R_2 = R_1 = 120k$$

4. Calculate R_3 from Equation (2.17.15).

   $R_3 = 2A_0 R_1 = 8R_1 = 8(120k) = 960k$

   Use $R_3 = 1$ Meg.

5. Calculate C_1 from Equation (2.17.12):

   $C_1 = \dfrac{Q}{2\pi f_0 A_0 R_1} = \dfrac{2}{(2\pi f_0)(4)(120k)}$

   $C_1 = \dfrac{6.63 \times 10^{-7}}{f_0}$

A table of standard values for C_1 vs. f_0 is given below.

TABLE 2.17.2

f_0 (Hz)	C_1
32	$0.022\mu F$
64	$0.011\mu F$
125	$0.0056\mu F$
250	$0.0027\mu F$
500	$0.0015\mu F$
1k	680 pF
2k	330 pF
4k	160 pF
8k	82 pF
16k	43 pF

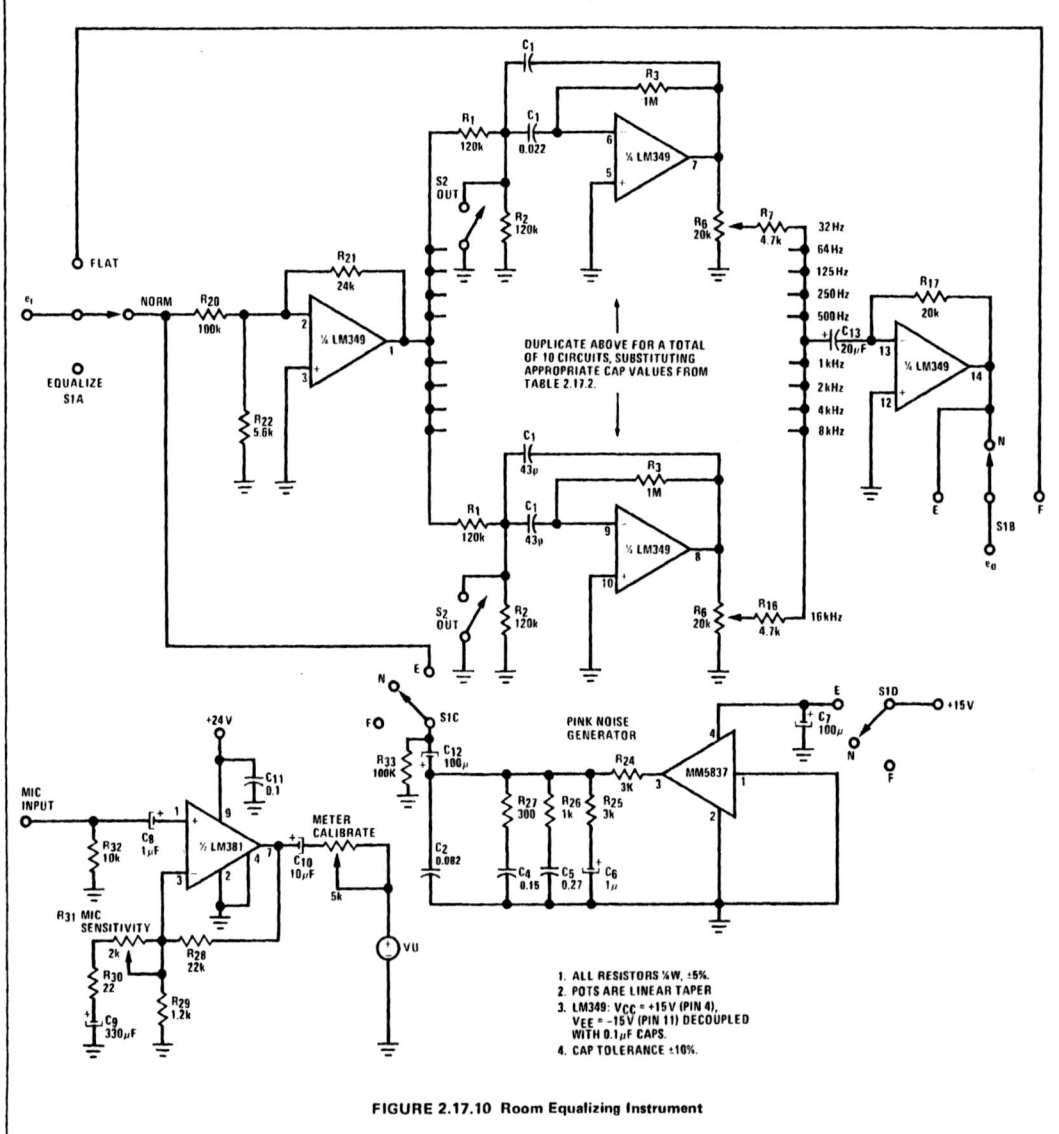

FIGURE 2.17.10 Room Equalizing Instrument

The complete room equalizing instrument appears as Figure 2.17.10. The input buffer and output summer are similar to those that appear in Figure 2.17.2, with some important differences. The input buffer acts as an active attenuator with a gain of 0.25 and the output summer has variable gain as a function of slider position. The purpose of these features is to preserve unity gain through a system that is really "cut-only" (since the gain of each filter section is fixed and the output is dropped across the potentiometers). The result is to create a boost and cut effect about the midpoint of the pot which equals unity gain. To see this, consider just one filter section, and let the input to the system equal 1V. The output of the buffer will be 0.25V and the filter output at the top of potentiometer R_6 will again be 1V (since $A_0 = 4$). The gain of the summer is given by $R_{17}/R_7 \approx 4$ when the slider of R_6 is at maximum, so the output will be equal to 4V, or +12dB relative to the input. With the slider at midposition the 4.7k summer input resistor R_7 effectively parallels 1/2 of R_6 for a net resistance from slider to ground of $4.7k \| 10k \approx 3.2k$. The voltage at the top of the pot is attenuated by the voltage divider action of the 10kΩ (top of pot to slider) and the 3.2kΩ (slider to ground). This voltage is approximately equal to 0.25V and is multiplied by 4 by the summer for a final output voltage of 1V, or 0dB relative to the input. With the slider at minimum there is no output from this section, but the action of the "skirts" of the adjacent filters tends to create -12dB cut relative to the input. So the net result is a ±12dB boost and cut effect from a cut only system.

The pink noise generator from Figure 2.17.7 is included as the noise source to each filter section only when switch S1 (3 position, 4 section wafer) is in the "Equalize" position. Power is removed from the pink noise generator during normal operation so that noise is not pumped back onto the supply lines. Switch S2 located on each filter section is used to ground the input during the equalizing process. The LM381 dual low noise preamplifier is used as the microphone amplifier to drive the VU meter. The second channel is added by duplicating all of Figure 2.17.9 with the exception of the pink noise generator which can be shared. Typical frequency response is given by Figure 2.17.11. While the system appears complex, a complete two-channel instrument is made with just 8 ICs (6-LM349, 1-LM381, and 1-MM5837).

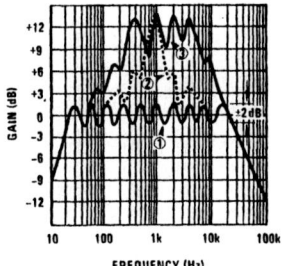

① ALL CONTROLS FLAT
② 1kHz BOOST, ALL OTHERS FLAT
③ 500Hz, 1kHz, 2kHz, 4kHz BOOST, ALL OTHERS FLAT

FIGURE 2.17.11 Typical Frequency Response of Room Equalizer

For detailed discussions about room equalization, the interested reader is directed to the references that follow this section.

REFERENCES

1. Davis, D., "Facts & Fallacies on Detailed Sound System Equalization," *AUDIO* reprint available from ALTEC, Anaheim, California.
2. Eargle, J., "Equalization in the Home," *AUDIO*, vol. 57, no. 11, November 1973, pp. 54-62.
3. Eargle, J., "Equalizing the Monitoring Environment," *Jour. Aud. Eng. Soc.*, vol. 21, no. 2, March 1973, pp. 103-107.
4. Engebretson, M. E., "One-Third Octave Equalization Techniques and Recommended Practices," Technical Letter No. 232, ALTEC, Anaheim, California.
5. Heinz, H. K., "Equalization Simplified," *Jour. Aud. Eng. Soc.*, vol. 22, no. 9, November 1974, pp. 700-703.
6. Queen, D., "Equalization of Sound Reinforcement Systems," *AUDIO*, vol. 56, no. 11, November 1972, pp. 18-26.
7. Thurmond, G. R., "A Self-Contained Instrument for Sound-System Equalization," *Jour. Aud. Eng. Soc.*, vol. 22, no. 9, November 1974, pp. 695-699.

2.18 MIXERS

2.18.1 Introduction

A microphone mixing console or "mixer" is an accessory item used to combine the outputs of several microphones into one or more common outputs for recording or public address purposes. They range from simple four input-one output, volume-adjust-only units to ultra-sophisticated sixteen channel, multiple output control centers that include elaborate equalization, selective channel reverb, taping facilities, test oscillators, multi-channel panning, automatic mix-down with memory and recall, individual VU meters, digital clocks, and even a built-in captain's chair. While appearing complex and mysterious, mixing consoles are more repetitious than difficult, being constructed from standard building-block modules that are repeated many times.

2.18.2 Six Input-One Output Mixer

A detailed analysis of all aspects of mixer design lies beyond the scope of this book; however, as a means of introduction to the type of design encountered Figure 2.18.1 is included to show the block diagram of a typical six input-one output mixer. Below each block, the section number giving design details is included in parentheses for easy cross reference.

Individual level and tone controls are provided for each input microphone, along with a choice of reverb. All six channels are summed together with the reverb output by the master summing amplifier and passed through the master level control to the octave equalizer. The output of the equalizer section drives the line amplifier, where monitoring is done via a VU meter.

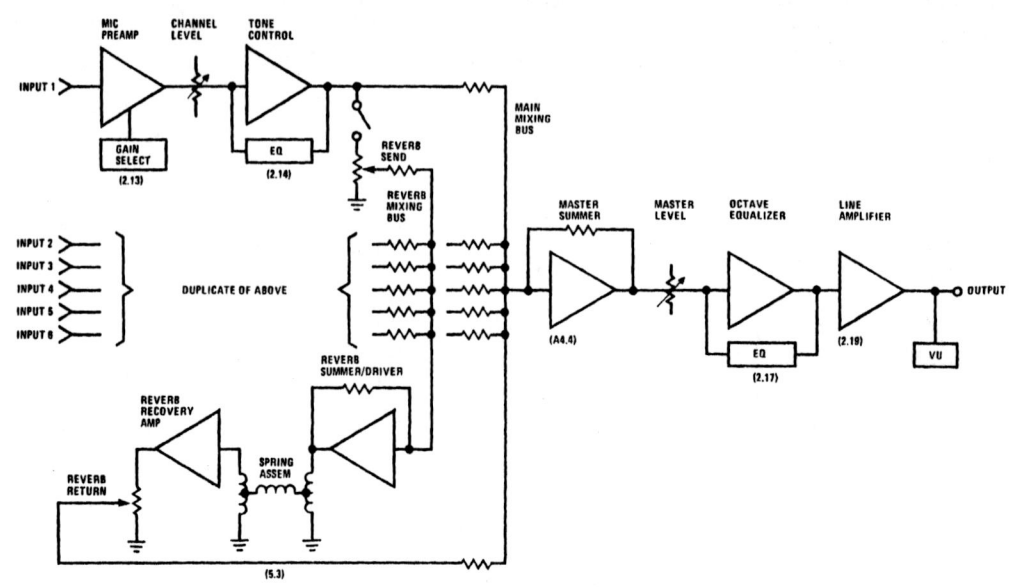

FIGURE 2.18.1 Six Input-One Output Microphone Mixing Console (Design details given in sections shown in parentheses.)

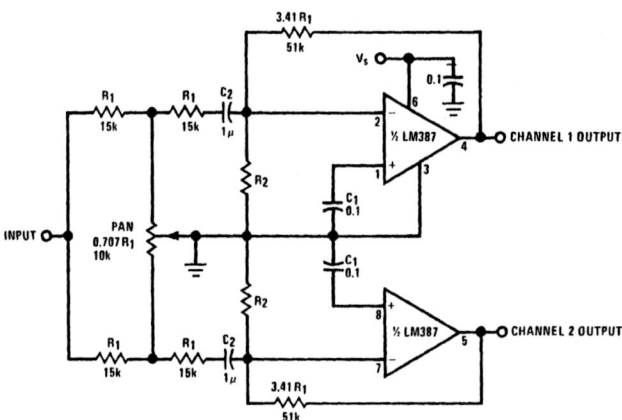

FIGURE 2.18.2 Two Channel Panning Circuit

Expansion of the system to any number of inputs requires only additional input modules, with the limiting constraint being the current driving capability of the summing amplifiers. (The summing amp must be capable of sourcing and sinking the *sum* of all of the input amplifiers driving the summing bus. For example, consider ten amplifiers, each driving a 10kΩ summing input resistor to a maximum level of 5V$_{RMS}$. The summing amplifier is therefore required to handle 5mA.) Expanding the number of output channels involves adding additional parallel summing busses and amplifiers, each with separate level, equalizer, and VU capabilities. Other features (test oscillator, pink noise generator, panning, etc.) may be added per channel or per console as required.

2.18.3 Two Channel Panning Circuit

Having the ability to move the apparent position of one microphone's input between two output channels often is required in recording studio mixing consoles. Such a circuit is called a panning circuit (short for panoramic control circuit) or a pan-pot. Panning is how recording engineers manage to pick up your favorite pianist and "float" the sound over to the other side of the stage and back again. The output of a pan circuit is required to have unity gain at each extreme of pot travel (i.e., all input signal delivered to one output channel with the other output channel zero) and -3dB output from each channel with the pan-pot centered. Normally panning requires two oppositely wound controls ganged together; however, the circuit shown in Figure 2.18.2 provides smooth and accurate panning with only one linear pot. With the pot at either extreme the effective negative input resistor equals $3.41R_1$ (see Appendix A3.1) and gain is unity. Centering the pot yields an effective input resistor on each side equal to $4.83R_1$ and both gains are -3dB. The net input impedance as seen by the input equals $0.6R_1$, independent of pan-pot position. Using standard 5% resistor values as shown in Figure 2.18.2, gain accuracies within 0.4dB are possible; replacing R_1 with 1% values (e.g., input resistors equal 14.3kΩ and feedback resistors equal 48.7kΩ) allows gain accuracies of better than 0.1dB. Biasing resistor R_2 is selected per section 2.8 as a function of supply voltage. Capacitor C_1 is used to decouple the positive input, while C_2 is included to prevent shifts in output DC level due to the changing source impedance.

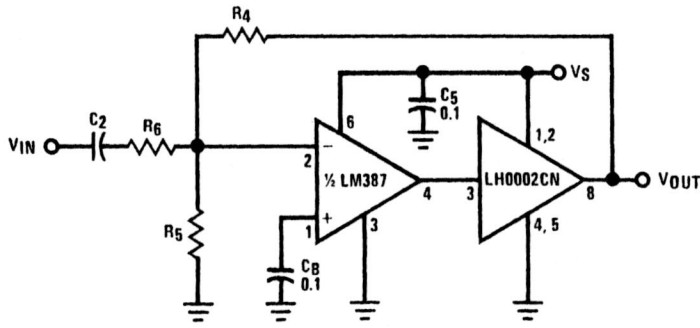

FIGURE 2.19.1 Preamp Current Booster

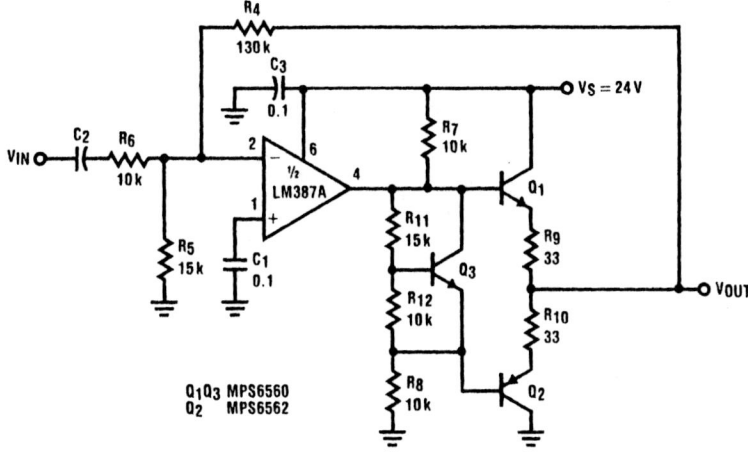

FIGURE 2.19.2 Discrete Current Booster Design

2.19 DRIVING LOW IMPEDANCE LINES

The output current and drive capability of a preamp may be increased for driving low impedance lines by incorporating a LH0002CN current amplifier within the feedback loop (Figure 2.9.1). Biasing and gain equations remain unchanged and are selected per section 2.8. Output current is increased to a maximum of ±100mA, allowing a LM387 to drive a 600Ω line to a full 24dBm when operated from a +36V supply. Insertion of the LH0002C adds less than 10 degrees additional phase shift at 15MHz, thereby not appreciably affecting the stability of the LM387 ($A_v \geq 10$).

Comparable performance can be obtained with the discrete design of Figure 2.19.2 for systems where parts count is not critical. Typical measured characteristics show a bandwidth of 15Hz – 250kHz at +10dBm output, with THD below 0.02% up to 20kHz. A maximum output level of +16dBm can be obtained before clipping.

2.20 NOISELESS AUDIO SWITCHING

2.20.1 Active Switching

As prices of mechanical switches continue to increase, solid state switching element costs have decreased to the point where they are now cost effective. By placing the switch on the PC board instead of the front panel, hum pickup and crosstalk are minimized, while at the same time replacing the complex panel switch assemblies.

The CMOS transmission gate is by far the cheapest solid state switching element available today, but it is plagued with spiking when switched, as are all analog switches. The switching spikes are only a few hundred nanoseconds wide, but a few volts in magnitude, which can overload following audio stages, causing audible pops. The switch spiking is caused by the switch's driver coupling through its capacitance to the load. Increasing the switch driver's transition time minimizes the spiking by reducing the transient current through the switch capacitance. Unfortunately, CMOS transmission gates do not have the drivers available, making them less attractive for audio use.

Discrete JFETs and monolithic JFET current mode analog switches such as AM97C11 have the switch element's input available. This allows the transition time of the drive to be tailored to any value, making noiseless audio switching possible. The current mode analog switches only need a simple series resistor and shunt capacitor to ground between the FET switch and the driver. (See Figure 2.10.1.)

Discrete JFETs may be used in place of the quad current mode switch; or, they can be used as voltage mode switches at a savings to the amplifier but at the expense of additional resistors and a diode.

Driver rise times shown in the figures, in the 1-10ms range, will result in coupled voltage spikes of only a few mV when used with the typical impedances found in audio circuits.

2.20.2 Mechanical Switching

A common mechanical switching arrangement for audio circuits involves a simple switch located after a coupling capacitor as diagrammed in Figure 2.20.3. For "pop" free switching the addition of a pull-down resistor, R_1, is essential. Without R_1 the voltage across the capacitor tends to float up and pops when contact is made again; R_1 holds the free end of the capacitor at ground potential, thus eliminating the problem.

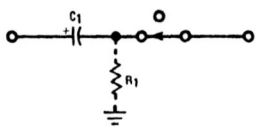

FIGURE 2.20.3 Capacitor Pull-Down Resistor

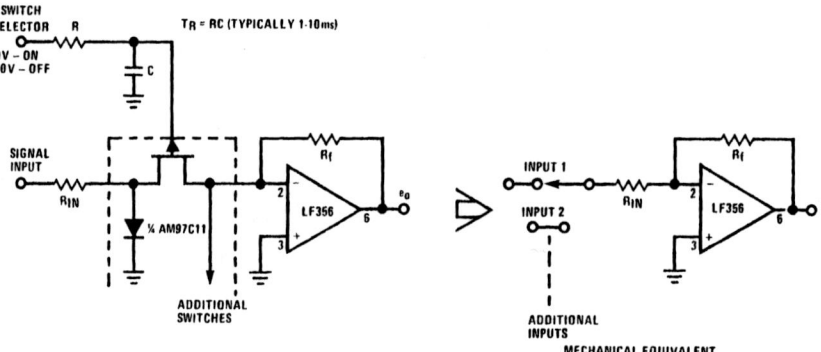

FIGURE 2.20.2 A Deglitched Current Mode Switch

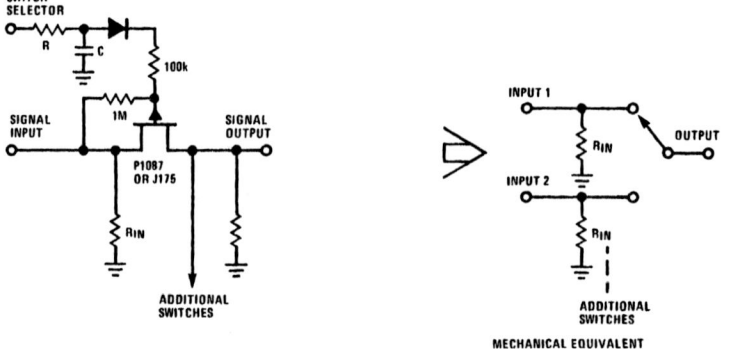

FIGURE 2.20.1 Deglitched Voltage Mode Switch

Section 3.0
AM, FM and FM Stereo

3.0 AM, FM and FM Stereo

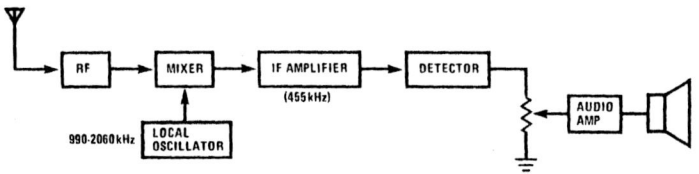

FIGURE 3.1.1 Superheterodyne Radio

3.1 AM RADIO

3.1.1 Introduction

Almost exclusively, the superheterodyne circuit reigns supreme in the design of AM broadcast radio. This circuit, shown in Figure 3.1.1, converts the incoming signal — 535kHz to 1605kHz — to an intermediate frequency, usually 262.5kHz or 455kHz, which is further amplified and detected to produce an audio signal which is further amplified to drive a speaker. Other types of receiver circuits include tuned RF (TRF) and regenerative.

In the tuned RF, the incoming signal is amplified to a relatively high level by a tuned circuit amplifier, and then demodulated.

Controlled positive feedback is used in the regenerative receiver to increase circuit Q and gain with relatively few components to obtain a satisfactory measure of performance at low cost.

Both the TRF and regenerative circuits have been used for AM broadcast, but are generally restricted to low cost toy applications.

3.1.2 Conversion of Antenna Field Strength to Circuit Input Voltage

Looking at Figure 3.1.1, the antenna converts incoming radio signals to electrical energy. Most pocket and table radios use ferrite loop antennas, while automobile radios are designed to work with capacitive whip antennas.

Ferrite Loop Antennas

The equivalent circuit of a ferrite rod antenna appears as Figure 3.1.2. Terms and definitions follow:

L = antenna inductance
C = tuning capacitor plus stray capacitance (20-150 pF typ.)
N_O = antenna turns ratio — primary to secondary
R_{IN} = circuit input impedance
R_p = equivalent parallel loss resistance (primarily a function of core material)
R_L = equivalent loading resistance
V_{IN} = volts applied to circuit
V_{ID} = volts induced to antenna
V_T = voltage transferred across tank
Q_u = unloaded Q of antenna coil
Q_L = loaded Q of antenna circuit
H_{eff} = effective height of antenna in meters
E = field strength in volts/meter

Necessary design equations appear below:

$$Q_u = \frac{R_p}{X_L} \quad (3.1.1)$$

$$Q_L = \frac{R_p \| R_L}{X_L} = \frac{R_T}{X_L} \quad (3.1.2)$$

$$R_L = N_O^2 R_{IN} \quad (3.1.3)$$

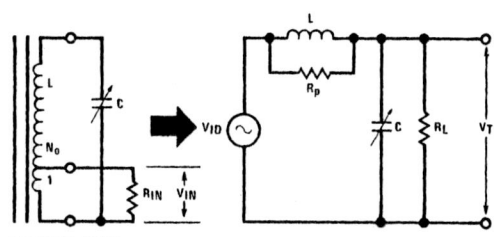

FIGURE 3.1.2 Ferrite Rod Antenna Equivalent Circuit

$$V_T = Q_L V_{ID} \quad (3.1.4)$$

$$V_{ID} = H_{eff} E \quad (3.1.5)$$

$$V_{IN} = \frac{V_T}{N_O} \quad (3.1.6)$$

The effective height of the antenna is a complex function of core and coil geometry, but can be approximated[1] by:

$$H_{eff} \approx \frac{2\pi \mu_r N_1 A}{\lambda} \quad (3.1.7)$$

where: N_1 = total number of turns
μ_r = relative permeability of antenna rod (primarily function of length)
A = cross sectional area of rod
λ = wavelength of received signal
$= \dfrac{3 \times 10^8 \text{m/sec}}{\text{freq (Hz)}}$

Noise voltage is calculated from the total Thevenin equivalent loading resistance, $R_T = R_p \| R_L$, using Equation (3.1.8):

$$e_n = \sqrt{4 K T \Delta f R_T} \qquad (3.1.8)$$

where: Δf = 3dB bandwidth of IF
T = temperature in °K
K = Boltzmann's constant
= 1.38×10^{-23} joules/°K

The signal-to-noise ratio in the antenna circuit can now be expressed as Equation (3.1.9):

$$S/N = \frac{V_T m}{e_n} = \frac{Q_L H_{eff} E m}{\sqrt{4 K T \Delta f R_T}} \qquad (3.1.9)$$

where: m = index of modulation

Example 3.1.1

Specify the turns ratio N_0, total turns N_1, effective height H_{eff}, and inductance required for an antenna wound onto a rod with the characteristics shown, designed to match an input impedance of 1 kΩ. Calculate the circuit input voltage resulting from a field strength of 100 μV/m with 20 dB S/N in the antenna circuit. Assume a 15-365 pF tuning capacitor set at 100 pF for an input frequency of 1 MHz.

Given: R_{IN} = 1 kΩ f_o = 1 MHz
E = 100 μV/m   rod dia. = 1.5 cm
S/N = 20 dB μ_r = 65 (rod length = 19 cm)
C = 100 pF   m = 0.3
Q_u = 200 Δf = 10 kHz

Calculate L, N_0, H_{eff}, N_1, V_{IN}

1. Since the circuit is "tuned," i.e., at resonance, then $X_L = X_C$, or

$$L = \frac{1}{C(2\pi f_o)^2} = \frac{1}{100 pF (2\pi \times 1 \times 10^6)^2}$$

$$= 2.53 \times 10^{-4} H$$

$$L \approx 250 \mu H$$

2. From Equation (3.1.1):

$$R_p = Q_u X_L = 200 \times 2\pi \times 1 MHz \times 250 \mu H$$

$$R_p \approx 314k$$

3. For matched conditions and using Equation (3.1.3):

$$R_p = R_L = N_o^2 R_{IN}$$

$$N_o = \sqrt{\frac{R_p}{R_{IN}}} = \sqrt{\frac{314k}{1k}} = 17.7$$

$$N_o \approx 18:1$$

4. From Equations (3.1.1) and (3.1.2):

$$Q_L = \frac{R_p \| R_L}{X_L} = \frac{R_p}{2 X_L} = \frac{Q_u}{2} \text{ since } R_p = R_L$$

$$Q_L = 100$$

5. Rearranging Equation (3.1.9) and solving for required H_{eff}:

$$H_{eff} = \frac{S/N \sqrt{4 K T \Delta f R_T}}{Q_L E m}$$

$$= \frac{10 \sqrt{(4)(1.38 \times 10^{-23})(300)(10 kHz)(157k)}}{(100)(100 \mu V/m)(0.3)}$$

$$= 1.7 cm$$

6. Rearranging Equation (3.1.7) and solving for N_1:

$$N_1 = \frac{H_{eff} \lambda}{2 \pi \mu_r A}$$

$$= \frac{(0.017 m)(3 \times 10^8 m/sec)}{(2\pi)(65)(1 \times 10^6 Hz)(\pi)(7.5 \times 10^{-3} m)^2} = 70.7$$

$$N_1 \approx 71 \text{ turns}$$

7. From Equation (3.1.5):

$$V_{ID} = H_{eff} E$$

$$= 0.017 m \times 100 \mu V/m$$

$$V_{ID} = 1.7 \mu V$$

8. Find V_T from Equation (3.1.4):

$$V_T = Q_L V_{ID}$$

$$= 100 \times 1.7 \mu V$$

$$V_T = 170 \mu V$$

9. Using Equation (3.1.6), find V_{IN}:

$$V_{IN} = \frac{V_T}{N_o} = \frac{170 \mu V}{18}$$

$$V_{IN} = 9.4 \mu V$$

Capacitive Automotive Antennas

A capacitive automobile radio antenna can be analyzed in a manner similar to the loop antenna. Figure 3.1.3 shows the equivalent circuit of such an antenna. C_1 is the capacitance of the vertical rod with respect to the horizontal ground plane, while C_2 is the capacitance of the shielded cable connecting the antenna to the radio. In order to obtain a useful signal output, this capacitance is tuned out with an inductor, L. Losses in the inductor and the input resistance of the radio form R_L. The signal appearing at the input stage of the radio is related to field strength:

$$V_T = V_{ID} Q_L \frac{C_1}{C_T} \qquad (3.1.10)$$

where: V_{ID} is defined by Equation (3.1.5)
Q_L is defined by Equation (3.1.2)
$C_T = C_1 + C_2$

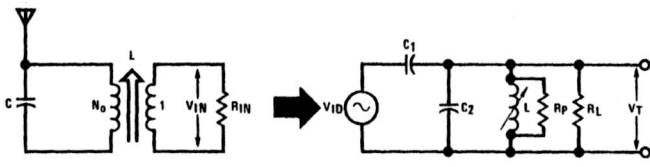

FIGURE 3.1.3 Capacitive Auto Antenna Equivalent Circuit

Similar to the ferrite rod antenna, the signal-to-noise ratio is given by:

$$S/N = \frac{H_{eff} \, E \, m \, Q_L \, (C_1/C_T)}{\sqrt{4KT\Delta f \, Q_L \, X_{CT}}} \quad (3.1.11)$$

The effective height of a capacitive vertical whip antenna can be shown[1] to equal Equation (3.1.12):

$$H_{eff} \approx \frac{h}{2} \quad (3.1.12)$$

where: h = antenna height in meters

Example 3.1.2

For comparison purposes, calculate the circuit input voltage, V_{IN}, for an automotive antenna operating in the same field as the previous example; assume same circuit input impedance of 1 kΩ and calculate the resultant S/N. Use the given data for a typical auto radio antenna extended two sections (1 meter).

Given: R_{IN} = 1 kΩ Δf = 10 kHz
        E = 100 μV/m C_1 = 10 pF
        Q_L = 80 C_T = 90 pF
        f_o = 1 MHz   m = 0.3

Calculate S/N, N_o, V_{IN}.

1. Calculate H_{eff} from Equation (3.1.12) and solve for X_{CT}

$$H_{eff} = \frac{h}{2} = 0.5 \, m$$

$$X_{CT} = \frac{1}{2\pi f C_T} = \frac{1}{2\pi \times 1\,MHz \times 90\,pF}$$

$$X_{CT} = 1768 \, \Omega$$

2. Rearranging Equation (3.1.11) and solving for S/N:

$$S/N = \frac{H_{eff} \, E \, m \, \frac{C_1}{C_T} \sqrt{Q_L}}{\sqrt{4KT\Delta F \, X_{CT}}}$$

$$S/N = \frac{(0.5)(100\mu V/m)(0.3)\frac{10pF}{90pF}\sqrt{80}}{\sqrt{(4)(1.38 \times 10^{-23})(300)(10k)(1768)}}$$

S/N = 27.55

S/N ≈ 29 dB

3. From Equations (3.1.10) and (3.1.5):

$$V_T = H_{eff} \, E \, Q_L \, \frac{C_1}{C_T}$$

$$= 0.5m \times 100\mu V/m \times 80 \times \frac{10pF}{90pF}$$

$$V_T = 444 \, \mu V$$

4. Since matching requires $R_p = R_L$, and resonance gives $X_{CT} = X_L$, then using Equation (3.1.2):

$$\frac{R_p}{2} = Q_L \, X_{CT}$$

$$R_p = 2 \times 80 \times \frac{1}{2\pi(1\,MHz)(90\,pF)} = 283k = R_L$$

5. Using Equation (3.1.3):

$$N_o = \sqrt{\frac{R_L}{R_{IN}}} = \sqrt{\frac{283k}{1k}} = 16.8$$

$$N_o \approx 17:1$$

6. From Equation (3.1.6):

$$V_{IN} = \frac{V_T}{N_o} = \frac{444\mu V}{17}$$

$$V_{IN} = 26.1 \, \mu V$$

7. From Equation (3.1.1):

$$Q_u = \frac{R_p}{X_{CT}} = 283k \times 2\pi \times 1\,MHz \times 90\,pF = 160$$

It is interesting to note that operating in the same field strength, the capacitive antenna will transfer approximately three times as much voltage to the input of the circuit, thus allowing the greater signal-to-noise ratio of 29 dB.

REFERENCES

1. Laurent, H. J. and Carvalho, C. A. B., "Ferrite Antennas for AM Broadcast Receivers," Application Note available from Bendix Radio Division of The Bendix Corporation, Baltimore, Maryland.

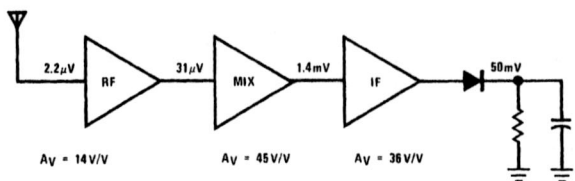

FIGURE 3.1.4 AM Radio Gain Stages

3.1.3 Typical AM Radio Gain Stages

The typical levels of Figure 3.1.4 give some idea of the gain needed in an AM radio. At the IF amplifier output, a diode detector recovers the modulation, and is generally designed to produce approximately $50\,mV_{RMS}$ of audio with m = 0.3. The gain required is therefore:

$$A_V = \frac{50\,mV}{2.2\,\mu V} = 23\,kV/V \text{ or } 87\,dB$$

3.2 LM3820 AM RECEIVER SYSTEM

The LM3820 is a 3 stage AM radio IC designed as an improved replacement for the LM1820. It consists of the following functional blocks:

RF Amplifier   IF Amplifier
Oscillator   AGC Detector
Mixer   Regulator

The RF amplifier section (Figure 3.2.1) consists of a cascode amplifier Q_2 and Q_3, whose geometries are specially designed for low noise operation from low source impedances. Q_2 is protected from overloads coupled via capacitive antennae by two back to back diodes. The cascode configuration has very low feedback capacitance to minimize stability problems, and a high output impedance to maximize gain. In addition, bias components (Q_1, etc.) are included. Biased at 5.6mA, the input stage is useful for frequencies in excess of 50mHz. Figure 3.2.2a shows the transconductance as a function of frequency.

Transistors Q_4 and Q_5 make up the local oscillator circuit. Positive feedback from the collector of Q_5 to the base of Q_4 is provided by the resistor divider R_9 and R_8. The oscillator frequency is set with a tuned circuit connected between pin 2 and V_{CC}. Transistors Q_4 and Q_5 are biased at 0.5mA each, so the transconductance of the differential pair is 10mmhos. For oscillation, the impedance at pin 2 must be high enough to provide a voltage gain greater than the loss associated with the resistor divider network R_9, R_8 and the input impedance of Q_4. Values of load impedance greater than 400Ω satisfy this condition, with values of $10k\Omega$ or greater being commonly used.

The differential pair Q_6 and Q_7 serve as a mixer, being driven with current from the oscillator. The input signal, applied to pin 1, is multiplied by the local oscillator frequency to produce a difference frequency at pin 14. This signal, the IF, is filtered and stepped down to match the input impedance of the IF amplifier.

Transistors Q_9 and Q_{10} form the IF amplifier gain stage. Again, a cascode arrangement is used for stability and high gain for a gm of 90mmhos.

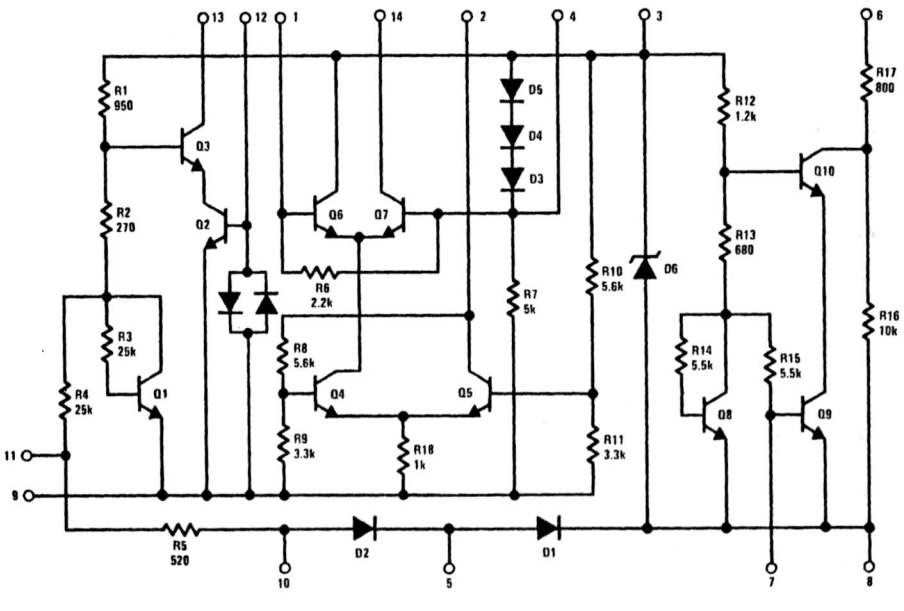

FIGURE 3.2.1 LM3820 Schematic Diagram

Basically, two ways exist for using the LM1820 in AM radio applications; these are illustrated in Figure 3.2.3. The mixer-IF-IF configuration (Figure 3.2.3a) results in an economical approach at some performance sacrifice because the mixer contributes excess noise at the antenna input, which reduces sensitivity. Since all gain is taken at the IF frequency, stability problems may be encountered if attention is not paid to layout.

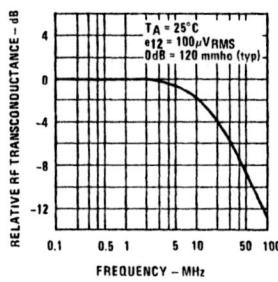

(a) RF Transconductance as a Function of Frequency

TABLE 3.2.1 Summary of Circuit Parameters

Parameter	RF Section	Mixer	IF
Input Resistance	1k	1.4k	1k
Input Capacitance	80 pF	8 pF	70 pF
Transconductance	120 mmhos	2.5 mmhos	90 mmhos
Input Noise Voltage, 6kHz Bandwidth	0.2 μV	0.5 μV	

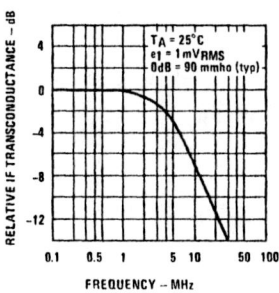

(b) IF Transconductance as a Function of Frequency

The RF-mixer-IF approach (Figure 3.2.3b) takes advantage of the low noise input stage to provide a high performance receiver for either automobile or high quality portable or table radio applications. Another approach which sacrifices little in performance, yet reduces cost associated with the three gang tuning capacitor, is to substitute a resistor for the tuned circuit load of the RF amplifier. The LM3820 has sufficient gain to allow for the mismatch and still provide good performance.

By appropriate impedance matching between stages, gain in excess of 120 dB is possible. This can be seen from Figure 3.2.3c, where the correct interstage matching values for maximum power gain are shown. The gain of the RF section is found from:

$$A_{V1} = \frac{V_1}{V_{IN}} = K_1 g_{m1} R_{L1} N K_2$$

where: N = turns ratio = $\sqrt{R_{sec}/R_{pri}}$

K_1 = 6dB loss @ output of RF amplifier due to matching 500k output impedance

K_2 = 6dB loss @ input to mixer due to matching 1.4k input impedance

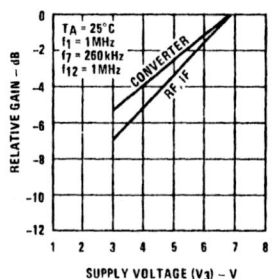

(c) Relative Gain as a Function of Supply Voltage (V_3)

FIGURE 3.2.2 LM3820 Performance Characteristics

An AGC detector is included on the chip. The circuit consists of diodes D_1 and D_2 which function as a peak to peak detector driven with IF signal from the output of the IF amplifier. As the output signal increases, a greater negative voltage is developed on pin 10 which diverts current away from the input transistor Q_2. This current reduction in turn reduces the gain of the input stage, effectively regulating the signal at the IF output.

A zener diode is included on the chip and is connected from V_{CC} to ground to provide regulation of the bias currents on the chip. However, the 3820 functions well at voltages below the zener regulating voltage as shown in Figure 3.2.2c. Table 3.2.1 summarizes circuit parameters.

For the values shown:

$$A_{V1} = \frac{1}{2}(120 \times 10^{-3})(500k)\sqrt{\frac{1.4k}{500k}} \cdot \frac{1}{2}$$

$$= 793.5 \approx 58 dB$$

Similarly, for the mixer:

$$A_{V2} = \frac{1}{2}(2.5 \times 10^{-3})(500k)\sqrt{\frac{1k}{500k}} \cdot \frac{1}{2}$$

$$= 14 \approx 23 dB$$

And for the IF:

$$A_{V3} = \frac{1}{2}(90 \times 10^{-3})(10k)\sqrt{\frac{5k}{10k}} \cdot \frac{1}{2}$$

$$= 159 \approx 44 dB$$

Total gain = $1.8 \times 16^6 \approx 125 dB$

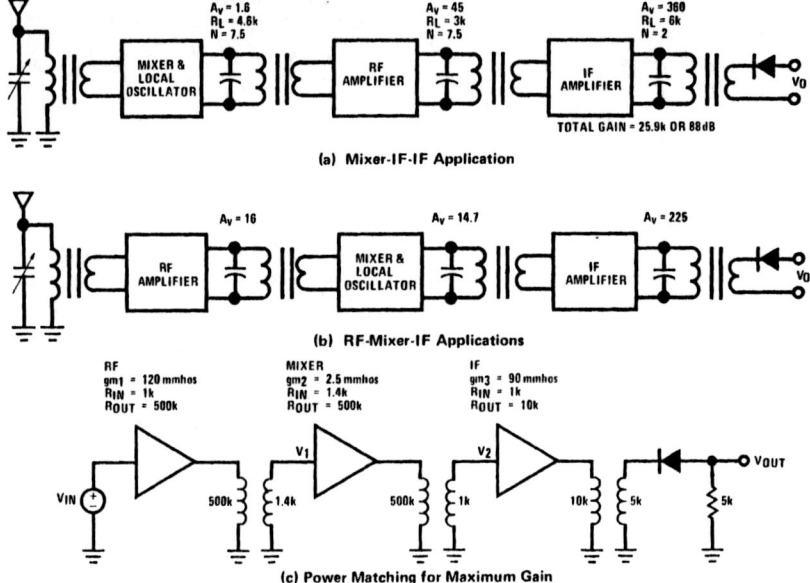

FIGURE 3.2.3 Circuit Configurations for AM Radios Using the LM3820

This much gain is undesirable from a performance standpoint, since it would result in 1.5V of noise to the diode detector due to the input noise, and it would probably be impossible to stabilize the circuit and prevent oscillation. From a design standpoint, it is desirable to mismatch the RF stage and mixer for less gain.

A capacitor tuned AM radio using the RF-mixer-IF configuration is shown in Figure 3.2.4. The RF amplifier is used with a resistor load to drive the mixer. A double tuned circuit at the output of the mixer provides selectivity, while the remainder of the gain is provided by the IF section, which is matched to the diode through a unity turns ratio transformer. A resistor from the detector to pin 10 bypasses the internal AGC detector in order to increase the recovered audio. The total gain in this design is 57k or 95dB from the base of the input stage to the diode detector.

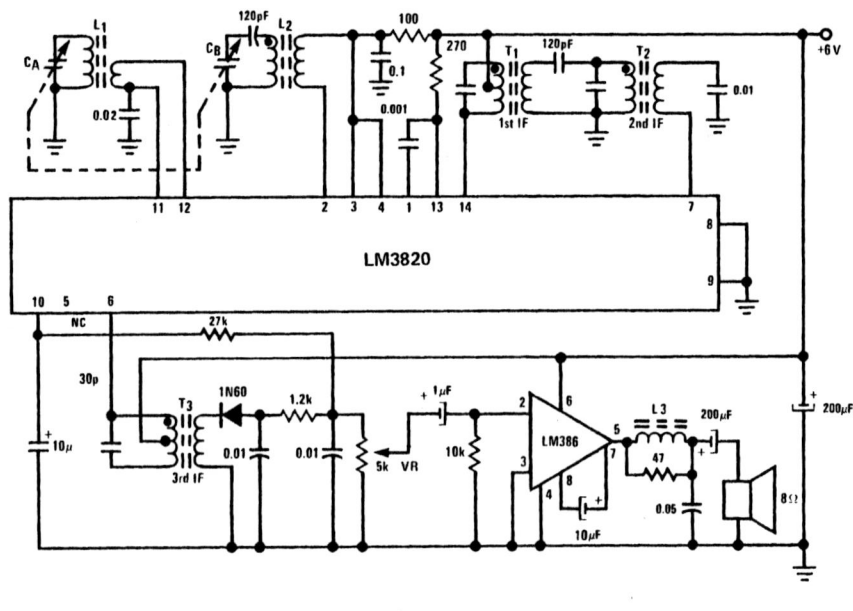

FIGURE 3.2.4 AM Radio Using RF-Mixer-IF

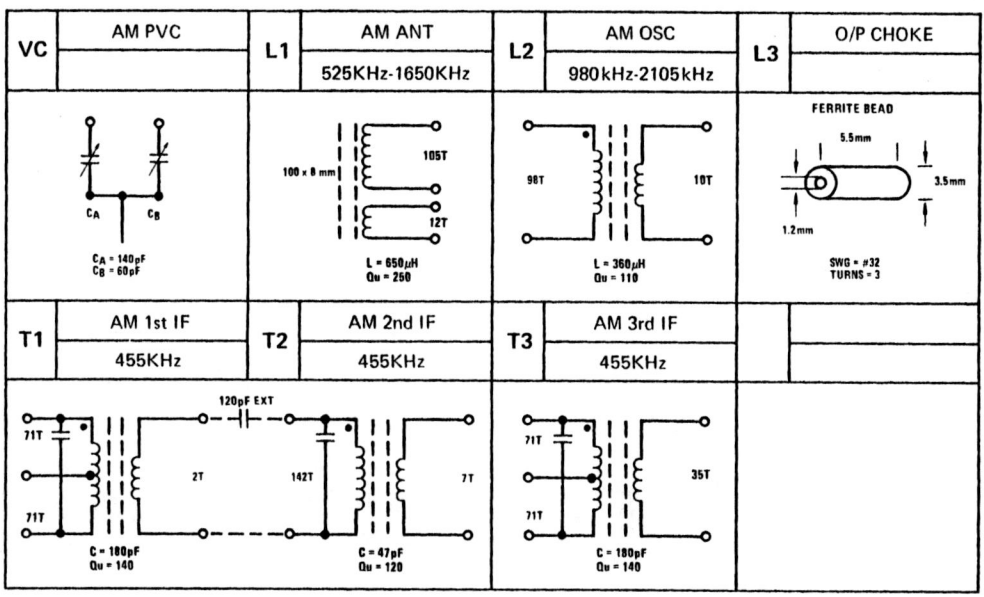

FIGURE 3.2.4 AM Radio Using RF-Mixer-IF continued

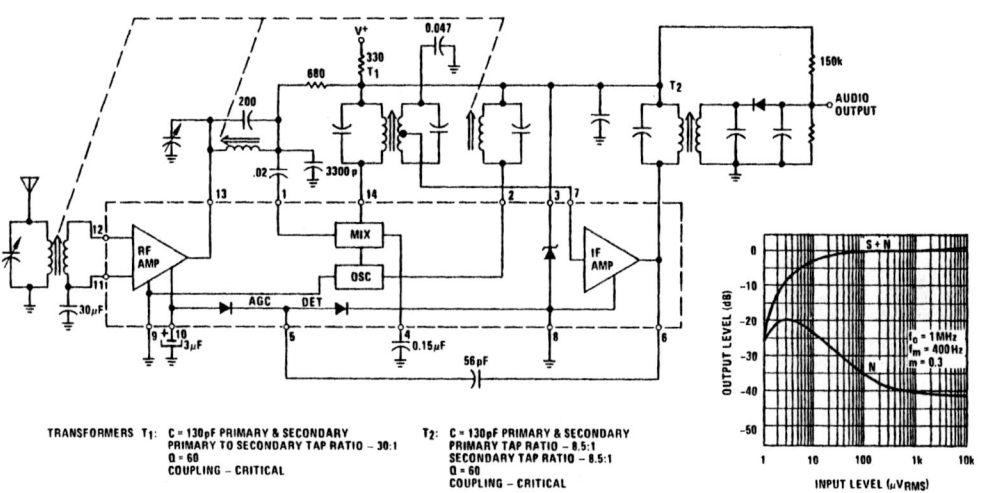

FIGURE 3.2.5 AM Auto Radio

A slug-tuned AM automobile radio design is shown in Figure 3.2.5. Tuning of both the input and the output of the RF amplifier and the mixer is accomplished with variable inductors. Better selectivity is obtained through the use of double tuned interstage transformers. Input circuits are inductively tuned to prevent microphonics and provide a linear tuning motion to facilitate push-button operation.

3.3 FM IF AMPLIFIERS AND DETECTORS

In the consumer field, two areas of application exist for FM IF amplifiers and detectors; in addition, applications exist in commercial two way and marine VHF FM radios:

TABLE 3.3.1 Application for FM-IF Amplifiers

Service	Frequency	Deviation	Input Limiting	Distortion
FM Broadcast	10.7 MHz	75 kHz	20 μV	0.5%
TV Sound	4.5 MHz	25 kHz	200 μV	1.5%
Two-Way Radio	various	5 kHz	5 μV	5%

The major requirement of an FM IF is good limiting characteristics, i.e., the ability to produce a constant output level to drive a detector regardless of the input signal level. This quality removes noise and amplitude changes that would otherwise be heard in the recovered signal.

3.4 THE LM3089 – TODAY'S MOST POPULAR FM IF SYSTEM

3.4.1 Introduction

LM3089 has become the most widely used FM IF amplifier IC on the market today. The major reason for this wide acceptance is the additional auxiliary functions not normally found in IC form. Along with the IF limiting amplifier and detector the following functions are provided:

1. A mute logic circuit that can mute or squelch the audio output circuit when tuning between stations.
2. An IF level or signal strength meter circuit which provides a DC logarithmic output as a function of IF input levels from $10\mu V$ to $100mV$ (four decades).
3. A separate AFC output which can also be used to drive a center-tune meter for precise visual tuning of each station.
4. A delayed AGC output to control front end gain.

The block diagram of Figure 3.4.1 shows how all the major functions combine to form one of the most complex FM IF amplifier/limiter and detector ICs in use today.

3.4.2 Circuit Description (Figure 3.4.2)

The following circuit description divides the LM3089 into four major subsections:

IF Amplifier
Quadrature Detector and IF Output
AFC, Audio and Mute Control Amplifiers
IF Peak Detectors and Drivers

IF Amplifier

The IF amplifier consists of three direct coupled amplifier-limiter stages Q_1-Q_{22}. The input stage is formed by a common emitter/common base (cascode) amplifier with differential outputs. The second and third IF amplifier stages are driven by Darlington connected emitter followers which provide DC level shifting and isolation. DC feedback via R_1 and R_2 to the input stage maintains DC operating point stability. The regulated supply voltage for each stage is approximately $5V$. The IF ground (pin 4) is used only for currents associated with the IF amplifiers. This aids in overall stability. Note that the current through R_9 and Z_1 is the only current on the chip directly affected by power supply variations.

Quadrature Detector and IF Output

FM demodulation in the LM3089 is performed accurately with a fully balanced multiplier circuit. The differential IF output switches the lower pairs Q_{34}, Q_{26} and Q_{39}, Q_{38}. The IF output at pin 8 is taken across 390Ω (R_{31}) and equals $300mV$ peak to peak. The upper pair-switching (Q_{35}, Q_{23}) leading by 90 degrees is through the externally connected quad coil at pin 9. The $5.6V$ reference at pin 10 provides the DC bias for the quad detector upper pair switching.

AFC, Audio and Mute Control Amplifiers

The differential audio current from the quad detector circuit is converted to a single ended output source for AFC by "turning around" the Q_{47} collector current to the collector of Q_{57}. Conversion to a voltage source is done externally

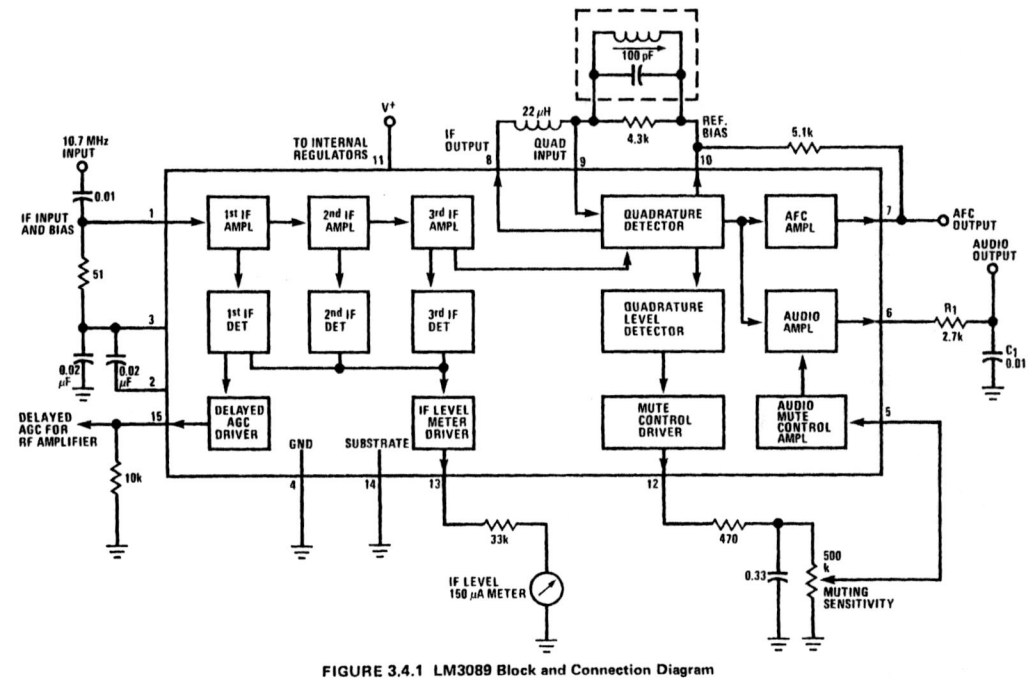

FIGURE 3.4.1 LM3089 Block and Connection Diagram

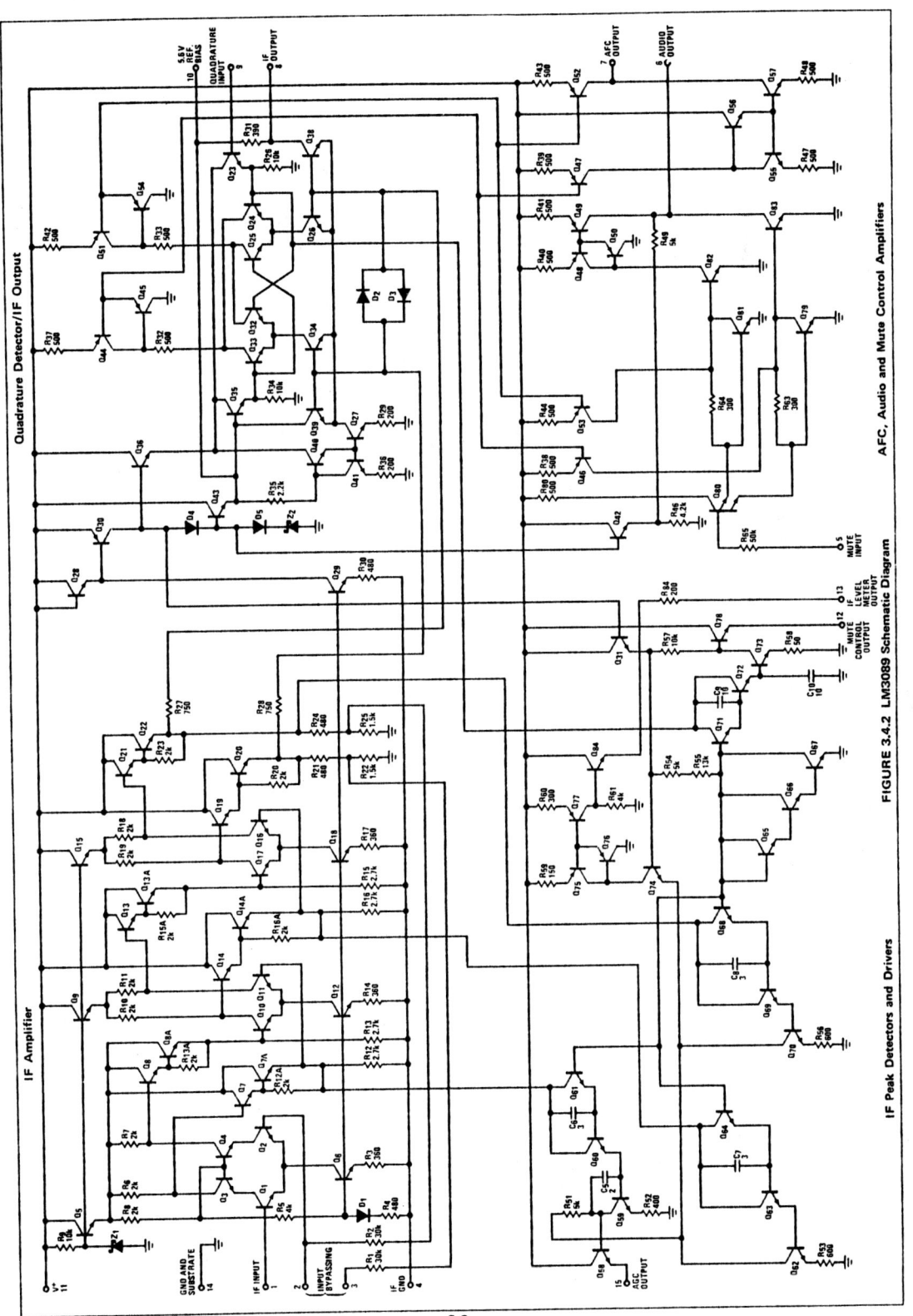

FIGURE 3.4.2 LM3089 Schematic Diagram

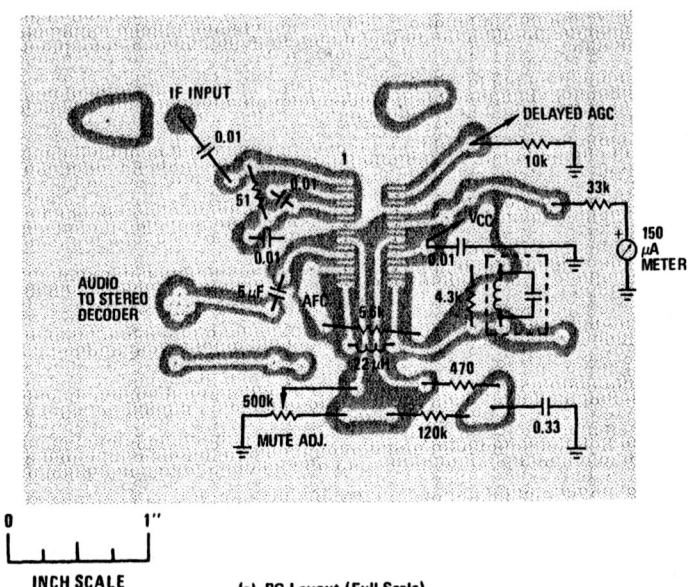

(a) PC Layout (Full Scale)

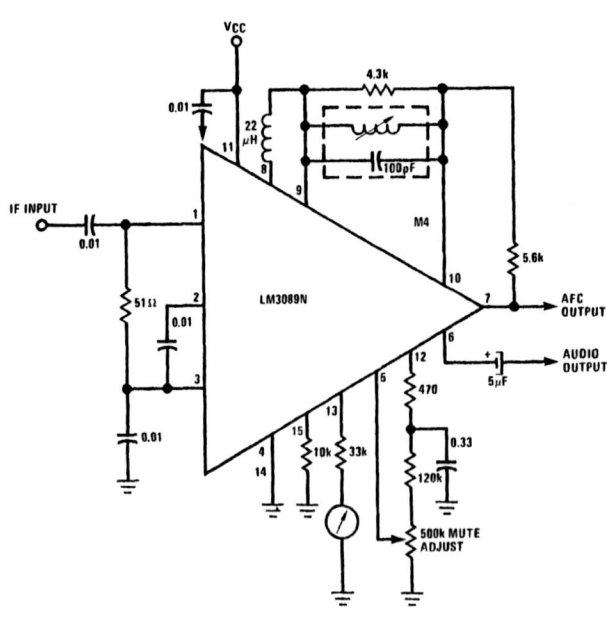

(b) Test Circuit

FIGURE 3.4.3 LM3089 Typical Layout & Test Circuit

by adding a resistor from pin 7 to pin 10. The audio amplifier stage operates in a similar manner as the AFC amplifier except that two "turn around" stages are used. This configuration allows the inclusion of muting transistor Q_{80}. A current into the base of Q_{80} will cause transistors Q_{79} and Q_{81} to saturate, which turns off the audio amplifier; the gain of the audio stage is set by internal resistor R_{49}. This 5kΩ resistor value is also the output impedance of the audio amplifier. When the LM3089 is used in mono receivers the 75μs de-emphasis (RC time constant) is calculated for a 0.01μF by including R_{49}. (RC = $[R_{49} + R_1][C_1]$, $R_1 = 75\mu s/0.01\mu F - 5k\Omega \approx 2.7k$, Figure 3.7.1.)

IF Peak Detectors and Drivers

Four IF peak or level detectors provide the delayed AGC, IF level and mute control functions. An output from the first IF amplifier drives the delayed AGC peak detector. Since the first IF amplifier is the last IF stage to go into limiting, Q_{60} and Q_{61} convert the first IF output voltage swing to a DC current (for IF input voltages between 10mV and 100mV). This changing current (0.1 to 1mA) is converted to a voltage across R_{51}. Emitter follower Q_{58} buffers this output voltage for pin 15. The top of resistor R_{51} is connected to a common base amplifier Q_{74} along with the output currents from the 2nd and 3rd stage IF peak detectors (which operate for IF input voltages between 10μV and 10mV). The output current from Q_{75} is turned around or mirrored by Q_{75}, Q_{76}, and Q_{77}, cut in half, then converted to a voltage across R_{61}. Emitter follower Q_{84} buffers this voltage for pin 13.

The fourth peak detector "looks" at the IF voltage developed across the quad coil. For levels above about 120mV at pin 9, Q_{73} will saturate and provide no output voltage at pin 12. Because the IF level at pin 9 is constant, as long as the last IF amplifier is in limiting, pin 12 will remain low. Sudden interruptions or loss of the pin 9 IF signal due to noise or detuning of the quad coil will allow the collector of Q_{73} to rise quite rapidly. The voltage at the collector of Q_{73} is buffered by Q_{78} for pin 12.

3.4.3 Stability Considerations

Because the LM3089 has wide bandwidth and high gain (> 80dB at 10.7MHz), external component placement and PC layout are critical. The major consideration is the effect of output to input coupling. The highest IF output signal will be at pins 8 and 9; therefore, the quad coil components should not be placed near the IF input pin 1. By keeping the input impedance low (< 500Ω) the chances of output to input coupling are reduced. Another and perhaps the most insidious form of feedback is via the ground pin connections. As stated earlier the LM3089 has two ground pins; the pin 4 ground should be used only for the IF input decoupling. The pin 4 ground is usually connected to the pin 14 ground by a trace under the IC. Decoupling of V_{CC} (pin 11), AGC driver (pin 15), meter driver (pin 13), mute control (pin 12) and in some cases the 5.6V REF (pin 10) should be done on the ground pin 14 side of the IC. The PC layout of Figure 3.4.3 has been used successfully for input impedances of 500Ω (1kΩ source/1kΩ load).

3.4.4 Selecting Quad Coil Components

The reader can best understand the selection process by example (see Figure 3.7.4):

Given: require quad coil bandwidth equal to 800kHz
f_o = 10.7 MHz
Q_u (unloaded) = 75

Find: L_{CH} and R_{EXT}

Find loaded Q of quad coil for required BW (Q_L)

$$Q_L = \frac{f_o}{BW} = \frac{10.7\,MHz}{0.8\,MHz} = 13.38$$

Find total resistance across quad coil for required BW (R_T)

$$R_T = Q_L X_{LI} = 13.38\,(2\pi f_o L_I) = 1981\Omega$$

Find reactance of coupling choke (XL_{CH})

$$XL_{CH} = \frac{R_T V_8}{V_9} = \frac{1981 \times 0.110}{0.15} = 1453\Omega$$

Find inductance of coupling choke (L_{CH})

$$L_{CH} = \frac{XL_{CH}}{2\pi f} = \frac{1453\Omega}{6.72 \times 10^7} = 22\mu H$$

Find parallel resistance of the unloaded quad coil (R_p)

$$R_p = X_{LI} Q_{UL} = 148\Omega \times 75 = 11.1k\Omega$$

Convert R_{31}, L_{CH} series to parallel resistance (R_{L31})

$$R_{L31} = \frac{(XL_{CH})^2}{R_{31}} + R_{31} = 5803\Omega$$

Find R_{EXT} for $R_T = R_p \| R_{L31} \| R_{EXT}$

$$R_{EXT} = \frac{1}{\frac{1}{R_T} - \frac{1}{R_p} - \frac{1}{R_{L31}}} = 4126$$

Use R_{EXT} = 4.3k.

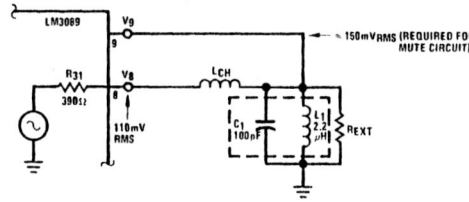

FIGURE 3.4.4 Quad Coil Equivalent Circuit

3.4.5 Typical Application of the LM3089

The circuit in Figure 3.4.5 illustrates the simplicity in designing an FM IF. The ceramic filters used in this application have become very popular in the last few years because of their small physical size and low cost. The filters eliminate all but one IF alignment step. The filters are terminated at the LM3089 input with 330Ω. Disc ceramic type capacitors with typical values of 0.01 to 0.02μF should be used for IF decoupling at pins 2 and 3.

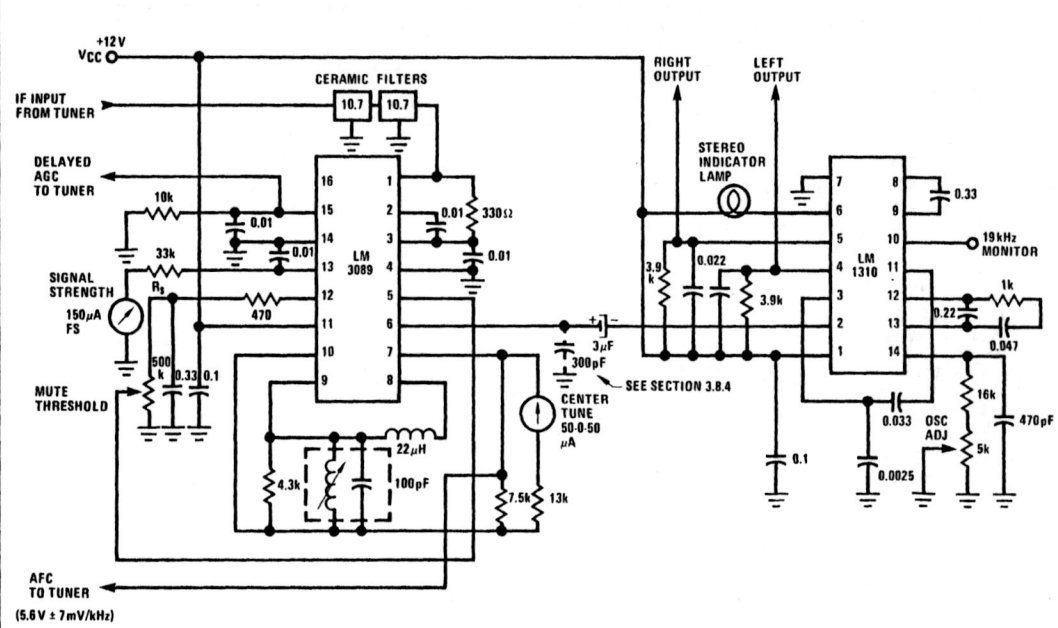

FIGURE 3.4.5 Typical Application of the LM3089

The AFC output at pin 7 can serve a dual purpose. In Figure 3.4.6 AFC sensitivity, expressed as mV/kHz, is programmed externally with a resistor from pin 7 to pin 10. A voltage reference other than pin 10 may be used as long as the pin 7 voltage stays less than 2V from the supply and greater than 2V from ground. The voltage change for a 5kΩ resistor will be ≈ 7.5mV/kHz or ≈ 1.5µA/kHz. The AFC output can also be used to drive a center tune meter. The full scale sensitivity is also programmed externally. The wide band characteristics of the detector and audio stage make the LM3089 particularly suited for stereo receivers. The detector bandwidth extends greater than 1MHz, therefore the phase delay of the composite stereo signal, especially the 38kHz side bands, is essentially zero.

The audio stage can be muted by an input voltage to pin 5. Figure 3.4.8 shows this attenuation characteristic. The voltage for pin 5 is derived from the mute logic detector pin 12. Figure 3.4.7 shows how the pin 12 voltage rises when the IF input is below 100µV. The 470Ω resistor and 0.33µF capacitor filter out noise spikes and allow a smooth mute transition. The pot is used to set or disable the mute threshold. When the pot is set for maximum mute sensitivity some competitors' versions of the LM3089 would cause a latch-up condition, which results in pin 12 staying high for all IF input levels. National's LM3089 has been designed such that this latch-up condition cannot occur.

The signal strength meter is driven by a voltage source at pin 13 (Figure 3.4.9). The value of the series resistor is determined by the meter used:

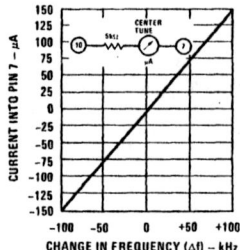

FIGURE 3.4.6 AFC (Pin 7) Characteristics vs. IF Input Frequency Change

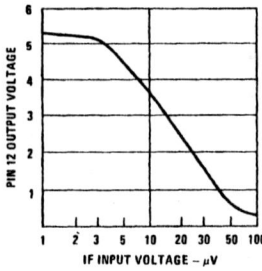

FIGURE 3.4.7 Mute Control Output (Pin 12) vs. IF Input Signal

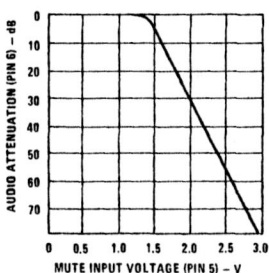

FIGURE 3.4.8 Typical Audio Attenuation (Pin 6) vs. Mute Input Voltage (Pin 5)

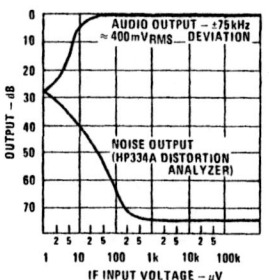

FIGURE 3.4.10 Typical (S+N)/N and IF Limiting Sensitivity vs. IF Input Signal

$$R_S = \frac{V_{MAX}(13)}{I_{FS}} = \frac{5V}{150\mu A} = 33k$$

The maximum current from pin 13 should be limited to approximately 2 mA. Short circuit protection has been included on the chip.

The delayed AGC (pin 15) is also a voltage source (Figure 3.4.9). The maximum current should also be limited to approximately 2 mA.

Figure 3.4.10 shows the typical limiting sensitivity (measured at pin 1) of the LM3089 when configured per Figure 3.4.3b and using PC layout of Figure 3.4.3a.

3.5 THE LM3189

3.5.1 Introduction

The LM3189 offers all the features of the LM3089 with improvements in performance in some areas, and increased flexibility in others. Since the major functions of the LM3189 are similar to the LM3089, the following sections will detail only the changes that have been made.

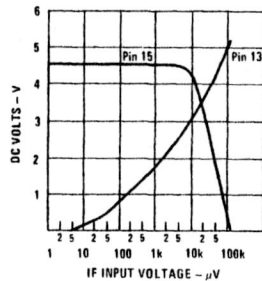

FIGURE 3.4.9 Typical AGC (Pin 15) and Meter Output (Pin 13) vs. IF Input Signal

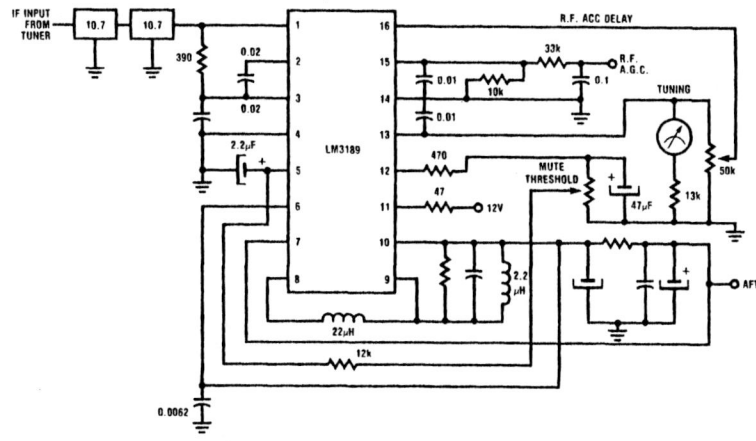

FIGURE 3.5.1 Typical Application of the LM3189

3.5.2 I.F. Amplifier

The input cascode stage has been optimized for low input capacitance and high gain for use with ceramic filters. An improvement in the I.F. amplifier noise performance has been accomplished by reducing the IC bandwidth. If the amplifier bandwidth is significantly higher than that needed to accommodate the operating I.F. frequency, out of band signals can be amplified and multiplied together in the non-linear stages to produce in-band noise components. The IC bandwidth has been decreased to about 15mHz and this will also help make the p.c.b. layout less sensitive. Nevertheless, attention to layout is still important and the I.F. amplifier ground (Pin 4) should be used only for decoupling the I.F. amplifier input.

3.5.3 R.f. a.g.c.

Instead of having a fixed r.f. a.g.c. delay threshold at the 10mV input signal level, the LM3189 allows the designer to select the a.g.c. threshold at any point between 200μV and 200mV, depending on the individual tuner requirements. A control voltage at the previously unused Pin 16 will determine the onset of r.f. a.g.c. action with a threshold level of 1.3V. This control voltage is obtained by a resistive divider connected to the signal strength meter Pin 13 as shown in Figure 3.5.1

3.5.4 Muting

Normally the muting circuit will operate by rectifying the signal that appears across the quad coil. Absence of a signal or noise "holes" in the carrier are peak detected and filtered to give the mute control voltage. The muting circuit of the LM3189 has been modified to include an early mute action when a strong signal with no noise "holes" is mistuned sufficiently. This is done to prevent dc shifts at the audio output from producing audible "thumps" in the loudspeaker.

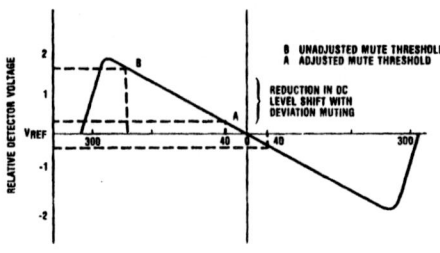

FIGURE 3.5.2 LM3189 Detector S Curve

Figure 3.5.2 shows a typical strong signal S curve for the LM3189 detector circuit. The dc voltage at the audio output will track the dc voltage level at the detector and, at center tuning, the output voltage will be the same as that held by the muting circuit between stations. However, when the signal is mistuned, the dc offset at the detector can reach as much as $\pm 2V_{DC}$ before the mute circuit operates and returns the audio output to the reference voltage level — thus producing an audible "thump" at the loudspeaker. To prevent this, two additional comparators are referenced to the AFC circuit control voltage such that the mute circuit will operate when a predetermined tuning deviation is reached, which results in much smaller dc offsets that can be adequately filtered. The degree of tuning deviation permitted before muting is set by the resistor connected between Pin 7 and Pin 10, with 15kΩ causing muting at ±40kHz. Because the muting control voltage changes only when the tuning is close to the proper point, Pin 12 can be used to indicate "on station" for automatic scanning tuning systems.

3.6 FM STEREO MULTIPLEX

3.6.1 Introduction

The LM1310/1800 is a phase locked loop FM stereo demodulator. In addition to separating left (L) and right (R) signal information from the detected IF output, this IC family features automatic stereo/monaural switching and a 100mA stereo indicator lamp driver. The LM1800 has the additional advantage of 45dB power supply rejection. Particularly attractive is the low external part count and total elimination of coils. A single inexpensive potentiometer performs all tuning. The resulting FM stereo system delivers high fidelity sound while still meeting the cost requirements of inexpensive stereo receivers.

Figures 3.6.1 and 3.6.2 outline the role played by the LM1310/1800 in the FM stereo receiver. The frequency domain plot shows that the composite input waveform contains L+R information in the audio band and L-R information suppressed carrier modulated on 38kHz. A 19kHz pilot tone, locked to the 38kHz subcarrier at the transmitter, is also included. SCA information occupies a higher band but is of no importance in the consumer FM receiver.

The block diagram (Figure 3.6.2) of the LM1800 shows the composite input signal applied to the audio frequency amplifier, which acts as a unity gain buffer to the decoder section. A second amplified signal is capacitively coupled to two phase detectors, one in the phase locked loop and the other in the stereo switching circuitry. In the phase locked loop, the output of the 76kHz voltage controlled oscillator (VCO) is frequency divided twice (to 38, then 19kHz), forming the other input to the loop phase detector. The output of the loop phase detector adjusts the VCO to precisely 76kHz. The 38kHz output of the first frequency divider becomes the regenerated subcarrier which demodulates L-R information in the decoder section. The amplified composite and an "in phase" 19kHz signal, generated in the phase locked loop, drive the "in phase" phase detector. When the loop is locked, the DC output voltage of this phase detector measures pilot amplitude. For pilot signals sufficiently strong to enable good stereo reception the trigger latches, applying regenerated subcarrier to the decoder and powering the stereo indicator lamp. Hysteresis, built into the trigger, protects against erratic stereo/monaural switching and the attendant lamp flicker.

In the monaural mode (electronic switch open) the decoder outputs duplicate the composite input signal except that the de-emphasis capacitors (from pins 3 and 6 to ground) roll off with the load resistors at 2kHz. In the stereo mode (electronic switch closed), the decoder demodulates the L-R information, matrixes it with the L+R information, then delivers buffered separated L and R signals to output pins 4 and 5 respectively.

Figure 3.6.3 is an equivalent schematic of an LM1800. The LM1310 is identical except the output turnaround circuitry (Q_{35}-Q_{38}) is eliminated and the output pins are connected to the collectors of Q_{39}-Q_{42}. Thus the LM1310 is essentially a 14 pin version of the LM1800, with load resistors returned to the power supply instead of ground. The National LM1800 is a pin-for-pin replacement for the UA758, while the LM1310 is a direct replacement for the MC1310.

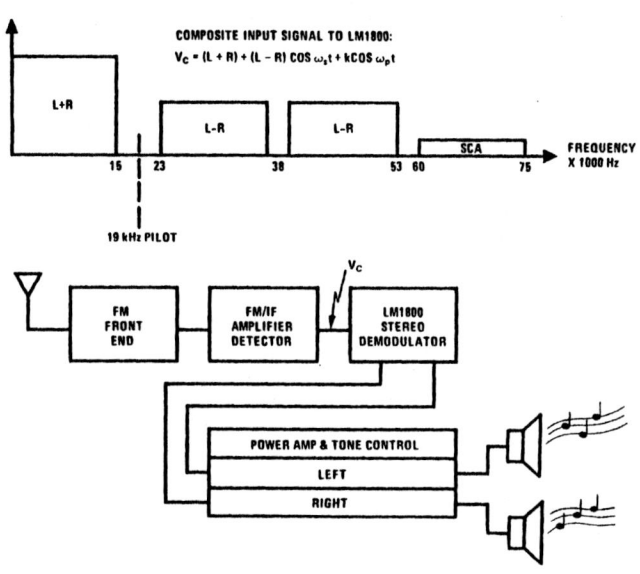

FIGURE 3.6.1 FM Receiver Block Diagram and Frequency Spectrum of LM1800 Input Signal

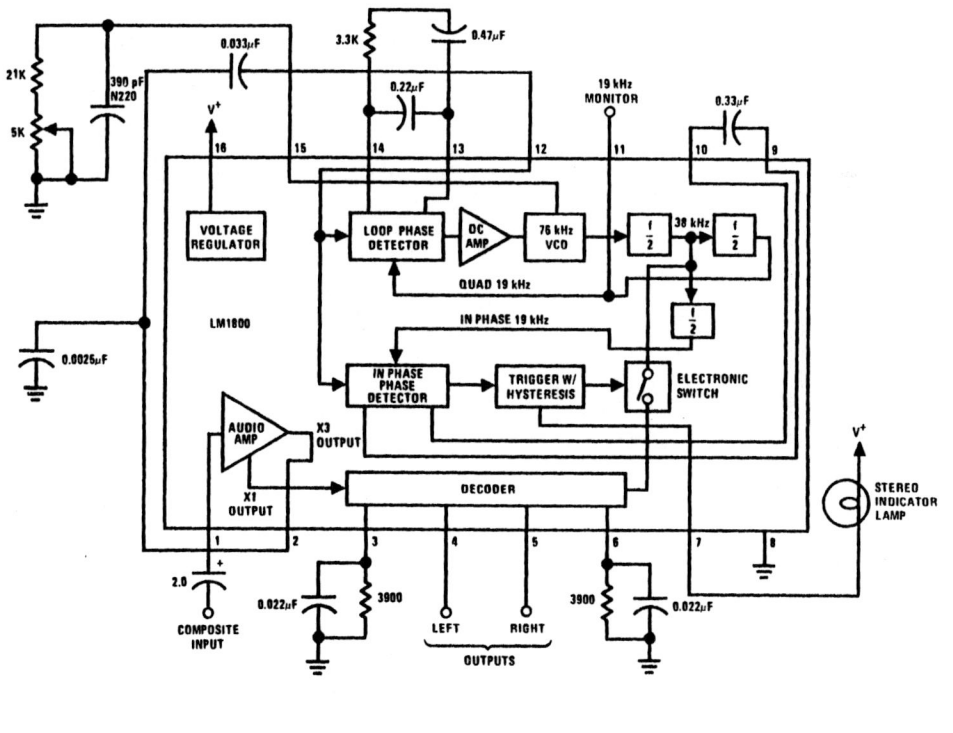

FIGURE 3.6.2 LM1800 Block Diagram

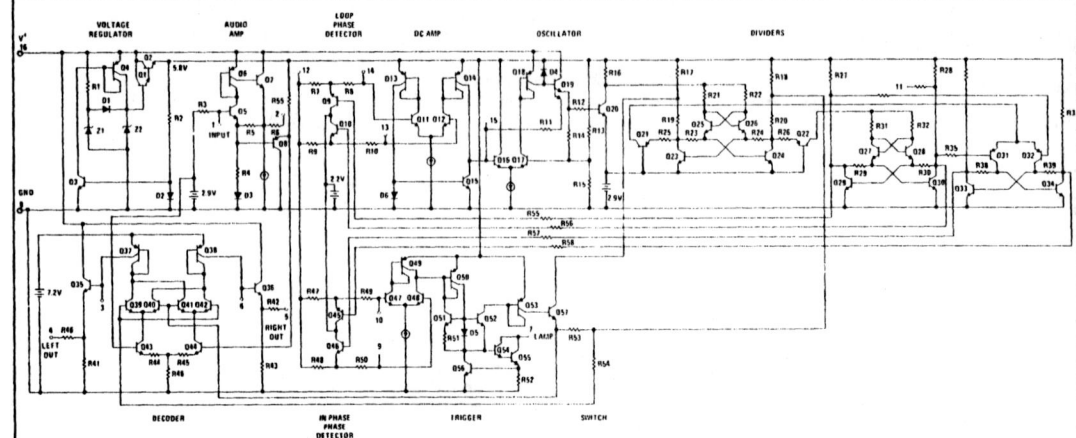

FIGURE 3.6.3 LM1800 Equivalent Schematic

3.6.2 LM1800 Typical Application

The circuit in Figure 3.6.4 illustrates the simplicity of designing an FM stereo demodulation system using the LM1800. R_3 and C_3 establish an adequate loop capture range and a low frequency well damped natural loop resonance. C_8 has the effect of shunting phase jitter, a dominant cause of high frequency channel separation problems. Recall that the 38 kHz subcarrier regenerates by phase locking the output of a 19 kHz divider to the pilot tone. Time delays through the divider result in the 38 kHz waveform leading the transmitted subcarrier. Addition of capacitor C_9 (0.0025 μF) at pin 2 introduces a lag at the input to the phase lock loop, compensating for these frequency divider delays. The output resistance of the audio amplifier is designed at 500 Ω to facilitate this connection.

The capture range of the LM1800 can be changed by altering the external RC product on the VCO pin. The loop gain can be shown to decrease for a decrease in VCO resistance ($R_4 + R_5$ in Figure 3.6.4). Maintaining a constant RC product, while increasing the capacity from 390 pF to 510 pF, narrows the capture range by about 25%. Although the resulting system has slightly improved channel separation, it is more sensitive to VCO tuning.

When the circuits so far described are connected in an actual FM receiver, channel separation often suffers due to imperfect frequency response of the IF stage. The input lead network of Figure 3.6.5 can be used to compensate for roll off in the IF and will restore high quality stereo sound. Should a receiver designer prefer a stereo/monaural switching point different from those programmed into the

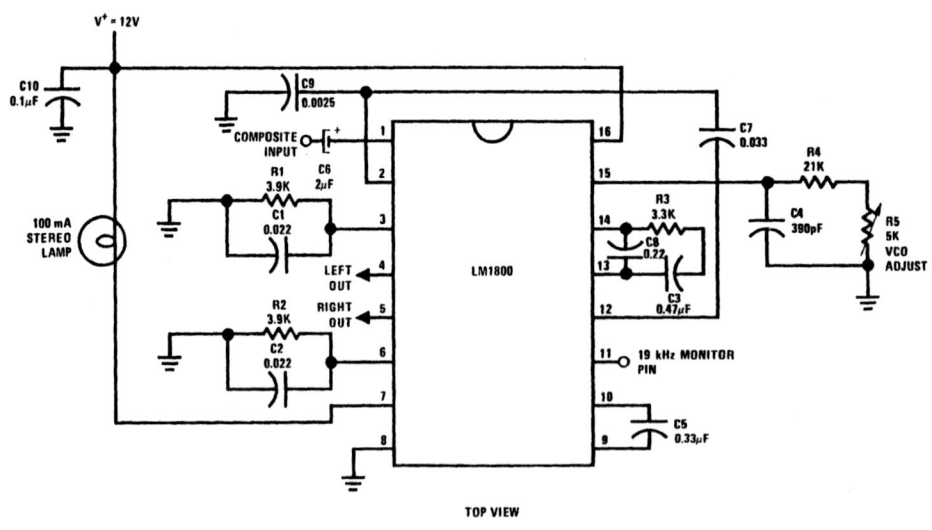

FIGURE 3.6.4 LM1800 Typical Application

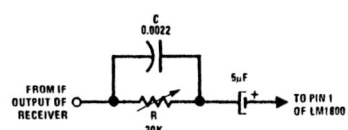

FIGURE 3.6.5 Compensation for Receiver IF Rolloff

LM1800 (pilot: 15mV$_{RMS}$ on, 6.0mV$_{RMS}$ off typical), the circuit of Figure 3.6.6 provides the desired flexibility.

The user who wants slightly increased voltage gain through the demodulator can increase the size of the load resistors (R_1 and R_2 of Figure 3.6.4), being sure to correspondingly change the de-emphasis capacitors (C_1 and C_2). Loads as high as 5600Ω may be used (gain of 1.4). Performance of the LM1800 is virtually independent of the supply voltage used (from 10 to 16V) due to the on-chip regulator.

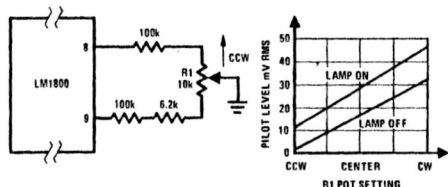

FIGURE 3.6.6 Stereo/Monaural Switch Point Adjustment

Although the circuit diagrams show a 100mA indicator lamp, the user may desire an LED. This presents no problem for the LM1800 so long as a resistor is connected in series to limit current to a safe value for the LED. The lamp or LED can be powered from any source (up to 18V), and need not necessarily be driven from the same supply as the LM1800.

3.6.3 LM1310 Typical Application

Figure 3.6.7 shows a typical stereo demodulator design using the LM1310. Capture range, lamp sensitivity adjustment and input lead compensation are all accomplished in the same manner as for the LM1800.

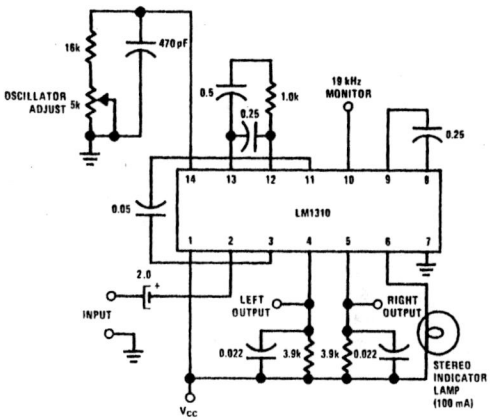

FIGURE 3.6.7 LM1310 Typical Application

3.6.4 Special Considerations of National's LM1310/1800

A number of FM stereo systems use the industry standard IF (LM3089) with an industry standard demodulator (LM1310/1800) as in Figure 3.8.8.

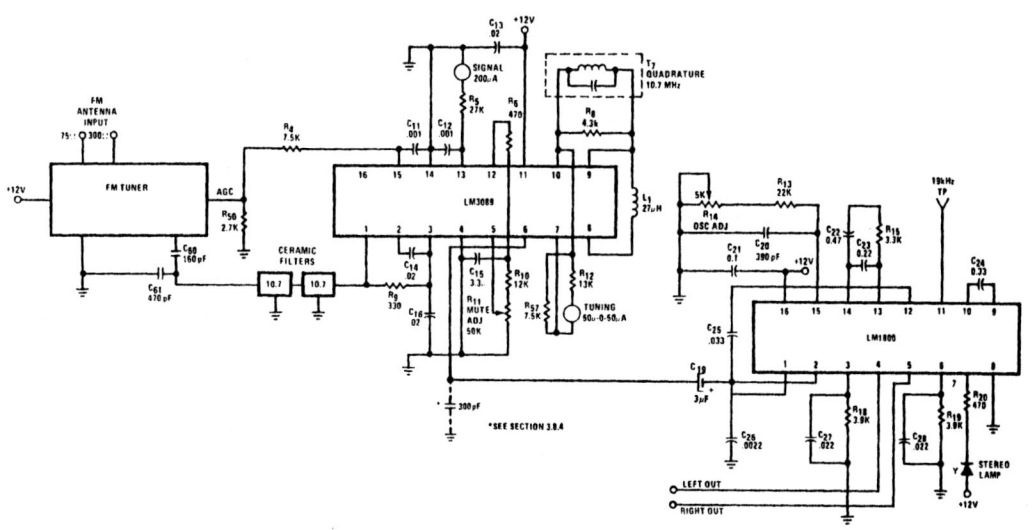

FIGURE 3.6.8 LM3089/LM1800 Application

The optional 300 pF capacitor on pin 6 of the LM3089 is often used to limit the bandwidth presented to the demodulator's input terminals. As the IF input level decreases and the limiting stages begin to come out of limiting, the detector noise bandwidth increases. Most competitive versions of the LM1310 would inadvertently AM detect this noise in their input "audio amplifier," resulting in decreased system signal-to-noise. They therefore require the 300 pF capacitor, which serves to eliminate this noise from the demodulator's input by decreasing bandwidth, and thus the system maintains adequate S/N.

The National LM1310 has been designed to eliminate the AM noise detection phenomenon, giving excellent S/N performance either with or without a bandlimited detected IF. Channel separation also is improved by elimination of the 300 pF capacitor since it introduces undesirable phase shift. The National LM1800 has the same feature, as do competitive 16 pin versions.

For systems demanding superior THD performance, the LM1800A is offered with a guaranteed maximum of 0.3%. Representing the industry's lowest THD value available in stereo demodulators, the LM1800A meets the tough requirements of the top-of-the-line stereo receiver market.

Utilization of the phase locked loop principle enables the LM1310/1800 to demodulate FM stereo signals without the use of troublesome and expensive coils. The numerous features available on the demodulator make it extremely attractive in a variety of home and automotive receivers.

3.7 MULTIPLEX WITH STEREO/MONAURAL BLEND

3.7.1 Introduction — Why Blend?

The signal to noise ratio of a strong, or local, stereo FM transmission is usually more than adequate. However, as many listeners to automotive radios will know, when the signal becomes weak, the S/N ratio in stereo is noticeably inferior to the S/N ratio of an equivalent strength monaural signal. Reference back to Figure 3.6.1 will show why this is the case. For a stereo broadcast a much wider frequency spectrum is used, in order to include the L-R channel information from 23 kHz to 53 kHz. When decoded, noise in this band will be translated down into the audio band, contributing to a higher noise level than if just L + R (or mono) were present.

Typical quieting curves for an FM stereo radio are shown in Figure 3.7.1, and it can be seen that for an S/N ratio of 50 dB, the stereo signal must be almost 20 dB greater than the mono signal. To prevent this degradation in S/N ratio the gain of the (L-R) channel in the decoder can be reduced as the r.f. signal strength decreases. Simultaneously, of course, there will be a corresponding reduction in stereo separation as the decoder gradually blends into a completely monaural signal output. This smooth loss of separation is much less noticeable than an abrupt switching into mono at a predetermined signal level. If an acceptable S/N ratio is 50 dB then the quieting curve to be followed is given by the dashed line in Figure 3.7.1. The required decrease in L-R gain is given by Figure 3.7.2 which also shows the change in stereo separation with signal level.

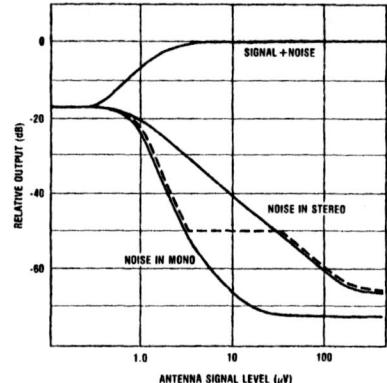

FIGURE 3.7.1 FM Radio S + N and N vs. Input Signal Level

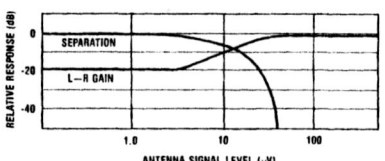

FIGURE 3.7.2 Change in Separation vs. Input Signal Level

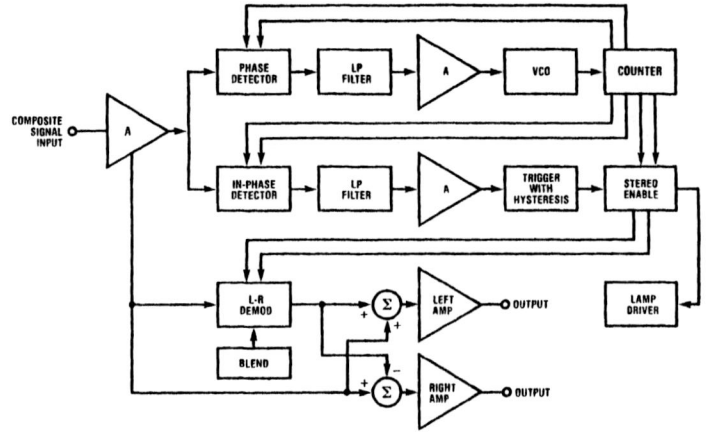

FIGURE 3.7.3 LM4500A Block Diagram

3.7.2 The LM4500A

The LM4500A is an improved stereo decoder with a new demodulation technique which minimizes subcarrier harmonics, and has a built-in blend circuit to optimize S/N ratios under weak FM signal conditions.

The block diagram of Figure 3.7.3 illustrates that the LM4500A has the same circuit functions as an LM1800, but with the addition of the blend circuit which operates on the L-R demodulator section. In this demodulation section both in-phase and antiphase components of the L-R signal are available and these can be gradually combined to finally produce complete cancellation of the L-R signal. The control voltage, which must be proportional to the r.f. signal strength, is obtained from the signal strength meter drive output of the FM IF. Usually a potentiometer adjustment will be needed to compensate for different Tuner/IF combinations. The change in separation with this control voltage is given by the curve of Figure 3.7.4

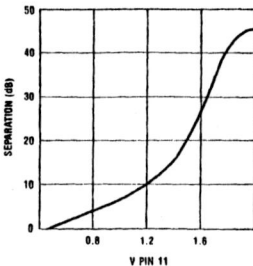

FIGURE 3.7.4 LM4500A Stereo Separation vs. Pin 11 Control Voltage

Not shown in the block diagram of Figure 3.7.3 are the different decoder switching waveforms used by the LM4500A. Conventional decoders, such as the LM1800, use square waves in-phase and anti-phase, which have a precise duty cycle, zero mean level and no even harmonics. While this provides excellent performance for the standard U.S. stereo broadcasts, problems with third harmonic radiation interference can occur in Europe where closer station spacing and the A.R.I. (Automotive Radio Information) signals are utilized. The third harmonic of the subcarrier is 114kHz, and an adjacent transmitter sideband can mix with this to produce audible components. Similarly, the third harmonic of the pilot carrier, at 57kHz, can mix with the A.R.I. system signal causing phase modulation of the V.C.O. and this results in intermodulation distortion.

The LM4500A avoids these problems by generating switching waveforms composed from square waves phase shifted such that their third harmonics are in antiphase and cancel out, Figure 3.7.5.

A complete schematic of the external components required for an LM4500A is shown in Figure 3.7.6 and this circuit exhibits at least 40dB stereo separation (optimized by P_2) and an 83dB S/N ratio. The subcarrier harmonics are typically better than 70dB down and the stereo T.H.D. is 0.07% with a 1.5V(p-p) composite signal level.

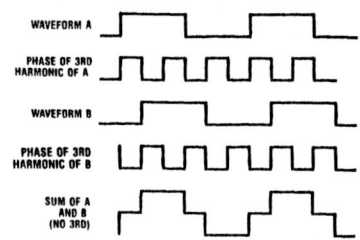

FIGURE 3.7.5 LM4500A Switching Waveform Generation

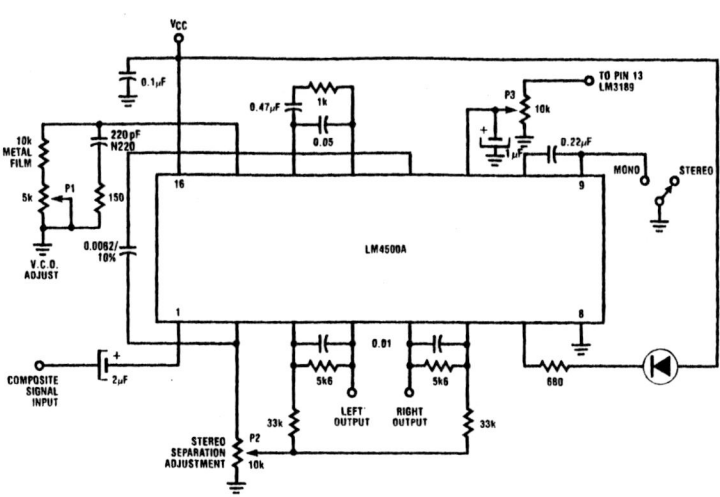

FIGURE 3.7.6 Typical Application of the LM4500A

3.7.3 The LM 1870

The LM1870 is another new stereo decoder IC from National that incorporates the variable blend feature. Instead of adding in-phase and antiphase components of the demodulated L-R signal, the LM1870 achieves stereo to mono blend before the demodulator, Figure 3.7.7. The composite input signal follows two paths, one of which has a flat, wideband frequency resonse. The other has a 2 pole low pass filter response and the output from both paths are summed in a multiplier circuit which is controlled by the r.f. signal strength. This control voltage is derived from the signal strength meter drive of the LM3189. Figure 3.7.8 shows the net result. As the r.f. signal level decreases, the h.f. portion of the composite signal containing the L-R information is decreased. At the same time the upper frequency response of the L + R signal is modified to further reduce the audible noise. Typical L+R and L-R response curves are shown in Figures 3.7.9 and 3.7.10.

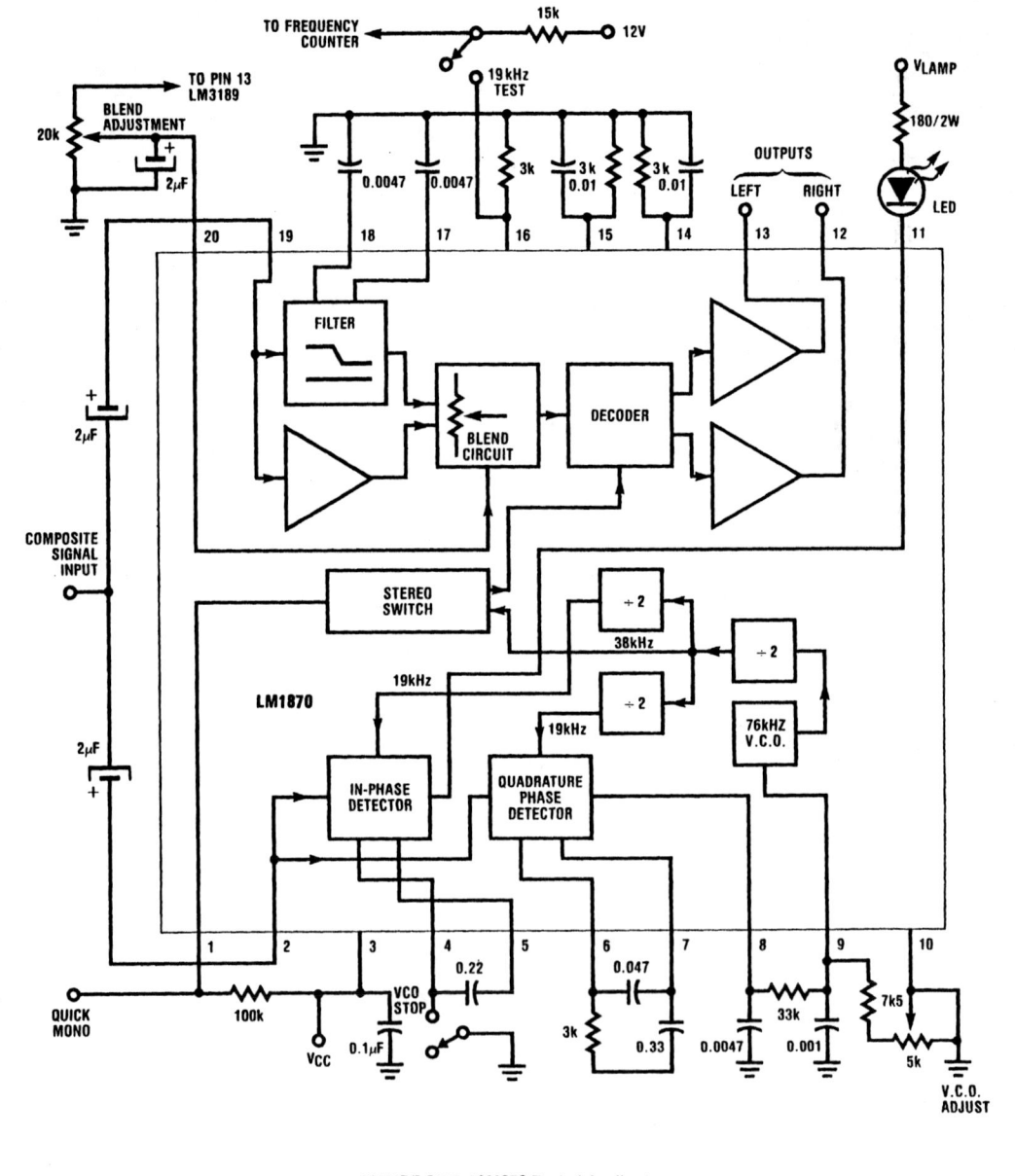

FIGURE 3.7.7 LM1870 Typical Application

The stereo performance of the LM1870 is very constant for small changes (<2%) in the free running frequency of the V.C.O. Low temperature coefficient components should be used for the oscillator capacitor and tuning resistors. Tuning the V.C.O. is done by adjusting the 5kΩ pot to obtain 19kHz ±20Hz with no signal input at Pin 2. 19kHz is available at Pin 16 if a resistor is connected from Pin 16 to the supply voltage. In normal operation, Pin 16 is connected via a resistor to ground which programs the blend characteristic, Figure 3.7.9.

Although the LM1870 outputs are low impedance and capable of sinking or sourcing 1mA, if the supply pin (Pin 3) is open or grounded, then both outputs are at a high impedance. This facilitates switching in AM-FM radios since the outputs do not have to be disconnected when the radio is in the AM mode.

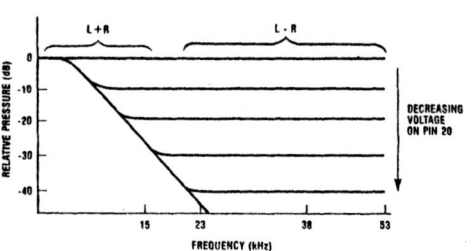

FIGURE 3.7.8 Response vs. Frequency of LM1870 Blend Circuit

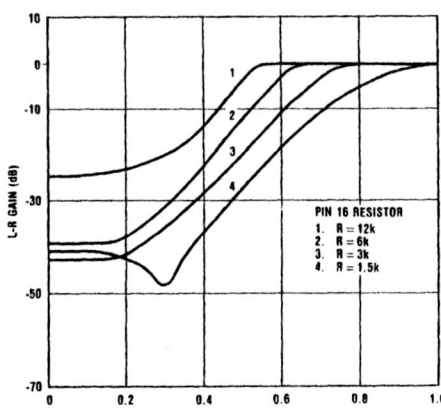

FIGURE 3.7.9 L-R Gain vs. Blend Control Voltage

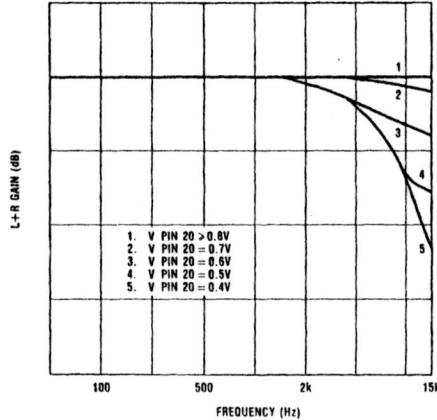

FIGURE 3.7.10 L+R Frequency Response

4.0 Power Amplifiers

4.1 INSIDE POWER INTEGRATED CIRCUITS

Audio power amplifiers manufactured using integrated circuit technology do not differ significantly in circuit design from traditional operational amplifiers. Use of current sources, active loads and balanced differential techniques predominate, allowing creation of high-gain, wide bandwidth, low distortion devices. Major design differences appear only in the class AB high current output stages where unique geometries are required and special layout techniques are employed to guarantee thermal stability across the chip.

The material presented in the following sections serves as a brief introduction to the design techniques used in audio power integrated circuits. Hopefully, a clearer understanding of the internal "workings" will result from reading the discussion, thus making application of the devices easier.

4.1.1 Frequency Response and Distortion

Most audio amplifier designs are similar to Figure 4.1.1. An input transconductance block ($gm = i_o/v_1$) drives a high gain inverting amplifier with capacitive feedback. To this is added an output buffer with high current gain but unity voltage gain. The resulting output signal is defined by:

$$v_o = v_1 \, gm \, X_C \qquad (4.1.1)$$

or, rewriting in terms of gain:

$$A_v = \frac{v_o}{v_1} = gm \, X_C = \frac{gm}{sC} = \frac{gm}{j\omega C} \qquad (4.1.2)$$

Setting Equation (4.1.2) equal to unity allows solution for the amplifier unity gain cross frequency:

$$A_v = 1 = \frac{gm}{j\omega C} = \frac{gm}{j2\pi f C} \qquad (4.1.3)$$

$$f_{UNITY} = \frac{gm}{2\pi C} \qquad (4.1.4)$$

Equation (4.1.2) indicates a single pole response resulting in a 20 dB/decade slope of the gain-frequency plot in Figure 4.1.1. There is, of course, a low frequency pole which is determined by the compensation capacitor and the resistance to ground seen at the input of the inverting amplifier. Usually this pole is below 100 Hz so it plays only a small role in determining amplifier performance in usual feedback arrangements.

For an amplifier of this type to be stable in unity gain feedback circuits, it is necessary to arrange gm and C so that the unity gain crossover frequency is about 1 MHz. This is, in short, due to a few other undesirable phase shifts that are difficult to avoid when using lateral PNP transistors in monolithic realizations of the transconductance as well as the buffer blocks. Figure 4.1.1 shows that if f_{UNITY} is 1 MHz then only 34 dB of gain is available at 20 kHz! Since most audio circuits require more gain, most IC audios are not compensated to unity. Evaluation of an IC audio amplifier will show stability troubles in loops fed back for less than 20 dB closed loop gain.

Consider for a moment the problem in audio designs with distortion (THD). The buffer of Figure 4.1.1 is essentially an emitter follower (NPN during positive half cycles and PNP during negative halves due to class B operation). As a result the load presented to the collector of the gain transistor is different depending on which half cycle the output is in. The buffer amplifier itself often contributes in the form of crossover distortion. Suppose for a moment that the amplifier were to be used open loop (i.e., without any AC feedback) and that the result was an output signal distorted 10% at 10 kHz. Further assume the open loop gain-frequency is as in Figure 4.1.2 so that the amplifier is running at 60 dB of gain. Now add negative feedback around the amplifier to set its gain at 40 dB and note that its voltage gain remains flat with frequency throughout the audio band. In this configuration there is 20 dB of loop gain (the difference between open loop gain and closed loop gain) which works to correct the distortion in the output waveform by about 20 dB, reducing it from the 10% open loop value to 1%. Further study of Figure 4.1.2 shows that there is more loop gain at lower frequencies which should, and does, help the THD at lower frequencies. The reduction in loop gain at high frequencies likewise allows more of the open loop distortion to show.

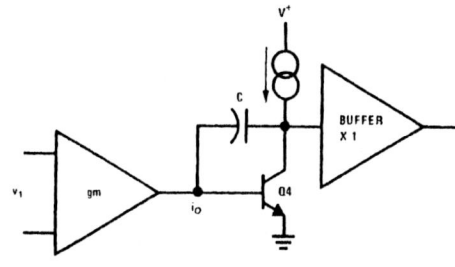

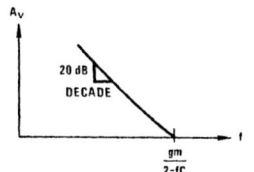

FIGURE 4.1.1 Audio Amp Small Signal Model

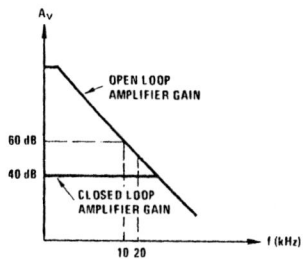

FIGURE 4.1.2 Feedback and "Loop Gain"

4.1.2 Slew Rate

Not only must IC audio amplifiers have more bandwidth than "garden variety" op amps, they must also have higher slew rates. Slew rate is a measure of the ability of an amplifier's large signal characteristics to match its own small signal responses. The transconductance block of Figure 4.1.1 delivers a current out for a given small signal input voltage. Figure 4.1.3 shows an input stage typically used in audio amplifiers. Even for large differential input voltage drives to the PNP bases, the current available can never surpass I. And this constant current (I) charging the compensation capacitor (C) results in a ramp at Q_1's collector. The slope of this ramp is defined as slew rate and usually is expressed in terms of volts per microsecond. Increasing the value of the current source does increase slew rate, but at the expense of increased input bias current and gm. Large gm values demand larger compensation capacitors which are costly in IC designs. The optimum compromise is to use large enough I to achieve adequate slew rate and then add emitter degeneration resistors to the PNPs to lower gm.

in proximity to an RF receiver. Among the stabilization techniques that are in use, with varying degrees of success are:

1. Placing an external RC from the output pin to ground to lower the gain of the NPN. This works pretty well and appears on numerous data sheets as an external cure.
2. Utilizing device geometry methods to improve the PNP's frequency response. This has been done successfully in the LM378 and LM379. The only problem with this scheme is that biasing the improved PNP reduces the usable output swing slightly, thereby lowering output power capability.
3. Addition of resistance in series with either the emitter or base of Q_3.
4. Making Q_3 a controlled gain PNP of unity, which has the added advantage of keeping gain more nearly equal for each half cycle.
5. Adding capacitance to ground from Q_3's collector.

These last three work sometimes to some degree at most current levels.

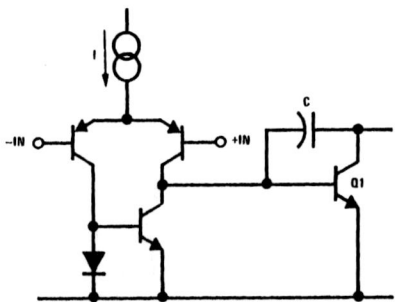

FIGURE 4.1.3 Typical gm Block

Slew rate can be calculated knowing only I and C:

$$\frac{\Delta V}{\Delta t} = \frac{I}{C} \qquad (4.1.5)$$

To more clearly understand why slew rate is significant in audio amplifiers, consider a 20kHz sine wave swinging $40V_{p-p}$, a worst case need for most of today's audios. The rate of change of voltage that this demands is maximum at zero crossing and is $2.5V/\mu s$. Equation (4.1.6) is a general expression for solving required slew rate for a given sinusoid. (See Section 1.2.1.)

$$\text{Slew rate} = \frac{\Delta V}{\Delta t} = \pi f V_{p-p} \qquad (4.1.6)$$

4.1.3 Output Stages

In the final analysis a buffer stage that delivers amperes of load current is the main distinction between audio and op amp designs. The classic class B is merely a PNP and NPN capable of huge currents, but since the IC designer lacks good quality PNPs, a number of compromises results. Figure 4.1.4b shows the bottom side PNP replaced with a composite PNP/NPN arrangement. Unfortunately, Q_2/Q_3 form a feedback loop which is quite inclined to oscillate in the 2-5MHz range. Although the oscillation frequency is well above the audible range, it can be troublesome when placed

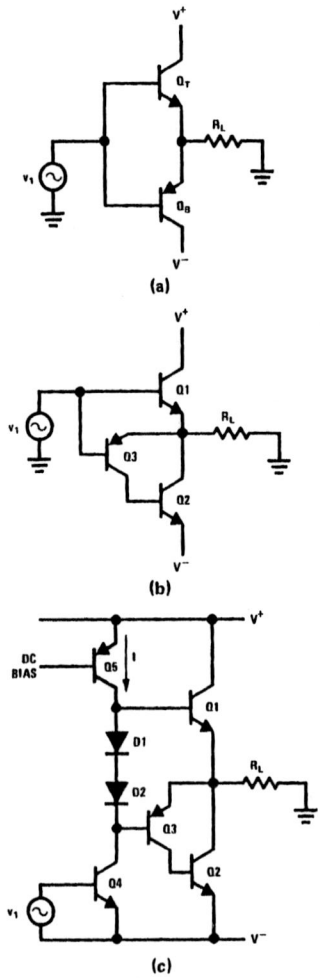

FIGURE 4.1.4 Basic Class B Output Drivers

Figure 4.1.5 illustrates crossover distortion such as would result from the circuit in Figure 4.1.4b. Beginning with Q_1 "on" and the amplifier output coming down from the top half cycle towards zero crossing, it is clear that the emitter of Q_1 can track its base until the emitter reaches zero volts. However, as the base voltage continues below 0.7 V, Q_1 must turn off; but Q_2/Q_3 cannot turn on until the input generator gets all the way to –0.7 V. Thus, there is a 1.4 V of dead zone where the output cannot respond to the input. And since the size of the dead zone is independent of output amplitude, the effect is more pronounced at low levels. Of course feedback works to correct this, but the result is still a somewhat distorted waveform — one which has an unfortunately distasteful sound. Indeed the feedback loop or the composite PNP sometimes rings as it tries to overcome the nonlinearity, generating harmonics that may disturb the receiver in radio applications. The circuit of Figure 4.1.4c adds "AB bias." By running current through D_1 and D_2, the output transistors are turned slightly "on" to allow the amplifier to traverse the zero volts region smoothly. Normally much of the power supply current in audio amplifiers is this AB bias current, running anywhere from 1 to 15 mA per amplifier.

The distortion components discussed so far have all been in terms of circuit nonlinearities and the loop gain covering them up. However, at low frequencies (below 100 Hz) thermal problems due to chip layout can cause distortion. In the audio IC, large amounts of power are dissipated in the output driver transistors causing thermal gradients across the die. Since a sensitive input amplifier shares the same piece of silicon, much care must be taken to preserve thermal symmetry to minimize thermal feedback.

Despite the many restrictions on audio IC designs, today's devices do a credible job, many boasting less than 1% THD from 20 Hz to 20 kHz — not at all a bad feat!

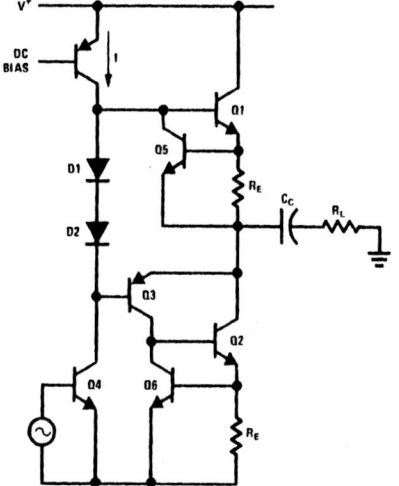

FIGURE 4.1.6 Simple Current Limit

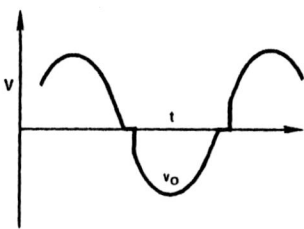

FIGURE 4.1.5 Crossover Distortion

Some amplifiers at high frequencies (say 10 kHz) exhibit slightly more crossover distortion when negative going than when positive going through zero. This is explained by the slow composite PNPs' (Q_2/Q_3) delay in turning "on." If the amplifier delivers any appreciable load current in the top half cycle, the emitter current of Q_1 causes its base-emitter voltage to rise and shut "off" Q_3 (since the voltage across D_1 and D_2 is fixed by I). Thus, fast negative going signals demand the composite to go from full "off" to full "on" — and they respond too slowly. As one might imagine, compensating the loop (Q_2 and Q_3) for stability even slows the switching time more. This problem makes very low distortion IC amplifiers (< 0.2%) difficult at the high end of the audio (20 kHz).

Another interesting phenomenon occurs when some IC amplifiers oscillate at high frequencies — their power supply current goes up and they die! This usually can be explained by positive going output signals where the fast top NPN transistor (Q_1) turns "on" before the sluggish composite turns "off," resulting in large currents passing straight down through the amplifier (Q_1 and Q_2).

4.1.4 Output Protection Circuitry

By the very nature of audio systems the amplifier often drives a transducer — or speaker — remote from the electronic components. To protect against inadvertent shorting of the speaker some audio ICs are designed to self limit their output current at a safe value. Figure 4.1.6 is a simple approach to current limiting: here Q_5 or Q_6 turns "on" to limit base drive to either of the output transistors (Q_1 or Q_2) when the current through the emitter resistors is sufficient to threshold an emitter base junction. Limiting is sharp on the top side since Q_5 has to sink only the current source (I). However, the current that Q_6 must sink is more nebulous, depending on the alpha holdup of Q_3, resulting in soft or mushy negative side limiting. Other connections can be used to sharpen the limiting action, but they usually result in a marginally stable loop that must be frequency compensated to avoid oscillation during limiting. The major disadvantage to the circuit of Figure 4.1.6 is that as much as 1.4 V is lost from loaded output swing due to voltage dropped across the two R_Es.

The improved circuit of Figure 4.1.7 reduces the values of R_E for limiting at the same current but is usable only in Darlington configurations. It suffers from the same negative side softness but only consumes about 0.4 V of output swing. There are a few other methods employed, some even consuming less than 0.4 V. Indeed it is further possible to add voltage information to the current limit transistor's base and achieve safe operating area protection. Care must be taken in such designs, however, to allow for a leading or lagging current of up to 60° to accommodate the variety of speakers on the market. However, the circuitry shown in Figures 4.1.6 and 4.1.7 is representative of the vast majority of audio ICs in today's marketplace.

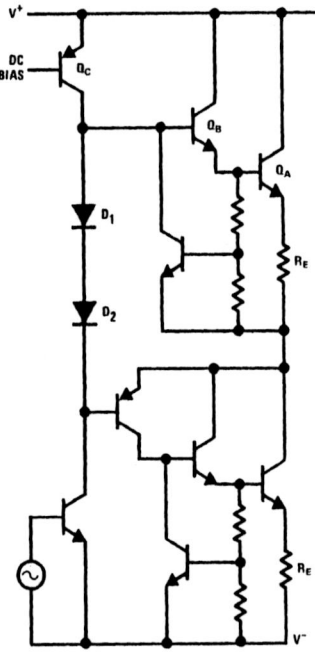

FIGURE 4.1.7 Improved Current Limit

Large amounts of power dissipation on the die cause chip temperatures to rise far above ambient. In audio ICs it is popular to include circuitry to sense chip temperature and shut down the amplifier if it begins to overheat. Figure 4.1.8 is typical of such circuits. The voltage at the emitter of Q_1 rises with temperature due both to the TC of the zener (Z_1) and Q_1's base-emitter voltage. Thus, the voltage at the junction of R_1 and R_2 rises while the voltage required to threshold Q_2's emitter-base junction falls with temperature. In most designs the resistor ratio is set to threshold Q_2 at about 165°C. The collector current of Q_2 is then used to disable the amplifier.

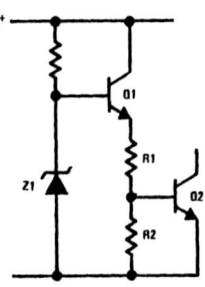

FIGURE 4.1.8 Typical Thermal Shutdown

The addition of thermal shutdowns in audio ICs has done much to improve field reliability. If the heat sinking is inadequate in a discrete design, the devices burn up. In a thermally protected IC the amplifier merely reduces drive to the load to maintain chip temperature at a safe value.

4.1.5 Bootstrapping

A look at the typical Class B output stage of Figure 4.1.4 shows that the output swings positive only until Q_5 saturates (even unloaded). At this point the output voltage swing lost across Q_1 is

$$V^+ - V_{OUTPEAK} = V_{SAT}(Q_5) + V_{BE}(Q_1) \approx 1.1V \quad (4.1.7)$$

Further, the output swings negative until Q_4 saturates when the output voltage swing loss is

$$V^- - V_{OUTPEAK} = V_{SAT}(Q_4) + V_{BE}(Q_3) \approx 0.9V \quad (4.1.8)$$

Despite the fact that there is no load current, the maximum possible output swing is about 2V less than the total supply voltage. While it is possible that with very high load currents the saturation voltages of Q_1 and Q_2 can exceed 1V each, most audio I/C's are limited by Equations (4.1.7) and (4.1.8). For battery operated systems in particular, this loss in output swing can seriously reduce the available output power to the load.

Larger positive swings can be obtained by utilizing "bootstrap" techniques (Figure 4.1.9). In the quiescent state the amplifier output is halfway between the supply voltages so that the capacitor is charged to a voltage given by,

$$V_{CBS} \approx \frac{V^+ - V^-}{2} \times \frac{R_2}{R_1 + R_2} \quad (4.1.9)$$

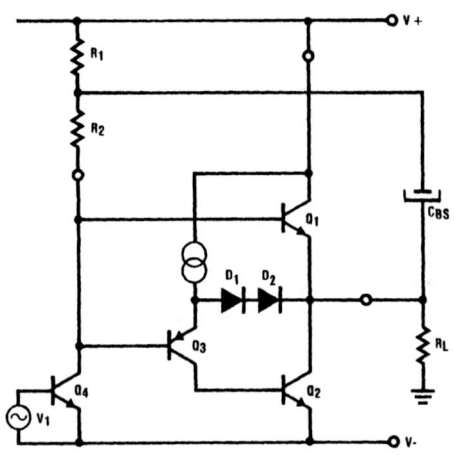

FIGURE 4.1.9 "Bootstrap" Output State

Less the diode drop of Q_1 base-emitter, this is the voltage across R_2. If C_{BS} is very large this voltage will remain constant, even as the output swings positive, so the voltage across R_2 (and consequently the current through R_2) will be maintained. In this way the bootstrap circuitry appears to be a current source to Q_4 and during extreme positive output excursions the base of Q_1 can be pulled above the supply rail, leaving the output swing limited only by Q_1 saturation voltage.

4-4

To improve the negative output swing the AB bias network of a current souce and diodes D_1 and D_2 is connected in the emitter circuit of Q_3. Now when Q_4 approaches saturation, the output voltage loss is given by,

$$V^- - V_{OUTPEAK} = V_C(Q_4) + V_{BE}(Q_3) - V_{D1} - V_{D2} \quad (4.1.7)$$

If Q_4 is allowed to saturate, this could be as low as

$$0.2V + 0.7V - 1.2V = -0.3V!$$

In practice the saturation voltage of Q_2 will define the lowest negative excursion — which will occur before Q_4 saturates.

4.2 DESIGN TIPS ON LAYOUT, GROUND LOOPS AND SUPPLY BYPASSING

Layout, grounding and power supply decoupling of audio power integrated circuits require the same careful attention to details as preamplifier ICs. All of the points discussed in Section 2.2 of this handbook apply directly to the use of power amplifiers and should be consulted before use.

The relevant sections are reproduced here for cross-reference and convenience:

Section 2.2.1 Layout
Section 2.2.2 Ground Loops
Section 2.2.3 Supply Bypassing
Section 2.2.4 Additional Stabilizing Tips

4.3 POWER AMPLIFIER SELECTION

National Semiconductor's line of audio power amplifiers consists of two major families: the "Duals" represented by LM1877/LM2877, LM378, LM378, LM1896/2896, and the "Monos" represented by ten products. Available power output ranges from miniscule 320mW battery operated devices to hefty 9.6W line operated systems. The power driver LM391 is capable of driving output stages, delivering 60W. Although most of the amplifiers are designed for single supply operation, all devices may be operated from split supplies where required. Tables 4.3.1 and 4.3.2 summarize the dual family for ease of selection, while Table 4.3.3 compares the mono devices.

TABLE 4.3.1 Dual Power Amplifier Characteristics

PARAMETER	LM1877/LM2877			LM378/LM379			LM1896/LM2896		
	MIN	TYP	MAX	MIN	TYP	MAX	MIN	TYP	MAX
Supply Voltage	6V	20V	24V	10V	24V	35V	3V	6V	10V/15V
Quiescent Supply Current ($P_{OUT} = 0W$)		25mA	50mA		15mA	65mA		15mA/25mA	
Open Loop Gain ($R_s = 0\Omega$, f = 1kHz)		70dB		66dB	90dB			100dB	
Input Impedance		4mΩ		3mΩ				100kΩ	
Channel Separation[1] Output Referred ($C_F = 50\mu F$, f = 1kHz)	–50dB	–70dB			–36dB[2]			–60dB	
Power Supply Ripple Rejection ($C_F = 50\mu F$, f = 120Hz)	–50dB	–68dB			–36dB[2]			–54dB	
Equivalent Input Noise ($R_s = 600\Omega$, BW 20Hz to 20kHz)		3μV			3μV			1.9μV	
THD f = 1kHz P_O = 1W/Channel P_O = 50mW/Channel		0.07% 0.1%			0.07% 0.25%			–/0.14% 0.09/0.27%	

1. $A_V = 34$dB
2. $C_F = 250\mu F$
3. LM1877/LM378/LM1896   14 Pin D.I.P. ⎫
     LM2877/LM2896   11 Pin S.I.P.   ⎬ Package Styles
     LM379   14 Pin 'S' Type Power D.I.P. ⎭

TABLE 4.3.2 Dual Amplifier Output Power

V_{CC} (VOLTS)	LOAD (OHMS)	TYPICAL OUTPUT POWER (WATTS) AT 10% T.H.D.					
		LM1877	LM2877	LM387	LM389	LM1896	LM2896
6V	4Ω					1.1W	
	8Ω					600mW	
	8Ω (Bridge)					2.2W	
9V	4Ω						2.5W
	4Ω (Bridge)						7.8W
	8Ω	500mW	500mW			1.3W	
	8Ω (Bridge)						5W
12V	4Ω		1.9W				2.5W
	8Ω	1.2W	1.2W	1.6W	1.6W		
	8Ω (Bridge)						9.0W
14V	8Ω	1.8W	1.8W	1.9W	1.9W		
18V	8Ω		3.6W	3W	3W		
	16Ω			1.8W	1.8W		
20V	8Ω		4.5W		3.8W		
	16Ω			2.4W	2.4W		
24V	8Ω				5.4W		
	16Ω			3.5W	3.6W		
30V	16Ω				5.5W		

1. Specification apply for $T_{TAB} = 25°C$. For operation at higher ambient temperatures, the IC must be derated based on the package/heatsink thermal resistance and a 150°C max junction temperature.

TABLE 4.3.3 Mono Power Amplifier Characteristics

V_{CC} (VOLTS)	DEVICE TYPE	OUTPUT POWER (WATTS) AT 10% THD[2]				GAIN (dB)	OUTPUT PROTECTION
		2Ω	4Ω	8Ω	16Ω		
3V	LM2000/1	480mW	280mW	160mW		ADJUSTABLE	NO
6V	LM383	1.9W	800mW	440mW	240mW	ADJUSTABLE	YES
	LM386/389		340mW	325mW	180mW	26-46dB	NO
	LM388		800mW	600mW	300mW	26-46dB	NO
	LM390		1.0W	650mW	325mW	26-46dB	NO
	LM2000/1	2.0W	1.2W	600mW		ADJUSTABLE	NO
9V	LM383	3.5W	2.1W	1.2W	630mW	ADJUSTABLE	YES
	LM386/389		350mW/-	700mW/520mW	500mW	26-46dB	NO
	LM388		1.8W	1.3W	650mW	26-46dB	NO
	LM390		2.0W	1.4W	700mW	26-46dB	NO
	LM2000	4.8W	2.8W	1.5W		ADJUSTABLE	NO
12V	LM380		2.4W	1.5W	500mW	34dB	NO
	LM383	6.4W	4.0W	2.3W	1.2W	ADJUSTABLE	YES
	LM386/389		350mW/-	820mW/-	1.6W/900mW	26-46dB	NO
	LM388		2.4W	2.2W	1.3W	26-46dB	NO
	LM2000	8.8W	5.0W	2.6W		ADJUSTABLE	NO
14V	LM380		3.3W	2.2W	1.0W	34dB	NO
	LM383	8.9W	5.6W	3.7W	1.7W	ADJUSTABLE	YES
	LM386			830mW	1.3W	26-46dB	NO
	LM388		3.0W	3.0W	1.8W	26-46dB	NO
16V	LM380			3.0W	1.6W	34dB	NO
	LM383	10.5W	7.0W			ADJUSTABLE	YES
	LM386				1.6W	26-46dB	NO
	LM388		3.6W	3.8W	2.3W	26-46dB	NO
18V	LM380			4.0W	2.2W	34dB	NO
	LM383		9.6W	5.5W	2.9W	ADJUSTABLE	YES
	LM384		4.2W	4.0W	2.2W		
22V	LM384		3.5W	5.7W	3.5W	34dB	NO
±22V	LM39IN-60		30W	20W		ADJUSTABLE	YES
±30V	LM39IN-80		60W	40W		ADJUSTABLE	YES

1. Specifications apply for $T_A = 25°C$. For operation at ambient temperatures > 25°C the IC must be derated based on the case style thermal resistance and a maximum 150°C junction temperature.
2. P_O increases by 19% at 5% THD and by 30% at 10% THD. Clipping occurs just before 3% THD is reached.

4.4 LM1877, LM1896, LM378, AND LM379 DUAL TWO TO SIX WATT POWER AMPLIFIERS

4.4.1 Introduction

The "Duals" are two channel power amplifiers capable of delivering up to 6 watts into 8 or 16Ω loads. They feature on-chip frequency compensation, output current limiting, thermal shutdown protection, fast turn-on and turn-off without "pops" or pulses of active gain, an output which is self-centering at $V_{CC}/2$, and a 5 to 20MHz gain-bandwidth product. Applications include stereo or multi-channel audio power output for phono, tape or radio use over a supply range of 6 to 35V, as well as servo amplifier, power oscillator and various instrument system circuits. Normal supply is single-ended; however, split supplies may be used without difficulty or degradation in power supply rejection.

4.4.2 Circuit Description of LM378 and LM379

The simplified schematic of Figure 4.4.1 shows the important design features of the amplifier. The differential input stage made up of Q_1-Q_4 uses a double (split) collector PNP Darlington pair having several advantages. The high base-emitter breakdown of the lateral PNP transistor is about 60V, which affords significant input over-voltage protection. The double collector allows operation at high emitter current to achieve good first stage f_t and minimum phase shift while simultaneously operating at low transconductance to allow internal compensation with a physically small capacitor C_1. (Unity gain bandwidth of an amplifier with pole-splitting compensation occurs where the first stage transconductance equals ωC_1.)

Further decrease of transconductance is provided by degeneration caused by resistors at Q_2 and Q_3 emitters, which also allow better large signal slew rate. The second collector provides bias current to the input emitter follower for increased frequency response and slew rate. Full differential input stage gain is provided by the "turnaround" differential to single-ended current source loads Q_5 and Q_6. The input common-mode voltage does not extend below about 0.5V above ground as might otherwise be expected from initial examination of the input circuit. This is because Q_7 is actually preceded by an emitter follower transistor not shown in the simplified circuit.

The second stage Q_7 operates common-emitter with a current source load for high gain. Pole splitting compensation is provided by C_1 to achieve unity gain bandwidth of about 10MHz. Internal compensation is sufficient with closed-loop gain down to about $A_v = 10$.

The output stage is a complementary common-collector class AB composite. The upper, or current sourcing section, is a Darlington emitter follower Q_{12} and Q_{13}. The lower, or current sinking, section is a composite PNP made up of Q_{14}, Q_{15}, and Q_9. Normally, this type of PNP composite has low f_t and excessive delay caused by the lateral PNP transistor Q_9. The usual result is poor unity gain bandwidth and probable oscillation on the negative half of the output waveform. The traditional fix has been to add an external series RC network from output to ground to reduce loop gain of the composite PNP and so prevent the oscillation. In the LM378/LM379 amplifiers, Q_9 is a field-aided lateral PNP to overcome these performance limitations and so reduce external parts count. There is no need for the external RC network, no oscillation is present on the negative half cycle, and bandwidth is better with this output stage. Q_{10} and Q_{11} provide output current limiting at

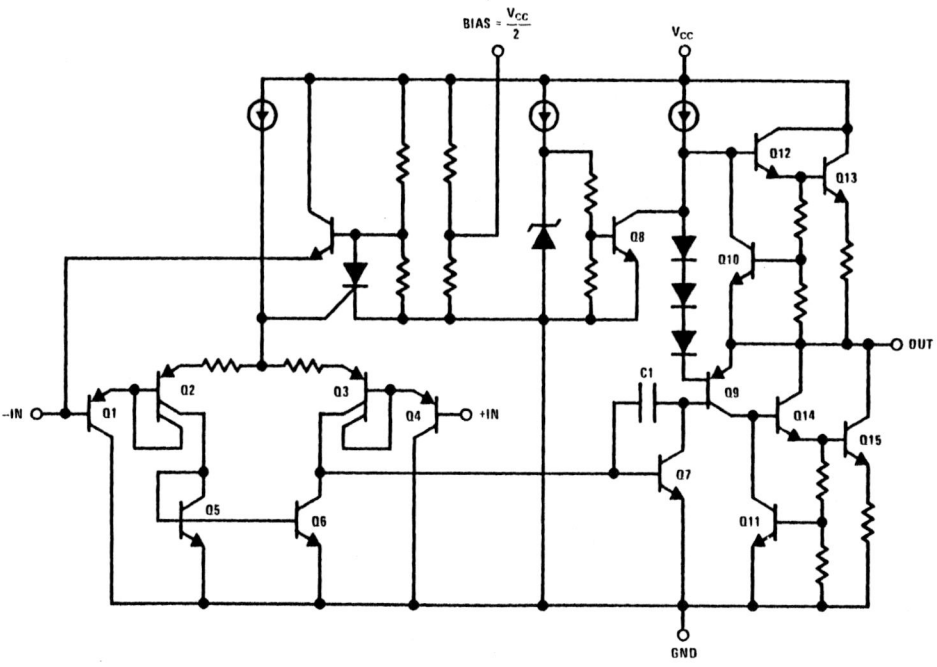

FIGURE 4.4.1 Simplified Schematic Diagram

about 1.3A, and there is internal thermal limiting protection at 150°C junction temperature. The output may be AC shorted without problem; and, although not guaranteed performance, DC shorts to ground are acceptable. A DC short to supply is destructive due to the thermal protection circuit which pulls the output to ground.

To achieve a stable DC operating point, it is desirable to close the feedback loop with unity DC gain. To achieve this simultaneously with a high AC gain normally requires a fairly large bypass capacitor, C_1, in Figure 4.4.2.

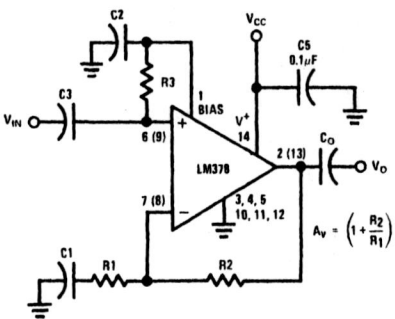

FIGURE 4.4.2 Non-Inverting Amplifier Connection

Establishing the initial charge on this capacitor results in a turn-on delay. An additional capacitor, C_2, is normally required to supply a ripple-free reference to set the DC operating point. To achieve good supply rejection X_{C_2} is normally made much smaller than a series resistor from the bias divider circuit (R_S in Figure 4.4.3). Where a supply rejection of 40dB is required with 40dB closed-loop gain, 80dB ripple attenuation is required of $R_S C_2$. The turn-on time can be calculated as follows:

$$PSRR = \frac{R_S - jX_{C_2}}{X_{C_2}} \approx \frac{R_S}{X_{C_2}} = \omega R C = \omega T$$

$$T = \frac{PSRR}{\omega} = \frac{80dB}{2\pi \, 120 \, Hz} = \frac{10^4}{754} = 13.3 \, sec$$

$$t_{ON} \approx \frac{T}{3} = 4.5 \text{ seconds to small signal operation}$$

$$t_{ON} \approx 3T = 40 \text{ seconds to full output voltage swing}$$

The 3T delay might normally be considered excessive! The LM378/379 amplifiers incorporate active turn-on circuitry to eliminate the long turn-on time. This circuitry appeared in Figure 4.4.1 as Q_{16} and an accompanying SCR; it is repeated and elaborated in Figure 4.4.3. In operation, the turn-on circuitry charges the external capacitors, bringing output and input levels to $V_{CC}/2$, and then disconnects itself leaving only the $V_{CC}/2$ divider R_B/R_B in the circuit.

The turn-on circuit operation is as follows. When power is applied, approximately $V_{CC}/2$ appears at the base of Q_{16}, rapidly charging C_1 and C_2 via a low emitter-follower output impedance and series resistors of 3k and 1k. This causes the emitters of the differential input pair to rise to $V_{CC}/2$, bringing the differential amp Q_3 and Q_4 into balance. This, in turn, drives Q_3 into conduction. Transistors

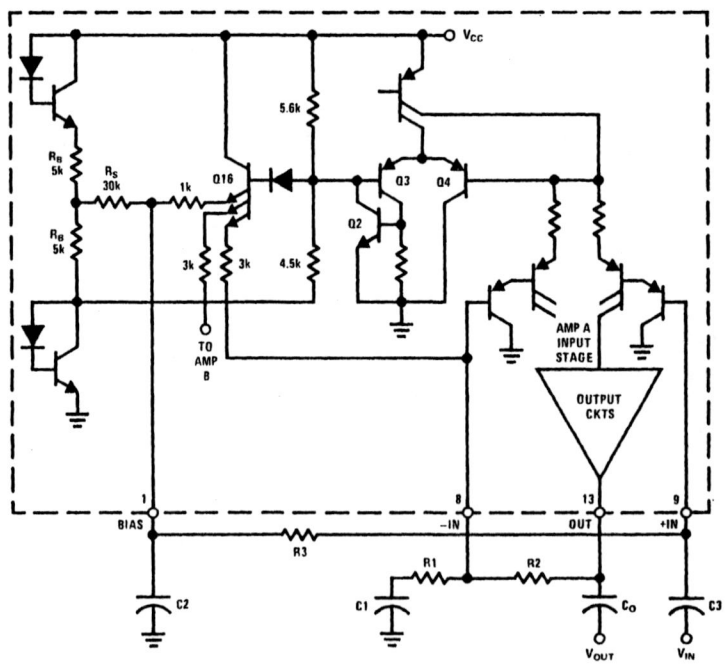

FIGURE 4.4.3 Internal Turn-On Circuitry

Q_2 and Q_3 form an SCR latch which then triggers and clamps the base of Q_{16} to ground, thus disabling the charging circuit. Once the capacitors are charged, the internal voltage divider R_B/R_B maintains the operating point at $V_{CC}/2$. Using $C_2 = 250\mu F$, the $t_{ON} = 3T \approx 0.3s$ and PSRR $\approx$ 75dB at 120Hz due to the 30k resistor R_S. Using $C_2 = 1000\mu F$, PSRR would be 86dB. The internal turn-on circuit prevents the usual "pop" from the speaker at turn-on. The turn-off period is also pop-free, as there is no series of pulses of active gain often seen in other similar amplifiers.

Note that the base of Q_4 is tied to the emitters of only one of the two input circuits. Should only one amplifier be in use, it is important that it be that with input at pins 8 and 9.

4.4.3 External Biasing Connection

The internal biasing is complete for the inverting gain connection of Figure 4.4.4 except for the external C_2 which provides power supply rejection. The bias terminal 1 may be connected directly to C_2 and the non-inverting input terminals 6 and 9. Normal gain-set feedback connections to the inverting inputs plus input and output coupling capacitors complete the circuitry. The output will Q up to $V_{CC}/2$ in a fraction of one second.

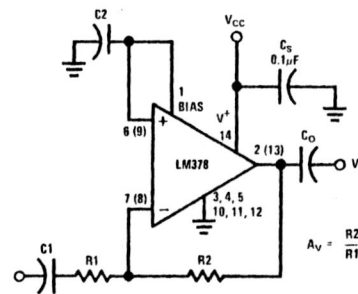

FIGURE 4.4.4 Inverting Amplifier Connection

The non-inverting circuit of Figure 4.4.2 is only slightly more complex, requiring the input return resistor R_3 from input to the bias terminal and additional input capacitor C_3. C_1 must remain in the circuit at the same or larger value than in Figure 4.4.4.

4.4.4 Circuit Description of LM1877 and LM2877

The LM1877 is a dual power amplifier designed to deliver 2W/channel continuously into 8Ω loads. It has an identical pin-out to the older LM377 and is intended as a direct replacement for that device in most applications.

The LM1877 differs internally in several respects from the LM378-LM379 series as shown by Figure 4.4.5 A differential input stage of NPN Darlington pairs is used and is optimized to give low equivalent input noise when the amplifier is driven from low impedance sources. Coupling to the second stage is through the current mirror Q_5. Note that Q_5 will hold the collector of Q_3 at 0.7V above $V_{CC}/2$ (Pin 1 bias level). This will limit the input voltage swing at Q_1 or Q_4 base to +700mV above $V_{CC}/2$. To accomodate input voltage swings that go higher than half supply (comparator or stereo amplifier applications) Pin 1 can be externally connected to Pin 14. The second stage is compensated internally for a unity gain bandwidth of 6mHz, which helps minimize the chance of rf radiation from the LM1877 into adjacent circuits (an AM radio input stage for example).

A large output swing capability is obtained by configuring the output stage and protection circuitry as shown in Section 4.1.4. Therefore an external R-C network from the output to ground is required to suppress oscillations that can occur during negative going signal swings as noted in Section 4.1.3.

Biasing for the amplifier stages is from a ΔV_{BE} reference voltage circuit (Q_{102}-Q_{109}) instead of from a zener, to allow operation with supplies as low as 6 volts. Q_{102}, Q_{103} and Q_{104} form the start-up circuit for the voltage reference by bleeding base current for Q_{105} (and hence Q_{108}) at turn-on. The double collector of Q_{108} will deliver equal currents to Q_{106} and Q_{107} which have a 4:1 ratio in emitter size. For transistors with a current density ratio of R, the difference in base-emitter voltage is given by,

$$\Delta V_{BE} = \frac{kT}{q} \log_e R$$

$$= 36mV \text{ for } R = 4 @ T = 300°K$$

In order for Q_{106} to have the same base voltage as Q_{107} (when it has the same current but one quarter the current density), this 36mV must appear across the 360Ω resistor in Q_{106} emitter. This sets the current level in the devices to 100μA so that a temperature compensated voltage of $0.7V + (100 \times 10^{-6} \times 5 \times 10^3)V = 1.2V$ appears at the base of Q_{109}. Once the circuit has started up, the current flowing in the 5k resistor in Q_{106} collector circuit wil cause Q_{105} to be shut off.

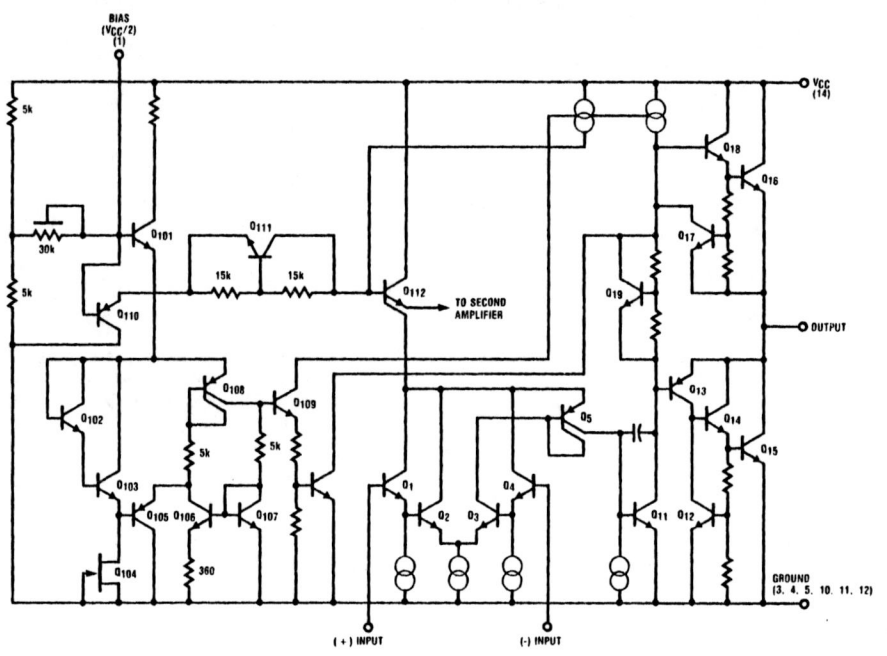

FIGURE 4.4.5 LM1887 Schematic Diagram (One Channel)

4.4.5 The LM1896 and LM2896

The newly introduced LM1896 is a dual power amplifier which has been optimized for maximum power output on low voltage supplies. As shown in Table 4.3.2., with a 6 volt supply, the LM1896 can deliver 1W/Ch into 4Ω or 2W into 8Ω when configured as a bridge amplifier. Good output swing capability is obtained by bootstrapping the output stages (Section 4.1.4) and a unique circuit design ensures low r.f. noise radiation — particularly important for obtaining high sensitivity and good S/N ratios in AM radios. Operation down to 3 volts and a low quiescent current drain of around 12mA make the LM1896 ideally suited for battery operated equipment requiring relatively high audio power output levels.

4.4.6 Stereo Amplifier Applications

The obvious and primary intended application is as an audio frequency power amplifier for stereo music systems. The amplifiers may be operated in either the non-inverting or the inverting modes of Figures 4.4.2 and 4.4.4. The inverting circuit has the lowest parts count so is most economical when driven by relatively low-impedance circuitry. Figure 4.4.6 shows the total parts count for such a stereo amplifier. The feedback resistor value of 1 meg in Figure 4.4.6 is about the largest practical value due to an input bias current max of approximately 1/2 μA (100nA typ). This will cause a −0.1 to 0.5V shift in DC output level, thus limiting peak negative signal swing. This output voltage shift can be corrected by the addition of series resistors (equal to the RF in value) in the + input lines. However, when this is done, a potential exists for high frequency instability due to capacitive coupling of the

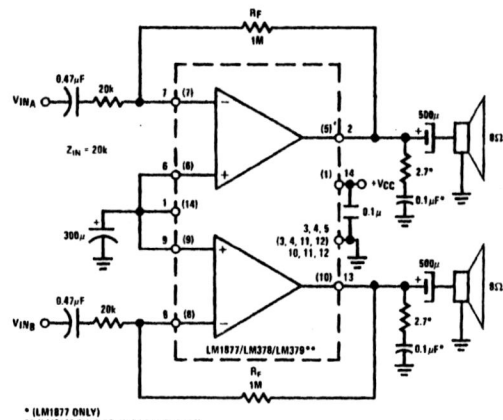

	LM1877	LM378	LM379
P_o =	2W/CH	3W/CH	4W/CH
e_i =	80mV MAX	98mV MAX	113mV MAX
A_V =	50	50	50
V_{CC} =	18V	24V	28V

FIGURE 4.4.6 Inverting Stereo Amplifier

output signal to the + input. Bypass capacitors could be added at + inputs to prevent such instability, but this increases the parts count equal to that of the non-inverting circuit of Figure 4.4.7, which has a superior input impedance. For applications utilizing high impedance tone and volume controls, the non-inverting connection will normally be used.

will typically be *double* the rated per channel *undistorted* power output into a resistive load. Since many of the smaller audio power amplifiers are rated at 10% THD, knowing that the output power at 10% is 30% larger than the undistorted power output enables a quick calculation to be made of the maximum amplifier dissipation,

$$P_{D(MAX)} \approx \frac{2 \times P_{MAX(RATE)}}{1.3}$$

or 1.5 times the output power at 10% THD.

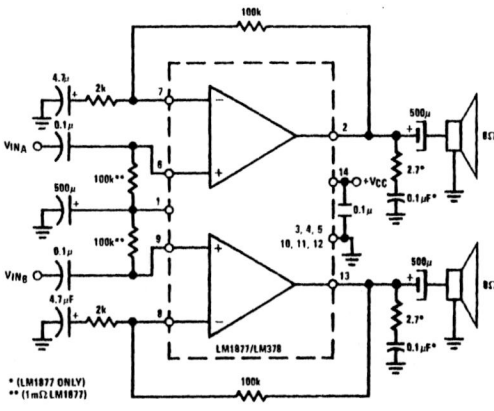

FIGURE 4.4.7 Non-Inverting Stereo Amplifier

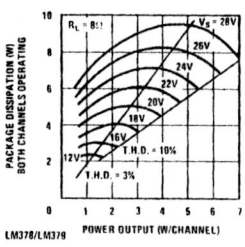

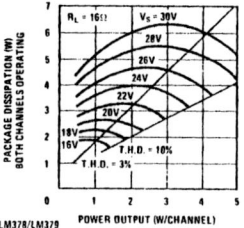

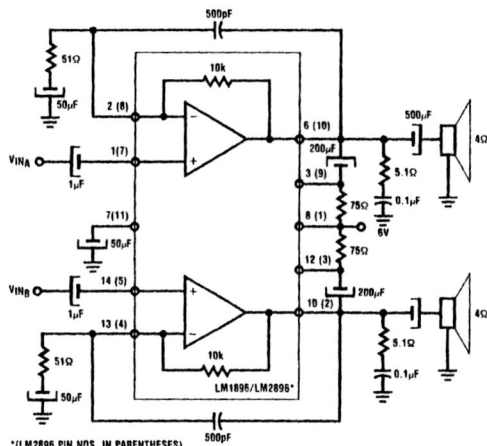

FIGURE 4.4.8 Low Voltage Stereo Amplifier 1.1 W/Ch

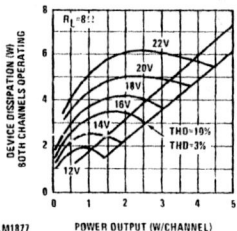

FIGURE 4.4.9 Device Dissipation for 8Ω and 16Ω Loads

4.4.7 Power Output per Channel (Both Channels Driven)

Figure 4.4.9 gives the package dissipation for the dual amplifiers with different supply voltages and 8Ω or 16Ω loads. The points at which 3% THD and 10% THD are reached are shown by the straight lines intersecting the curves. At 3% THD the output waveform has noticeable clipping while at 10% THD severe clipping of the output is occurring. It is also worth noting that the maximum amplifier power dissipation

Figure 4.4.10 gives the power derating curves for the dual amplifiers. Used in conjunction with Figure 4.4.9, the derating curves will indicate the heatsink requirements for continuous operation at any output power level and ambient temperature. It should be obvious from these curves that in most cases continuous or rms power at the rated output can require substantial heatsinking. Although the LM379 can be effectively heatsinked because of the low thermal resistance of the "S" Package style, with practical heatsinks the LM378 and

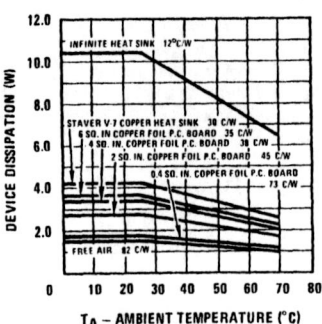

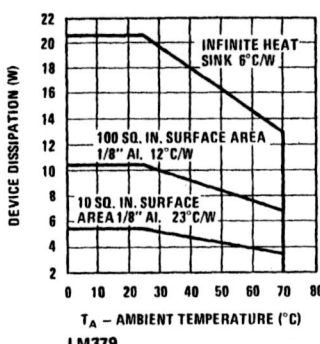

FIGURE 4.4.10 Dual Amplifier Maximum Dissipation vs. Ambient Temperature

LM1877 are limited to about 2 Watts per channel output at elevated temperatures. This can be illustrated by the use of these curves to select a suitable heatsink for a 2 Watt/Channel amplifier driving an 8Ω load from an unregulated 18 Volt power supply. Operation without thermal shutdown is required at a maximum ambient temperature of 55°C.

Solution:

1. Unloaded supply voltage at high line = 18×1.1 = 19.8 Volts.
   Amplifier maximum supply voltage rating must be ≥ 20 Volts.
2. When delivering the rated output the supply will sag by about 15% at 2 Watts, V_S = 15.7V.
3. From Figure 4.4.9, for the LM1877 with this supply voltage, the amplifier can deliver 2 Watts before clipping and 2.5 Watts at 10% THD.
4. From the same curve the peak device power dissipation is 3.2 Watts.
*5. Figure 4.4.10 shows that for dissipating 3.2 Watts at 55°C, a Staver V7-1 heatsink is needed.

From the above calculations there doesn't seem to be much point in publishing curves for an amplifier driving 8Ω with regulated supplies above 16 volts or unregulated supplies above 18 volts.

Also, a designer appears to be prevented from using higher supply voltages to provide a safety margin from clipping at rated outputs or power outputs in excess of 2.5 Watts. This is not always true. Usually for speech or music there is a 30dB ratio between the R.M.S. and peak power levels. It is possible to design the heatsink for power levels 20dB below the rated maximum, anticipating that the heatsink thermal capacity is adequate to carry through peak power levels. In any case, the dual amplifiers have thermal shutdown circuitry to protect the device if sustained peak power levels cause the junction temperature to increase above 150°C.

Where higher power levels must be sustained, the alternative is to use the Single-in-line Package style (S.I.P.), Figure 4.4.11. The S.I.P. not only permits more compact p.c.b. layouts to be obtained, but the large tab allows easier heatsinking. In this package the LM2877 is electrically equivalent to the LM1877, and the LM2896 is the S.I.P. version of the LM1896. Figure 4.4.12 shows the substantially better thermal performance of the S.I.P. The power output levels of the previous example can be handled by less than a $2 \times 2 \times 1/16''$ piece of aluminum. If the LM2877 is bolted to a typical chassis, then 5.5 watts can be dissipated at 55°C ambient temperature for output power levels in excess of 3W/Ch!

For the LM379S custom heatsinks are easily fabricated from sheet copper of aluminum and are bolted to the package tab. Power outputs of over 4W/Ch are possible, although the designer should watch out for the LM379 current limit specification. On the data sheet this is given as 1.5A measured at 25°C. As the I/C warms up, this current limit will decrease to between 1A and 1.25A. These peak currents correspond to 0.7 to 0.88 A_{RMS}, which will limit the output power into 8Ω to 4W or 6.2W respectively.

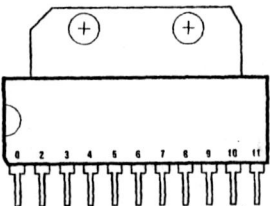

FIGURE 4.4.11 Molded Single-In-Line Package (NT)

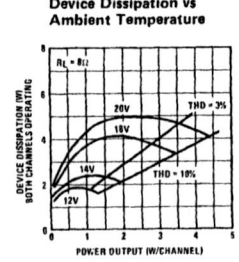

FIGURE 4.4.12 LM2887NT Power Dissipation and Temperature Derating Curves

4.4.8 Stabilization

The LM378/379 series amplifiers are internally stabilized so external compensation capacitors are not required. The high gain × BW provides a bandwidth greater than 50kHz, as seen in Figure 4.4.13. These amplifiers are, however, not intended for closed loop gain below 10. The typical Bode plot of Figure 4.4.14 shows a phase margin of 70° for gain of 5.6 (15dB), which is stable. At unity gain the phase margin is less than 30°, or marginally stable. This margin may vary considerably from device to device due to variation in gain × BW.

4.4.9 Layout

Ground and power connections must be adequate to handle the 1 to 2A peak supply and load currents. Ground loops can be especially troublesome because of these high currents. The load return line should be connected directly to the ground pins of the package on one side and/or the input and feedback ground lines should be connected directly to the ground pins (possibly on the other side of the package). The signal ground should not be connected so as to intercept any output signal voltage drop due to resistance between IC ground and load ground.

4.4.10 Split Supply Operation

The use of split power supplies offers a substantial reduction in parts count for low power stereo systems using dual power amplifiers. Split supply operation requires only redefinition of the ground pins for use with the negative power supply. The only precaution necessary is to observe that when thermal shutdown occurs the output is pulled down to the negative supply, instead of ground. Both supplies require bypassing with $0.1\mu F$ ceramic or $0.47\mu F$ mylar capacitors to ground.

Single supply operation (Figure 4.4.15) requires 6 resistors and 9 capacitors (excluding power supply parts) and uses the typical power supply shown. The same circuit using split supplies (Figure 4.4.16) requires only 4 resistors and 4 capacitors. This approach allows direct coupling of the amplifier to the speakers since the output DC level is approximately zero volts (offset voltages will be less than 25mV), thereby eliminating the need for large coupling capacitors and their associated degradation of power, distortion and cost. Since the input bias voltages are zero volts, the need for bias resistors and the bias-pin supply bypassing capacitor are also eliminated. Input capacitors are omitted to allow bias currents from the positive inputs to flow directly through the volume pots to ground.

Normally with split supply operation, the current loading of each supply is fairly symmetrical. Nevertheless, care should be taken that at turn-on both supplies increase from zero to full value at the same rate or within a couple volts of each other. If this is not so, referring the non-inverting inputs to ground instead of to pin 1 can cause a latch-up state to occur.

4.4.11 Unity Gain Power Buffers

Occasionally system requirements dictate the need for a unity gain power buffer, i.e., a current amplifier rather than a voltage amplifier. The peak output currents greater than one amp of the LM378/379 family make them a logical choice for this application.

Internal compensation limits stable operation to gains greater than 10 (20dB), thereby requiring additional components if unity gain operation is to be used. Stable unity gain inverting amplifiers (Figure 4.4.17) require only one additional resistor from the negative input to ground, equal in value to one tenth the feedback resistor. A discussion of this technique may be found in Section 2.8.4.

Non-inverting unity gain stability (Figure 4.4.18) can be achieved without additional components by judicious selection of the existing feedback elements. Writing the gain function of Figure 4.4.16 including the frequency dependent term of C_2 yields:

$$A_v = 1 + \frac{R_1}{R_2 + X_{C_2}}$$

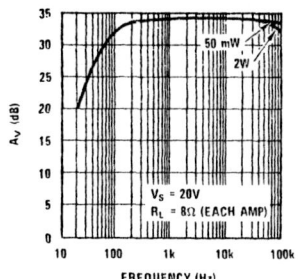

FIGURE 4.4.13 Frequency Response of the Stereo Amp of Figure 4.4.5

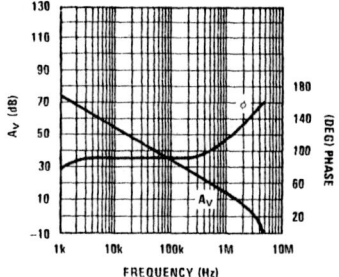

FIGURE 4.4.14 Open Loop Bode Plot (Approximately Worst Case)

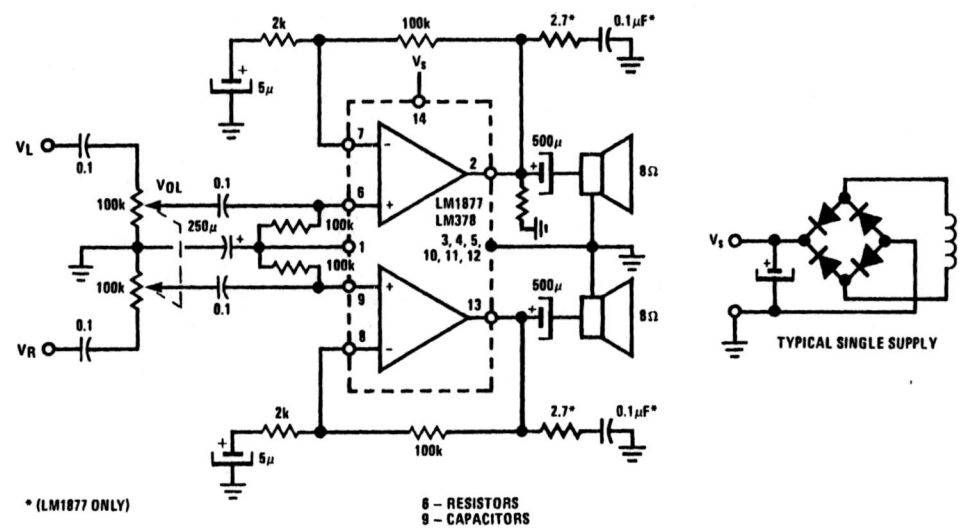

FIGURE 4.4.15 Non-Inverting Amplifier Using Single Supply

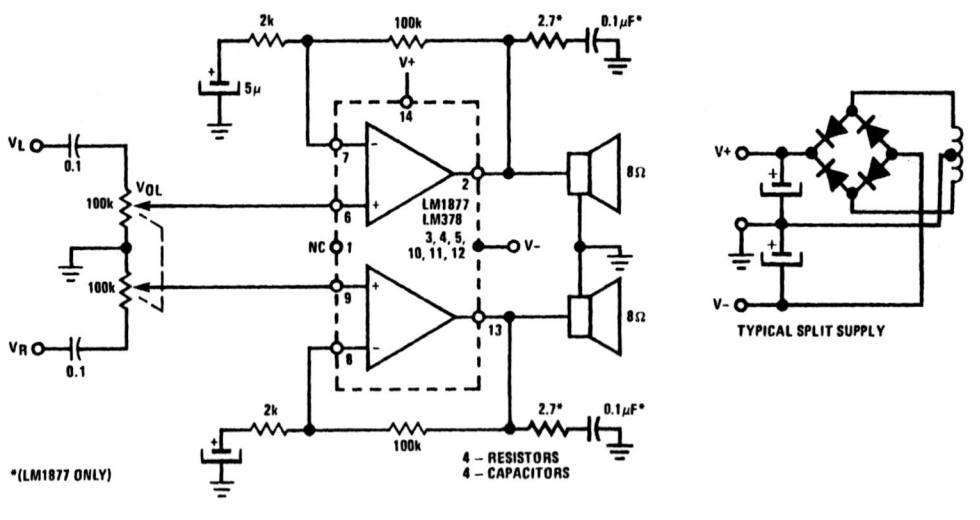

FIGURE 4.4.16 Non-Inverting Amplifier Using Split Supply

Satisfaction of unity gain *circuit* performance over the audio band and gain greater than 10 *amplifier* performance at high frequencies can be accomplished by making the frequency dependent term small (relative to one) over the audio band and allowing it to dominate the gain expression beyond audio. Rewriting the gain term using the Laplace variable S (The variable S is a complex frequency.) results in Equation (4.4.1):

$$A_V = 1 + \frac{R_1}{R_2 + \frac{1}{SC_2}} = \frac{S(R_1+R_2)C_2 + 1}{SR_2C_2 + 1}$$

$$\approx \frac{SR_1C_2 + 1}{SR_2C_2 + 1} \qquad (4.4.1)$$

Zero at $f_z = \dfrac{1}{2\pi R_1 C_2}$ (4.4.2)

Pole at $f_p = \dfrac{1}{2\pi R_2 C_2}$ (4.4.3)

Examination of Equation (4.4.1) shows it to have a frequency response zero at f_z (Equation (4.4.2)) and a pole at f_p (Equation (4.4.3)). By selecting f_z to fall at the edge of the audio spectrum (20 kHz as shown) and f_p prior to hitting the open loop response (340 kHz as shown) the frequency response of Figure 4.4.19 is obtained. This response satisfies the unity gain requirements, while allowing the gain to raise beyond audio to insure stable operation.

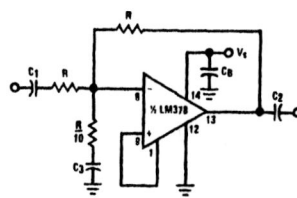

FIGURE 4.4.17 Inverting Unity Gain Amplifier

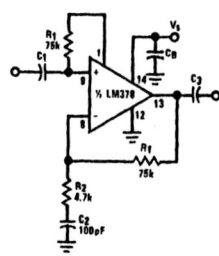

FIGURE 4.4.18 Non-Inverting Unity Gain Amplifier

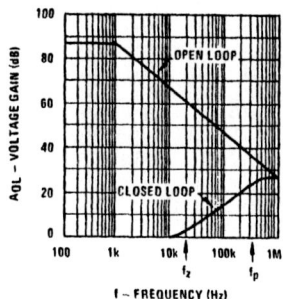

FIGURE 4.4.19 Frequency Response of Non-Inverting Unity Gain Amplifier

4.4.12 Bridge Amplifiers

The dual amplifiers are equally useful in the bridge configuration to drive floating loads, which may be loudspeakers, servo motors or whatever. Double the power output can be obtained in this connection, and output coupling capacitors are not required. Load impedance may be either 8 or 16Ω in the bridge circuit of Figure 4.4.20. Response of this circuit is 20Hz to 160kHz as shown in Figure 4.4.21 and distortion is 0.1% midband at 4W, rising to 0.5% at 10kHz and 50mW output (Figure 4.4.22). The higher distortion at low power is due to a small amount of crossover notch distortion which becomes more apparent at low powers and high frequencies. The circuit of Figure 4.4.23 is similar except for higher input impedance. In Figure 4.4.23 the signal drive for the inverting amplifier is derived from the feedback voltage of the non-inverting amplifier. Resistors R_1 and R_3 are the input and feedback resistors for A_2, whereas R_1 and R_2 are the feedback network for A_1. So far as A_1 is concerned, R_2 sees a virtual ground at the (−) input to A_2; therefore, the gain of A_1 is $(1 + R_2/R_1)$. So far as A_2 is concerned, its input signal is the voltage appearing at the (−) input to A_1. This equals that at the (+) input to A_1. The driving point impedance at the (−) input to A_1 is very low even though R_2 is 100k. A_1 can be considered a unity gain amplifier with internal $R = R_2 = 100k$ and $R_L = R_1 = 2k$. Then the effective output resistance of the unity gain amplifier is:

$$R_{OUT} = \frac{R_{INTERNAL}}{A_{OL}/A\beta} = \frac{100k}{600/1} = 167\Omega$$

Layout is critical if output oscillation is to be avoided. Even with careful layout, capacitors C_1 and C_2 may be required to prevent oscillation. With the values shown, the amplifier will drive a 16Ω load to 4W with less than 0.2% distortion midband, rising to 1% at 20kHz (Figure 4.4.24). Frequency response is 27Hz to 60kHz as shown in Figure 4.4.25. The low frequency roll off is due to the double poles C_3R_3 and C_4R_1.

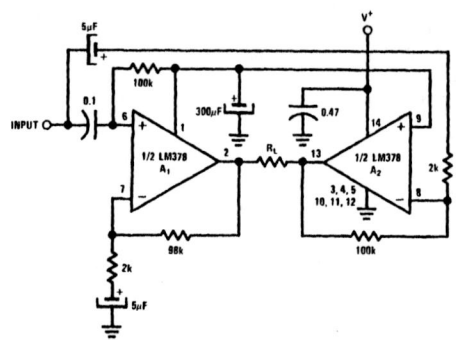

FIGURE 4.4.20 4-Watt Bridge Amplifier

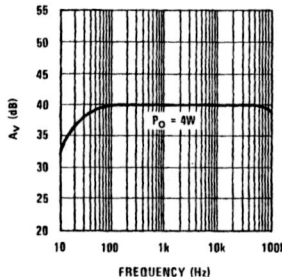

FIGURE 4.4.21 Frequency Response, Bridge Amp of Figure 4.4.20

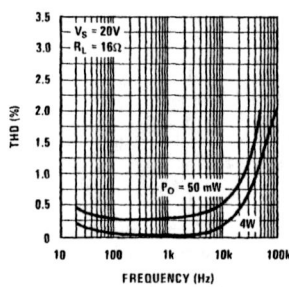

FIGURE 4.4.22 Distoration for Bridge Amp of Figure 4.4.20

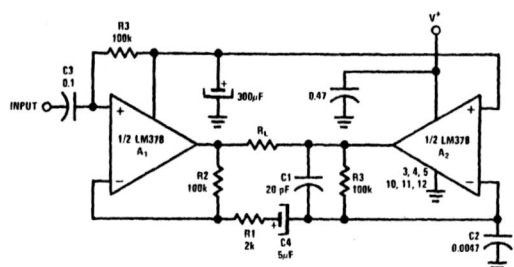

FIGURE 4.4.23 4-Watt Bridge Amplifier with High Input Impedance

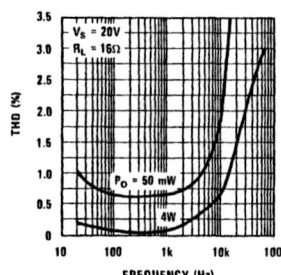

FIGURE 4.4.24 Distoration for Bridge Amp of Figure 4.4.23

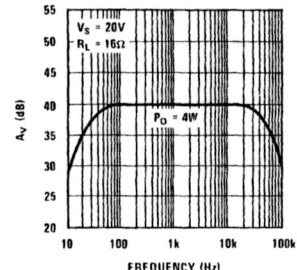

FIGURE 4.4.25 Frequency Response, Bridge Amp of Figure 4.4.23

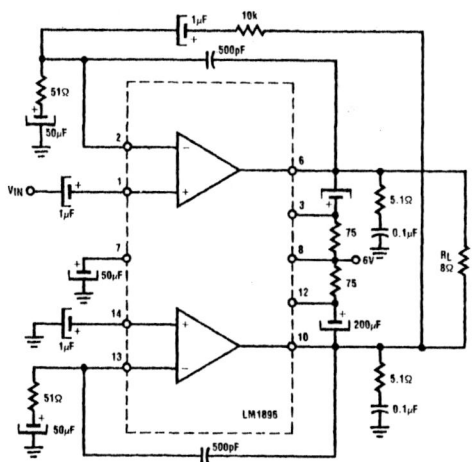

FIGURE 4.4.26 2-Watt Bridge Amplifier with 6V Supply

4.4.13 Power Oscillator

One half of an LM378 may be connected as an oscillator to deliver up to 2W to a load. Figure 4.4.27 shows a Wien bridge type of oscillator with FET amplitude stabilization in the negative feedback path. The circuit employs internal biasing and operates from a single supply. C_3 and C_6 allow unity gain DC feedback and isolate the bias from ground. Total harmonic distortion is under 1% to 10kHz, and could possibly be improved with careful adjustment of R_5. The FET acts as the variable element in the feedback attenuator R_4 to R_6. Minimum negative feedback gain is set by the resistors R_4 to R_6, while the FET shunts R_6 to increase gain in the absence of adequate output signal. The peak detector D_2 and C_8 senses output level to apply control bias to the FET. Zener diode D_1 sets the output level although adjustment could be made if R_9 were a potentiometer with R_8 connected to the slider. Maximum output level with the values shown is $5.3 V_{RMS}$ at 60 Hz. C_7 and the attenuator R_7 and R_8 couple 1/2 the signal of the FET drain to the gate for improved FET linearity and low distortion. The amplitude control loop could be replaced by an incandescent lamp in non-critical circuits (Figure 4.4.25), although DC offset will suffer by a factor of about 3 (DC gain of the oscillator). R_{10} matches R_3 for improved DC stability, and the network R_{11}, C_9 increases high frequency gain for improved stability. Without this RC, oscillation may occur on the negative half cycle of output waveform. A low inductance capacitor, C_5, located directly at the supply leads on the package is important to maintain stability and prevent high frequency oscillation on negative half cycle of the output waveform. C_5 may be $0.1\mu F$ ceramic, or $0.47\mu F$ mylar. Layout is important; especially take care to avoid ground loops as discussed in the section on amplifiers. If high frequency instability still occurs, add the R_{12}, C_{10} network to the output.

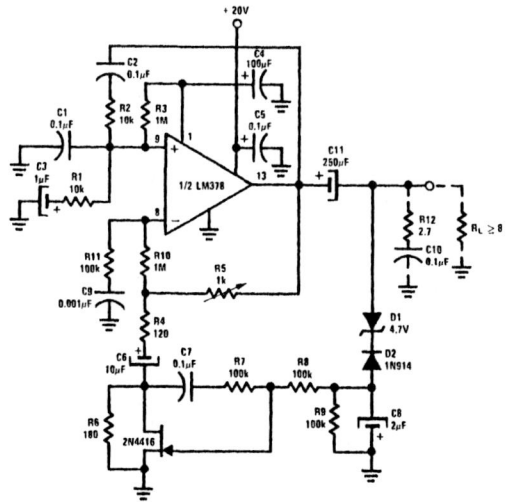

FIGURE 4.4.27 Wien Bridge Power Oscillator

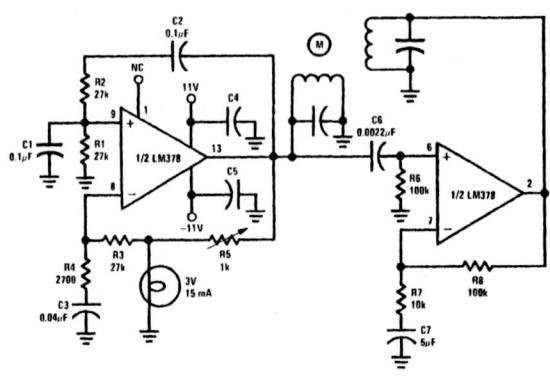

FIGURE 4.4.28 Two-Phase Motor Drive

4.4.14 Two-Phase Motor Drive

Figure 4.4.28 shows the use of the LM378 to drive a small 60 Hz two phase servo motor up to 3W per phase. Applications such as a constant (or selectable) speed phonograph turntable drive are adequately met by this circuit. A split supply is used to simplify the circuit, reduce parts count, and eliminate several large bypass capacitors. An incandescent lamp is used in a simple amplitude stabilization loop. Input DC is minimized by balancing DC resistance at (+) and (−) amplifier inputs ($R_1 = R_3$ and $R_6 = R_8$). High frequency stability is assured by increasing closed-loop gain from approximately 3 at 60 Hz to about 30 above 40 kHz with the network consisting of R_3, R_4 and C_3. The interstage coupling $C_6 R_6$ network shifts phase by 85° at 60 Hz to provide the necessary two phase motor drive signal. The gain of the phase shift network is purposely low so that the buffer amplifier will operate at a gain of 10 for adequate high frequency stability. As in other circuits, the importance of supply bypassing, careful layout, and prevention of output ground loops is to be stressed. The motor windings are tuned to 60 Hz with shunt capacitors. This circuit will drive 8Ω loads to 3W each.

4.4.15 Proportional Speed Controller

A low cost proportional speed controller may be simply designed using a LM378 amplifier. For use with 12-24 V_{DC} motors at continuous currents up to several hundred milliamps, this circuit allows remote adjustment of angular displacements in a drive shaft. Typical applications include rooftop rotary antennas and motor-controlled valves.

Proportional control (Figure 4.4.29) results from an error signal developed across the Wheatstone bridge comprised of resistors R_1, R_2 and potentiometers P_1, P_2. Control P_1 is

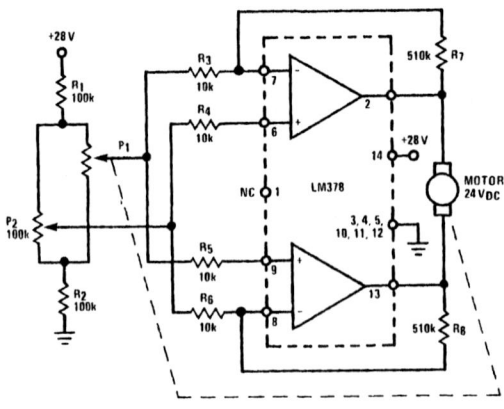

FIGURE 4.4.29 Proportional Speed Controller

mechanically coupled to the motor shaft as depicted by the dotted line and acts as a continuously variable feedback sensor. Setting position control P_2 creates an error voltage between the two inputs which is amplified by the LM378 (wired as a difference bridge amplifier); the magnitude and polarity of the output signal of the LM378 determines the speed and direction of the motor. As the motor turns, potentiometer P_1 tracks the movement, and the error signal, i.e., difference in positions between P_1 and P_2, becomes smaller and smaller until ultimately the system stops when the error voltage reaches zero volts.

Actual gain requirements of the system are determined by the motor selected and the required range. Figure 4.4.26 demonstrates the principle involved in proportional speed control and is not intended to specify final resistor values.

4.4.16 Complete Systems

The dual power amplifiers are useful in table or console radios, phonographs, tape players, intercoms, or any low to medium power music systems.

Figures 4.4.30 through 4.4.32 describe the complete electronic section of a 2-channel sound system with inputs for AM radio, stereo FM radio, phono, and tape playback. Figure 4.4.30 combines the power amplifier pair with loudness, balance, and tone controls. The tone controls allow boost or cut of bass and/or treble. Transistors Q_1 and Q_2 act as input line amplifiers with the triple function of (1) presenting a high impedance to the inputs, especially ceramic phono; (2) providing an amplified output signal to a tape recorder; and (3) providing gain to make up for the loss in the tone controls. Feedback tone controls of the Baxandall type employing transistor gain could be used; but then, with the same transistor count, the first two listed functions of Q_1Q_2 would be lost. It is believed that this circuit represents the lowest parts count for the complete system. Figure 4.4.31 is the additional circuitry for input switching and tape playback amplifiers. The LM382 with capacitors as shown provides for NAB tape playback compensation. For further information on the LM382 or the similar LM381 and LM387, refer to Section 2.0.

Figure 4.4.32 shows the relationship between signal source impedance and gain or input impedance for the amplifier stage Q_1Q_2. Stage gain may be set at a desired value by choice of either the source impedance or insertion of resistors in series with the inputs (as R_1 to R_4 in Figure 4.4.31). Gain is variable from -15 to $+24\,dB$ by choice of series R from 0 to 10 meg. Gain required for $e_{IN} = 100$ to $200\,mV$ (approximate value of recovered audio from FM stereo or AM radio) is about 18 to 21 dB overall for 2W into an 8Ω speaker at 1 Hz or 21 to 24 dB for 4W.

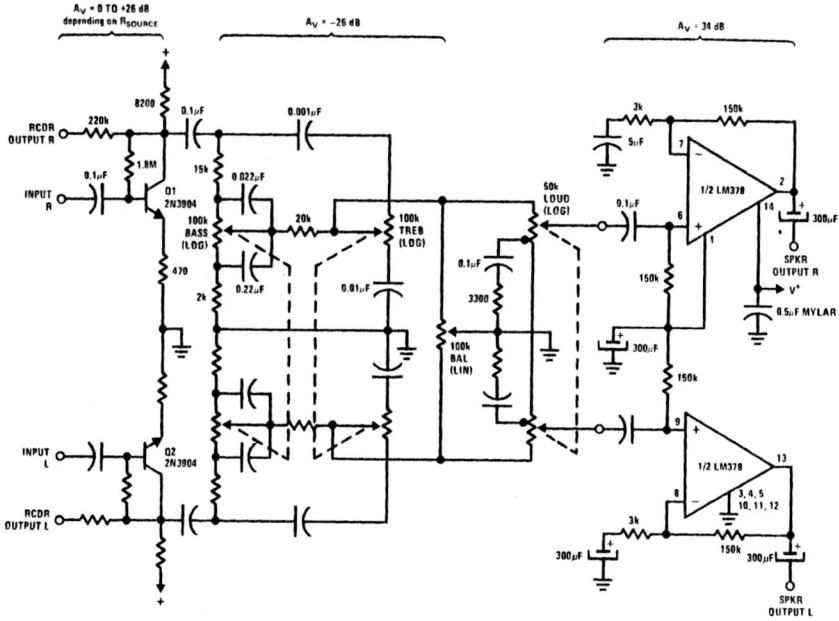

FIGURE 4.4.30 Two-Channel Power Amplifier and Control Circuits

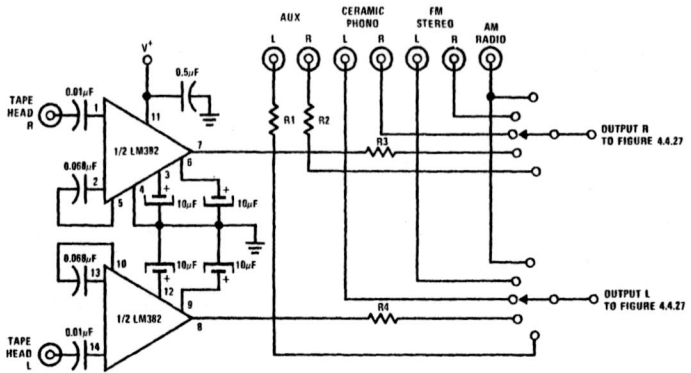

FIGURE 4.4.31 Two-Channel Tape-Playback Amplifier and Signal Switching

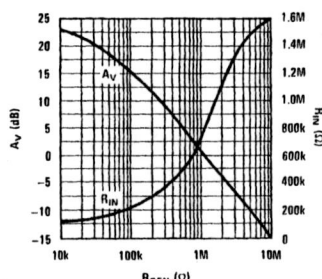

FIGURE 4.4.32 A_V and R_{IN} for Input Stage of Figure 4.4.26

4.4.17 Rear Channel Ambience Amplifier

The rear channel "ambience" circuit of Figure 4.4.33 can be added to an existing stereo system to extract a difference signal (R − L or L − R) which, when combined with some direct signal (R or L), adds some fullness, or "concert hall realism" to reproduction of recorded music. Very little power is required at the rear channels, hence an LM1877 will suffice for most "ambience" applications. The inputs are merely connected to the existing speaker output terminals of a stereo set, and two more speakers are connected to the ambience circuit outputs. Note that the rear speakers should be connected in opposite phase to those of the front speakers, as indicated by the +/− signs on the diagram of Figure 4.4.33.

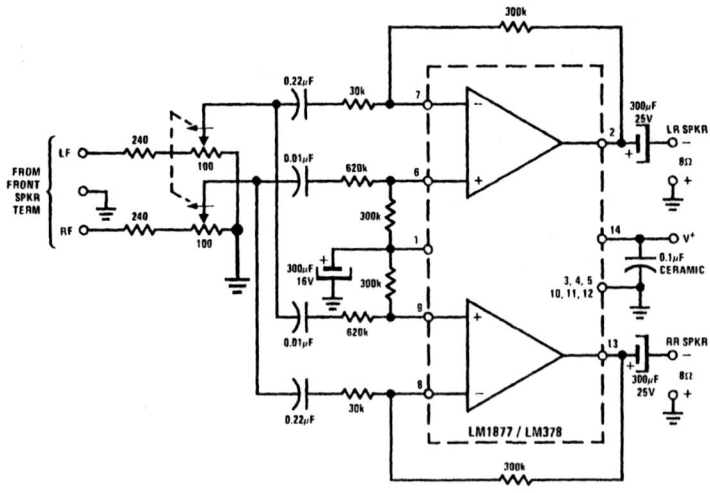

FIGURE 4.4.33 Rear Speaker Ambience (4-Channel) Amplifier

4.4.18 Ceramic Cartridge Stereo Phonograph

Ceramic cartridges, with a high output level of several hundred millivolts, can be used with the LM1877 as the only active gain element to provide a complete and inexpensive 2W/Ch stereo phonograph system. A suitable circuit is shown in Figure 4.4.34 where the cartridge is loaded directly with the 500kΩ gain control potentiometers. The LM1877 is configured in the non-inverting mode to minimize loading on the cartridge at maximum volume settings and a simple bass tone control circuit is added in the feedback network (see Section 2.14.7). Response of the tone control circuit is shown in Figure 4.4.35. At midband and higher frequencies the capacitors can be considered as short circuits, which gives a midband gain

$$A_V = \left(\frac{10k + 1k}{1k}\right) \times \left(\frac{510k + 51k}{51k}\right) = 121 \text{ or } 42dB.$$

For a typical ceramic cartridge output of 200mV to 300mV, this gain is more than adequate to ensure clipping at the speaker output with moderate gain control settings. The amplifier is capable of delivering 2 watts continuously in both channels at the 10% distortion level into 8Ω loads on a 14V supply. With a 16V supply, 2.5 watts continuous is available (See Figure 4.4.9).

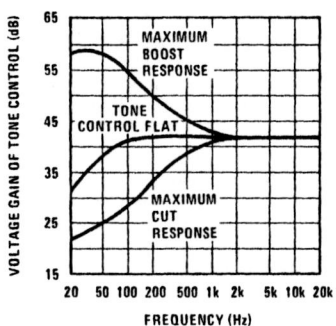

FIGURE 4.4.35 Frequency Response of Bass Tone Control

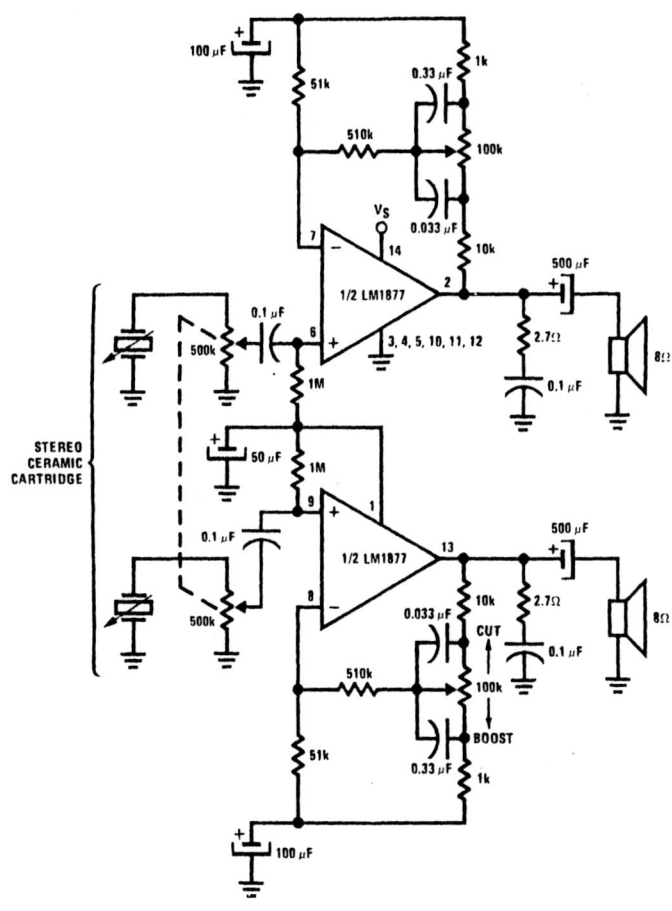

FIGURE 4.4.34 Stereo Phongraph Amplifier with Bass Tone Control

4.5 LM380 AUDIO POWER AMPLIFIER

4.5.1 Introduction

Most of the mono power amplifiers listed in Table 4.3.3 derive from the LM380 design; therefore, a detailed discussion of the internal circuitry will be presented as a basis for understanding each of the devices. Subsequent sections will describe only the variations on the LM380 design responsible for each unique part.

The LM380 is a power audio amplifier intended for consumer applications. It features an internally fixed gain of 50 (34 dB) and an output which automatically centers itself at one half of the supply voltage. A unique input stage allows inputs to be ground referenced or AC coupled as required. The output stage of the LM380 is protected with both short circuit current limiting and thermal shutdown circuitry. All of these internally provided features result in a minimum external parts count integrated circuit for audio applications.

4.5.2 Circuit Description

Figure 4.5.1 shows a simplified circuit schematic of the LM380. The input stage is a PNP emitter-follower driving a PNP differential pair with a slave current-source load. The PNP input is chosen to reference the input to ground, thus enabling the input transducer to be directly coupled.

The second stage is a common emitter voltage gain amplifier with a current-source load. Internal compensation is provided by the pole-splitting capacitor C. Pole-splitting compensation is used to preserve wide power bandwidth (100 kHz at 2W, 8Ω). The output is a quasi-complementary pair emitter-follower.

The output is biased to half the supply voltage by resistor ratio R_2/R_1. Simplifying Figure 4.5.1 still further to show the DC biasing of the output stage results in Figure 4.5.2, where resistors R_1 and R_2 are labeled R. Since the transistor operates with effectively zero volts base to collector, the circuit acts as a DC amplifier with a gain of one half (i.e., $A_v = R/[R + R]$) and an input of V^+; therefore, the output equals $V^+/2$.

The amplifier AC gain is internally fixed to 34 dB (or 50 V/V). Figure 4.5.3 shows this to be accomplished by the internal feedback network R_2-R_3. The gain is twice that of the ratio R_2/R_3 due to the slave current-source (Q_5, Q_6) which provides the full differential gain of the input stage.

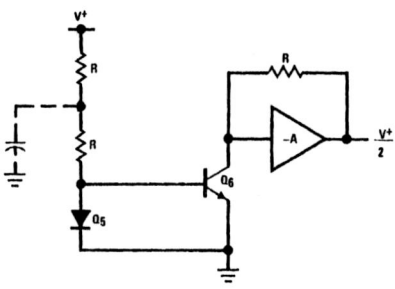

FIGURE 4.5.2 LM380 DC Equivalent Circuit

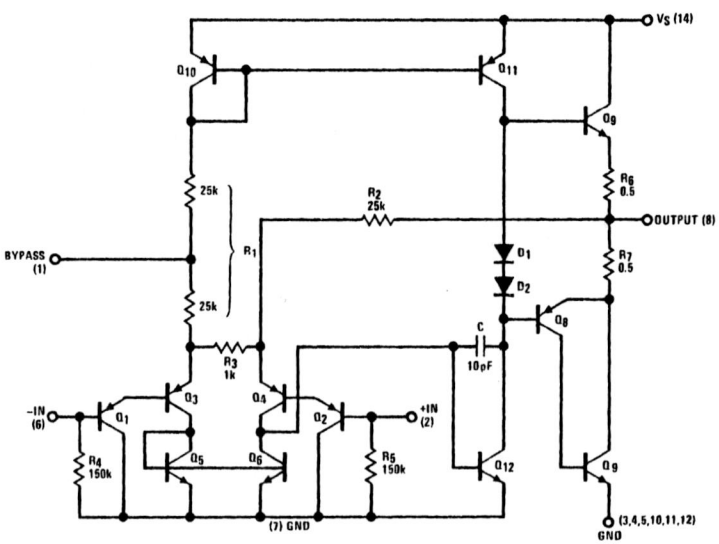

FIGURE 4.5.1 LM380 Simplified Schematic

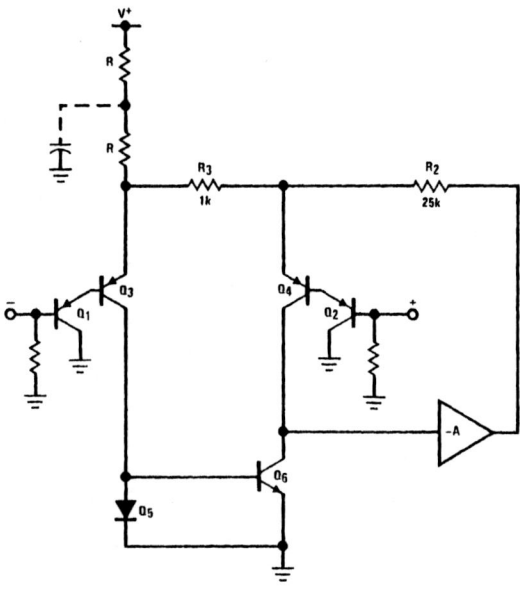

FIGURE 4.5.3 LM380 AC Equivalent Circuit

A gain difference of one exists between the negative and positive inputs, analogous to inverting and non-inverting amplifiers. For example, an inverting amplifier with input resistor equal to 1k and a 50k feedback resistor has a gain of 50 V/V, while a non-inverting amplifier constructed from the same resistors has a gain of 51 V/V. Driving the inverting terminal of the LM380, therefore, results in a gain of 50, while driving the non-inverting will give a gain of 51.

4.5.3 General Operating Characteristics

The output current of the LM380 is rated at 1.3 A peak. The 14 pin dual-in-line package is rated at 35°C/W when soldered into a printed circuit board with 6 square inches of 2 ounce copper foil (Figure 4.5.4). Since the device junction temperature is limited to 150°C via the thermal shutdown circuitry, the package will support 2.9W dissipation at 50°C ambient or 3.6W at 25°C ambient.

Figure 4.5.4a shows the maximum package dissipation vs. ambient temperature for various amounts of heat sinking. (Dimensions of the Staver V7 heat sink appear as Figure 4.5.4b.)

Figures 4.5.5a, -b, and -c show device dissipation versus output power for various supply voltages and loads.

The maximum device dissipation is obtained from Figure 4.5.4 for the heat sink and ambient temperature conditions under which the device will be operating. With this maximum allowed dissipation, Figures 4.5.5a, -b, and -c show the maximum power supply allowed (to stay within dissipation limits) and the output power delivered into 4, 8 or 16Ω loads. The three percent total harmonic distortion line is approximately the onset of clipping.

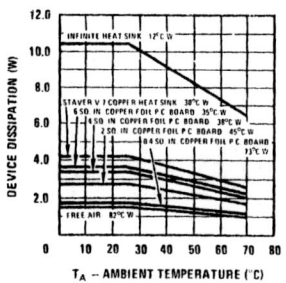

FIGURE 4.5.4a Device Dissipation vs. Maximum Ambient Temperature

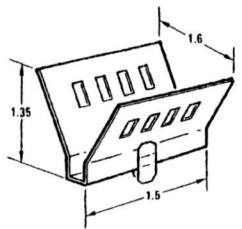

* – Staver Co.
Bayshore, N.Y.

FIGURE 4.5.4b Staver* "V7" Heat Sink

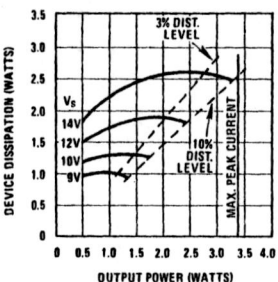

FIGURE 4.5.5a Device Dissipation vs. Output Power – 4Ω Load

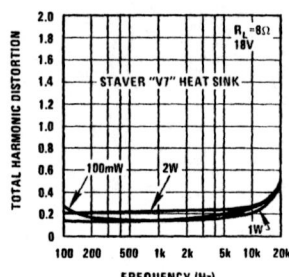

FIGURE 4.5.6 Total Harmonic Distortion vs. Frequency

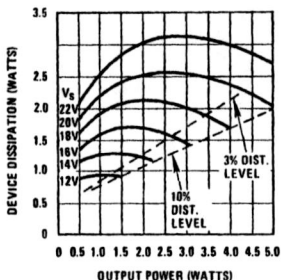

FIGURE 4.5.5b Device Dissipation vs. Output Power – 8Ω Load

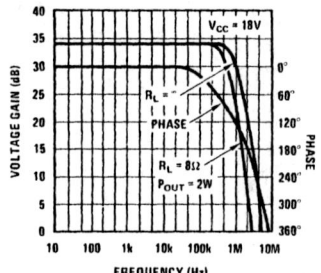

FIGURE 4.5.7 Output Voltage Gain vs. Frequency

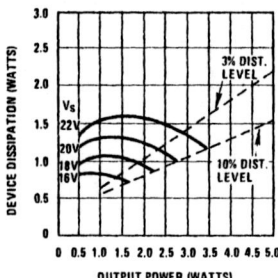

FIGURE 4.5.5c Device Dissipation vs. Output Power – 16Ω Load

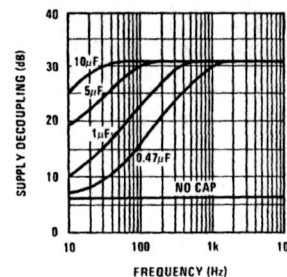

FIGURE 4.5.8 Supply Decoupling vs. Frequency

Figure 4.5.6 shows total harmonic distortion vs. frequency for various output levels, while Figure 4.5.7 shows the power bandwidth of the LM380.

Power supply decoupling is achieved through the AC divider formed by R_1 (Figure 4.5.1) and an external bypass capacitor. Resistor R_1 is split into two 25kΩ halves providing a high source impedance for the integrator. Figure 4.5.8 shows supply decoupling vs. frequency for various bypass capacitors.

4.5.4 Biasing

The simplified schematic of Figure 4.5.1 shows that the LM380 is internally biased with the 150kΩ resistance to ground. This enables input transducers which are referenced to ground to be direct-coupled to either the inverting or non-inverting inputs of the amplifier. The unused input may be either: (1) left floating, (2) returned to ground through a resistor or capacitor, or (3) shorted to ground. In most applications where the non-inverting input is used, the inverting input is left floating. When the inverting input is used and the non-inverting input is left floating, the amplifier may be found to be sensitive to board layout since stray coupling to the floating input is positive feedback. This can be avoided by employing one of three alternatives: (1) AC grounding the unused input with a small capacitor. This is preferred when using high source impedance transducers. (2) Returning the unused input to ground through a resistor. This is preferred when using moderate to low DC source impedance transducers and when output offset from half supply voltage is critical. The resistor is made equal to the resistance of the input transducer, thus maintaining balance in the input differential amplifier and minimizing output offset. (3) Shorting the unused input to ground. This is used with low DC source impedance transducers or when output offset voltage is non-critical.

4.5.5 Oscillation

The normal power supply decoupling precautions should be taken when installing the LM380. If V_S is more than 2" to 3" from the power supply filter capacitor it should be decoupled with a $0.1\mu F$ disc ceramic capacitor at the V_S terminal of the IC.

The R_C and C_C components in Figure 4.5.9 and throughout this section suppress a 5 to 10MHz small amplitude oscillation which can occur during the negative swing into a load which draws high current. The oscillation is of course at too high a frequency to pass through a speaker, but it should be guarded against when operating in an RF sensitive environment.

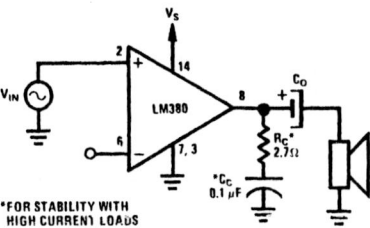

FIGURE 4.5.9 Oscillation Suppression Components

4.5.6 RF Precautions — See Section 2.3.10

4.5.7 Inverting Amplifier Application

With the internal biasing and compensation of the LM380, the simplest and most basic circuit configuration requires only an output coupling capacitor as seen in Figure 4.5.10.

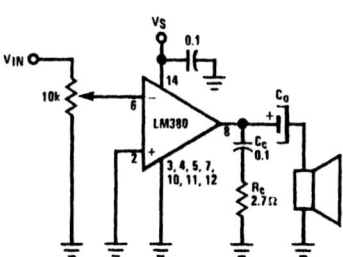

FIGURE 4.5.10 Minimum Component Configuration

4.5.8 Ceramic Phono Amplifier

An application of this basic configuration is the phonograph amplifier where the addition of volume and tone controls is required. Figure 4.5.11 shows the LM380 with a voltage divider volume control and high frequency roll-off tone control.

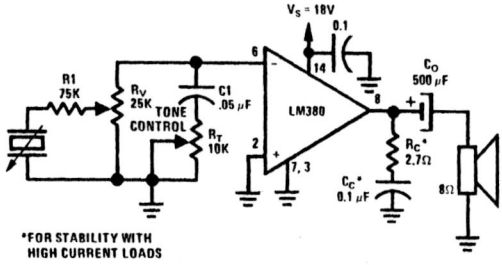

FIGURE 4.5.11 Ceramic Phono Amp

4.5.9 Common Mode Volume and Tone Controls

When maximum input impedance is required or the signal attenuation of the voltage divider volume control is undesirable, a "common mode" volume control may be used as seen in Figure 4.5.12.

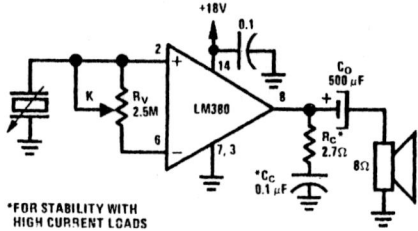

FIGURE 4.5.12 "Common Mode" Volume Control

With this volume control the source loading impedance is only the input impedance of the amplifier when in the full-volume position. This reduces to one half the amplifier input impedance at the zero volume position. Equation (4.5.1) describes the output voltage as a function of the potentiometer setting.

$$V_{OUT} = 50 V_{IN} \left(1 - \frac{150 \times 10^3}{k_1 R_V + 150 \times 10^3}\right)_{0 \leq k_1 \leq 1} \quad (4.5.1)$$

This "common mode" volume control can be combined with a "common mode" tone control as seen in figure 4.5.13.

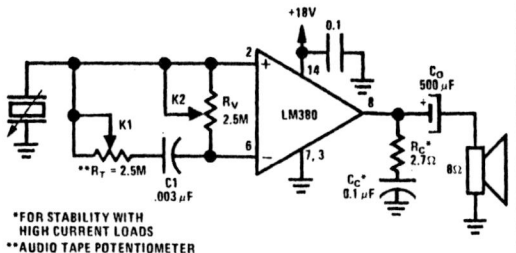

FIGURE 4.5.13 "Common Mode" Volume and Tone Control

This circuit has a distinct advantage over the circuit of Figure 4.5.10 when transducers of high source impedance are used, in that the full input impedance of the amplifier is realized. It also has an advantage with transducers of low source impedance, since the signal attenuation of the input voltage divider is eliminated. The transfer function of the circuit of Figure 4.5.13 is given by:

$$\frac{V_{OUT}}{V_{IN}} = 50 \left(1 - \frac{150k}{150k + \dfrac{k_1 R_T k_2 R_v + \dfrac{k_2 R_v}{j2\pi f C_1}}{k_1 R_T + k_2 R_v + \dfrac{1}{j2\pi f C_1}}}\right)$$

$$0 \leq K_1 \leq 1$$
$$0 \leq K_2 \leq 1$$

(4.5.2)

Figure 4.5.14 shows the response of the circuit of Figure 4.5.13.

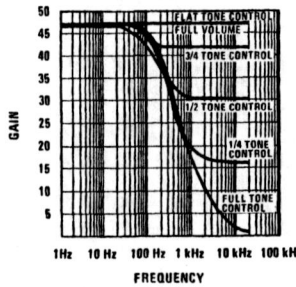

FIGURE 4.5.14 Tone Control Response

4.5.10 Bridge Amplifier

Where more power is desired than can be provided with one amplifier, two amps may be used in the bridge configuration shown in Figure 4.5.15.

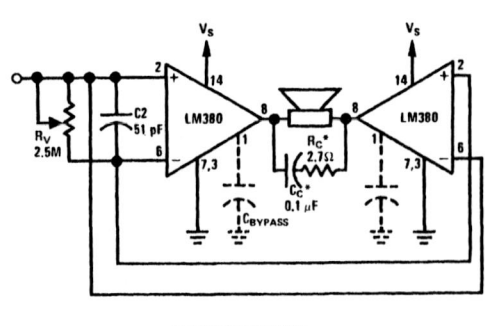

*FOR STABILITY WITH HIGH CURRENT LOADS

FIGURE 4.5.15 Bridge Configuration

This provides twice the voltage swing across the load for a given supply, thereby increasing the power capability by a factor of four over the single amplifier. However, in most cases the package dissipation will be the first parameter limiting power delivered to the load. When this is the case, the power capability of the bridge will be only twice that of the single amplifier. Figures 4.5.16a and -b show output power vs. device package dissipation for both 8 and 16Ω loads in the bridge configuration. The 3% and 10% harmonic distortion contours double back due to the thermal limiting of the LM380. Different amounts of heat sinking will change the point at which the distortion contours bend.

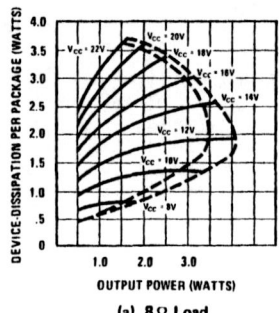

(a) 8Ω Load

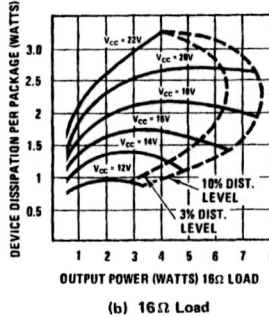

(b) 16Ω Load

FIGURE 4.5.16

The quiescent output voltage of the LM380 is specified at 9 ± 1 volts with an 18 volt supply. Therefore, under the worst case condition, it is possible to have two volts DC across the load.

With an 8Ω speaker this is 0.25A which may be excessive. Three alternatives are available: (1) care can be taken to match the quiescent voltages, (2) a non-polar capacitor may be placed in series with the load, or (3) the offset balance controls of Figure 4.5.17 may be used.

The circuits of Figures 4.5.15 and 4.5.17 employ the "common mode" volume control as shown before. However, any of the various input connection schemes discussed previously may be used. Figure 4.5.18 shows the bridge configuration with the voltage divider input. As discussed in the "Biasing" section the undriven input may be AC or DC grounded. If V_s is an appreciable distance from the power supply (> 3") filter capacitor it should be decoupled with a 1μF tantalum capacitor.

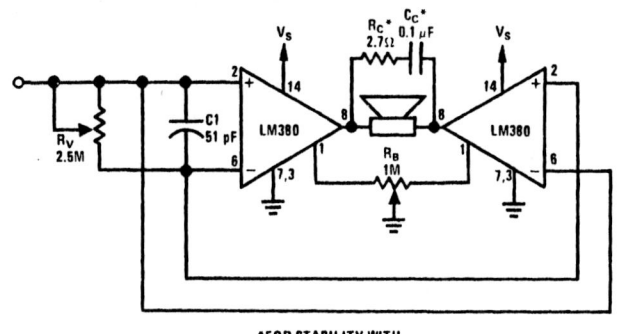

*FOR STABILITY WITH
HIGH CURRENT LOADS

FIGURE 4.5.17 Quiescent Balance Control

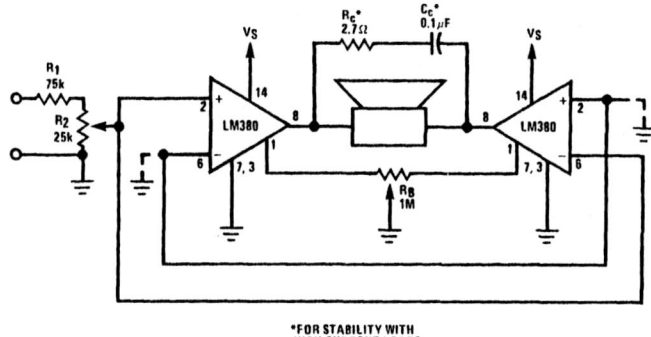

*FOR STABILITY WITH
HIGH CURRENT LOADS

FIGURE 4.5.18 Voltage Divider Input

4.5.11 Intercom

The circuit of Figure 4.5.19 provides a minimum component intercom. With switch S_1 in the talk position, the speaker of the master station acts as the microphone with the aid of step-up transformer T_1.

A turns ratio of 25 and a device gain of 50 allows a maximum loop gain of 1250. R_v provides a "common mode" volume control. Switching S_1 to the listen position reverses the role of the master and remote speakers.

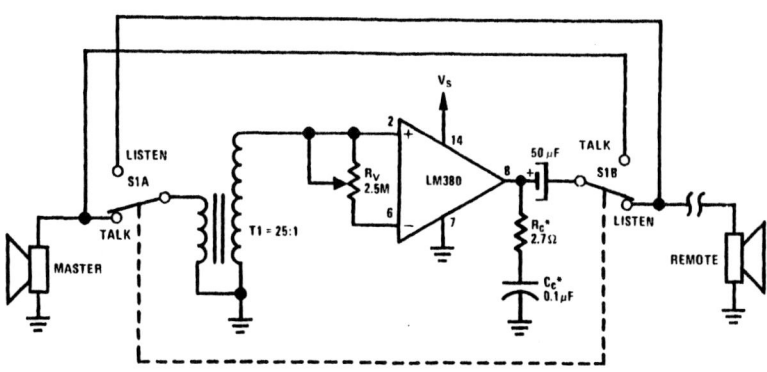

FIGURE 4.5.19 Intercom

4.5.12 Low Cost Dual Supply

The circuit shown in Figure 4.5.20 demonstrates a minimum parts count method of symmetrically splitting a supply voltage. Unlike the normal R, C, and power zener diode technique the LM380 circuit does not require a high standby current and power dissipation to maintain regulation.

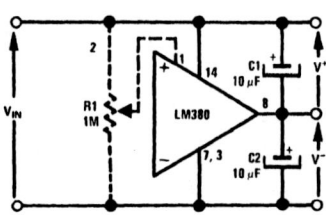

FIGURE 4.5.20 Dual Supply

With a 20 V input voltage (±10 V output) the circuit exhibits a change in output voltage of approximately 2% per 100 mA of unbalanced load change. Any balanced load change will reflect only the regulation of the source voltage V_{IN}.

The theoretical plus and minus output tracking ability is 100% since the device will provide an output voltage at one half of the instantaneous supply voltage in the absence of a capacitor on the bypass terminal. The actual error in tracking will be directly proportional to the imbalance in the quiescent output voltage. An optional potentiometer may be placed at pin 1 as shown in Figure 4.5.20 to null output offset. The unbalanced current output for the circuit of Figure 4.5.20 is limited by the power dissipation of the package.

In the case of sustained unbalanced excess loads, the device will go into thermal limiting as the temperature sensing circuit begins to function. For instantaneous high current loads or short circuits the device limits the output current to approximately 1.3 A until thermal shutdown takes over or until the fault is removed.

4.5.13 High Input Impedance Circuit

The junction FET isolation circuit shown in Figure 4.5.21 raises the input impedance to 22 MΩ for low frequency input signals. The gate to drain capacitance (2 pF maximum for the PN4221 shown) of the FET limits the input impedance as frequency increases.

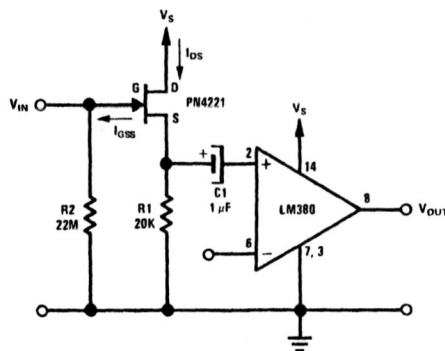

FIGURE 4.5.21 High Input Impedance

At 20 kHz the reactance of this capacitor is approximately $-j4\,M\Omega$, giving a net input impedance magnitude of 3.9 MΩ. The values chosen for R_1, R_2 and C_1 provide an overall circuit gain of at least 45 for the complete range of parameters specified for the PN4221.

When using another FET device the relevant design equations are as follows:

$$A_v = \left(\frac{R_1}{R_1 + \frac{1}{gm}}\right) \quad (50) \qquad (4.5.3)$$

$$gm = gm_0 \left(1 - \frac{V_{GS}}{V_p}\right) \qquad (4.5.4)$$

$$V_{GS} = I_{DS} R_1 \qquad (4.5.5)$$

$$I_{DS} = I_{DSS}\left(1 - \frac{V_{GS}}{V_p}\right)^2 \qquad (4.5.6)$$

The maximum value of R_2 is determined by the product of the gate reverse leakage I_{GSS} and R_2. This voltage should be 10 to 100 times smaller than V_p. The output impedance of the FET source follower is:

$$R_o = \frac{1}{gm} \qquad (4.5.7)$$

so that the determining resistance for the interstage RC time constant is the input resistance of the LM380.

4.5.14 Power Voltage-to-Current Converter

The LM380 makes a low cost, simple voltage-to-current converter capable of supplying constant AC currents up to 1 A over variable loads using the circuit shown in Figure 4.5.22.

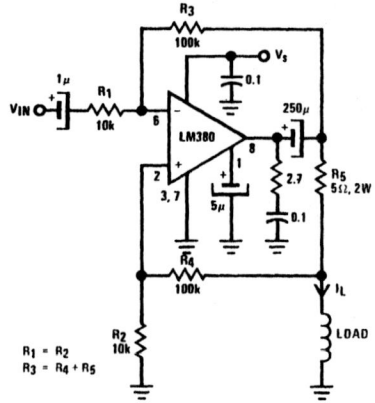

FIGURE 4.5.22 Power Voltage-to-Current Converter

Current through the load is fixed by the gain setting resistors R_1-R_3, input voltage, and R_5 per Equation (4.5.8).

$$I_L = -\frac{R_3 V_{IN}}{R_1 R_5} \qquad (4.5.8)$$

For AC signals the minus sign of Equation (4.5.8) merely shows phase inversion. As shown, Figure 4.5.22 will deliver

$1/2 A_{RMS}$ to the load from an input signal of $250 mV_{RMS}$, with THD less than 0.5%. Maximum current variation is typically 0.5% with a load change from 1-5Ω.

Flowmeters, or other similar uses of electromagnets, exemplify application of Figure 4.5.22. Interchangeable electromagnets often have different impedances but require the same constant AC current for proper magnetization. The low distortion, high current capabilities of the LM380 make such applications quite easy.

4.5.15 Muting

Muting, or operating in a squelched mode may be done with the LM380 by pulling the bypass pin high during the mute, or squelch period. Any inexpensive, general purpose PNP transistor can be used to do this function as diagrammed in Figure 4.5.23.

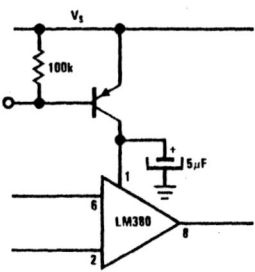

FIGURE 4.5.23 Muting the LM380

During the mute cycle, the output stage will be switched off and will remain off until the PNP transistor is turned off again. Muting attach and release action is smooth and fast.

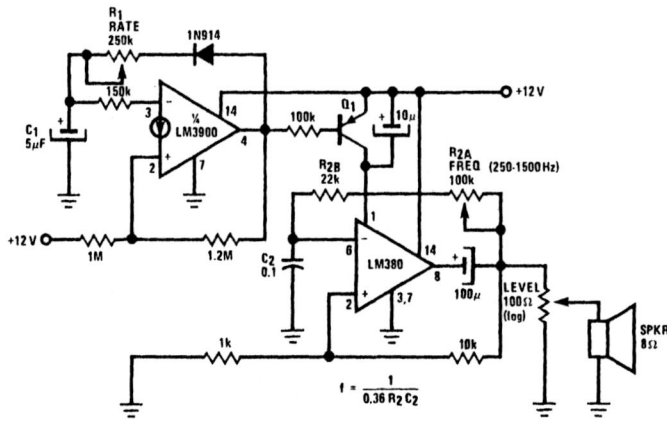

FIGURE 4.5.24 Siren with Programmable Frequency and Rate Adjustment

4.5.16 Siren

Use of the muting technique described in section 4.5.15 allows the LM380 to be configured into a siren circuit with programmable frequency and rate adjustment (Figure 4.5.24. The LM380 operates as an astable oscillator with frequency determined by R_2-C_2. Adding Q_1 and driving its base with the output of an LM3900 wired as a second astable oscillator acts to gate the output of the LM380 on and off at a rate fixed by R_1-C_1. The design equations for the LM3900 astable are given in detail in application note AN-72, page 20, and should be consulted for accurate variation of components. For experimenting purposes (i.e., playing around), changing just about any component will alter the siren effect.

4.6 LM384 AUDIO POWER AMPLIFIER

4.6.1 Introduction

Higher allowed operating voltage, thus higher output power, distinguishes the LM384 from the LM380 audio amplifier.

Typical power levels of 7.5W (10% THD) into 8Ω are possible when operating from a supply voltage of 26V. All other parameters remain as discussed for the LM380. The electrical schematic is identical to Figure 4.5.1.

4.6.2 General Operating Characteristics

Package power dissipation considerations regarding heatsinking are the same as the LM380 (Figure 4.5.4). Device dissipation versus output power curves for 4, 8 and 16Ω loads appear as Figures 4.6.1-4.6.3.

Figure 4.6.4 shows total harmonic distortion vs. output power, while total harmonic distortion vs. frequency for various output levels appears as Figure 4.6.5.

A typical 5W amplifier (V_s = 22V, R_L = 8Ω, THD = 10%) is shown by Figure 4.6.6. Note the extreme simplicity of the circuit. For applications where output ripple and small, high-frequency oscillations are not a problem, all capacitors except the 500μF output capacitor may be eliminated — along with the 2.7Ω resistor. This creates a complete amplifier with only *one* external capacitor and *no* resistors.

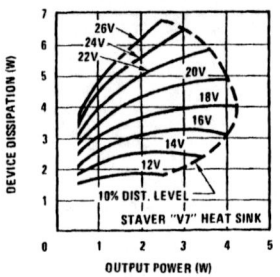

FIGURE 4.6.1 Device Dissipation vs. Output Power — 4 Ω Load

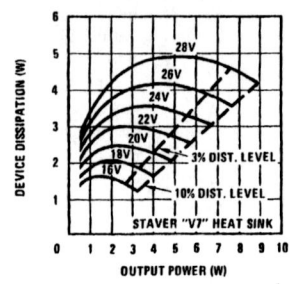

FIGURE 4.6.2 Device Dissipation vs. Output Power — 8 Ω Load

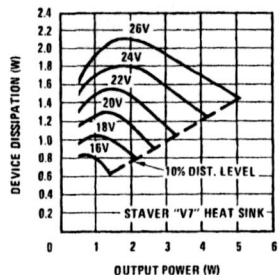

FIGURE 4.6.3 Device Dissipation vs. Output Power — 16 Ω Load

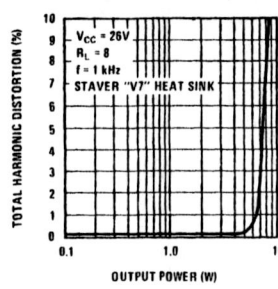

FIGURE 4.6.4 Total Harmonic Distortion vs. Output Power

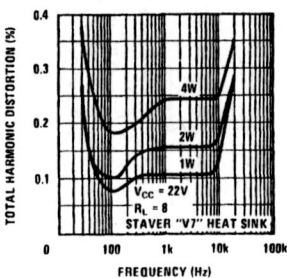

FIGURE 4.6.5 Total Harmonic Distortion vs. Frequency

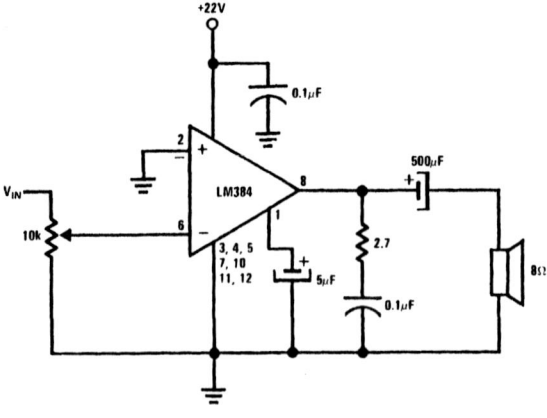

FIGURE 4.6.6 Typical 5W Amplifier

4.7 LM386 LOW VOLTAGE AUDIO POWER AMPLIFIER

4.7.1 Introduction

The LM386 is a power amplifier designed for use in low voltage consumer applications. The gain is internally set to 20 to keep external part count low, but the addition of an external resistor and capacitor between pins 1 and 8 will increase the gain to any value up to 200.

The inputs are ground referenced while the output is automatically biased to one half the supply voltage. The quiescent power drain is only 24 mW when operating from a 6 V supply, making the LM386 ideal for battery operation.

Comparison of the LM386 schematic (Figure 4.7.1) with that of the LM380 (Figure 4.5.1) shows them to be essentially the same. The major difference is that the LM386 has two gain control pins (1 and 8), allowing the internally set gain of 20 V/V (26 dB) to be externally adjusted to any value up to 200 V/V (46 dB). Another important difference lies in the LM386 being optimized for low current drain, battery operation.

4.7.2 General Operating Characteristics

Device dissipation vs. output power curves for 4, 8 and 16 Ω loads appear as Figures 4.7.2-4.7.4. Expected power output as a function of typical supply voltages may be noted from these curves. Observe the "Maximum Continuous Dissipation" limit denoted on the 4 and 8 Ω curves as a dashed line. The LM386 comes packaged in the 8-pin mini-DIP leadframe having a thermal resistance of 187°C/W, junction to ambient. There exists a maximum allowed junction temperature of 150°C, and assuming ambient temperature equal to 25°C, then the maximum dissipation permitted is 660 mW (P_{DMAX} = [150°C − 25°C]/[187°C/W]). Operation at increased ambient temperatures means derating the device at a rate of 187°C/W. Note from Figure 4.7.3 that operation from a 12V supply limits continuous output power to a maximum of 250 mW for allowed limits of package dissipation. It is therefore important that the power supply voltage be picked to optimize power output vs. device dissipation.

Figure 4.7.5 gives a plot of voltage gain vs. frequency, showing the wideband performance characteristic of the LM386. Both gain extremes are shown to indicate the narrowing effect of the higher gain setting.

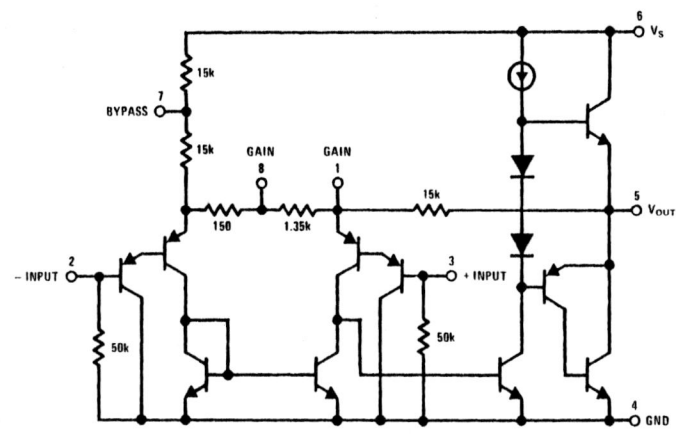

FIGURE 4.7.1 LM386 Simplified Schematic

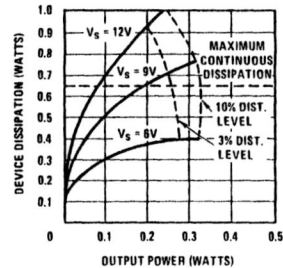

FIGURE 4.7.2 Device Dissipation vs. Output Power − 4 Ω Load

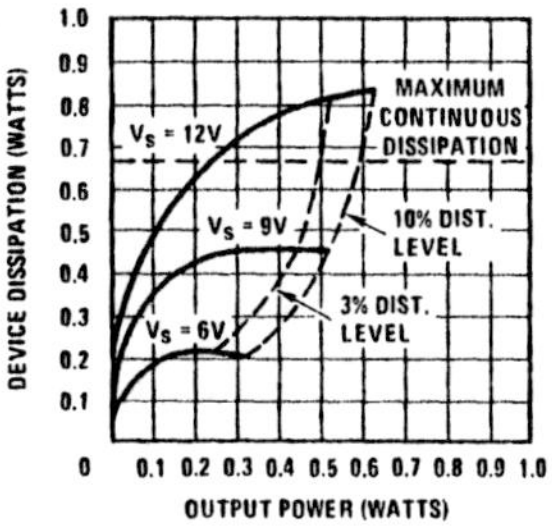

FIGURE 4.7.3 Device Dissipation vs. Output Power − 8 Ω Load

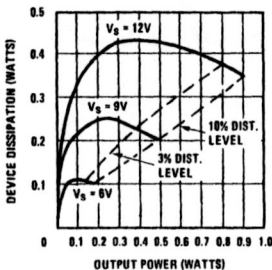

FIGURE 4.7.4 Device Dissipation vs. Output Power — 16 Ω Load

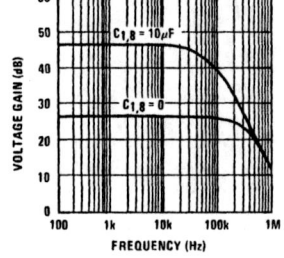

FIGURE 4.7.5 Voltage Gain vs. Frequency

4.7.3 Input Biasing

The schematic (Figure 4.7.1) shows that both inputs are biased to ground with a 50 kΩ resistor. The base current of the input transistors is about 250 nA, so the inputs are at about 12.5 mV when left open. If the DC source resistance driving the LM386 is higher than 250 kΩ it will contribute very little additional offset (about 2.5 mV at the input, 50 mV at the output). If the DC source resistance is less than 10 kΩ, then shorting the unused input to ground will keep the offset low (about 2.5 mV at the input, 50 mV at the output). For DC source resistances between these values we can eliminate excess offset by putting a resistor from the unused input to ground, equal in value to the DC source resistance. Of course all offset problems are eliminated if the input is capacitively coupled.

When using the LM386 with higher gains (bypassing the 1.35 kΩ resistor between pins 1 and 8) it is necessary to bypass the unused input, preventing degradation of gain and possible instabilities. This is done with a 0.1 µF capacitor or a short to ground depending on the DC source resistance on the driven input.

4.7.4 Gain Control

Figure 4.7.6 shows an AC equivalent circuit of the LM386, highlighting the gain control feature. To make the LM386 a more versatile amplifier, two pins (1 and 8) are provided for gain control. With pins 1 and 8 open the 1.35 kΩ resistor sets the gain at 20 (26 dB). If a capacitor is put from pin 1 to 8, bypassing the 1.35 kΩ resistor, the gain will go up to 200 (46 dB).

If a resistor (R_3) is placed in series with the capacitor, the gain can be set to any value from 20 to 200. Gain control can also be done by capacitively coupling a resistor (or FET) from pin 1 to ground. When adding gain control with components from pin 1 to ground, the *positive* input (pin 3) should always be driven, with the negative input (pin 2) appropriately terminated per Section 4.7.3.

Gains less than 20 dB should not be attempted since the LM386 compensation does not extend below 9 V/V (19 dB).

4.7.5 Muting

Similar to the LM380 (Section 4.5.15), the LM386 may be muted by shorting pin 7 (bypass) to the supply voltage. The LM386 may also be muted by shorting pin 1 (gain) to ground. Either procedure will turn the amplifier off without affecting the input signal.

4.7.6 R.F. Precautions

In AM radio applications in particular, r.f. interference caused by radiated wideband noise voltage at the speaker terminals needs to be considered. The pole splitting compensation used in monolithic audio power amplifiers to preserve a wide power bandwidth capability means that there will be plenty of excess

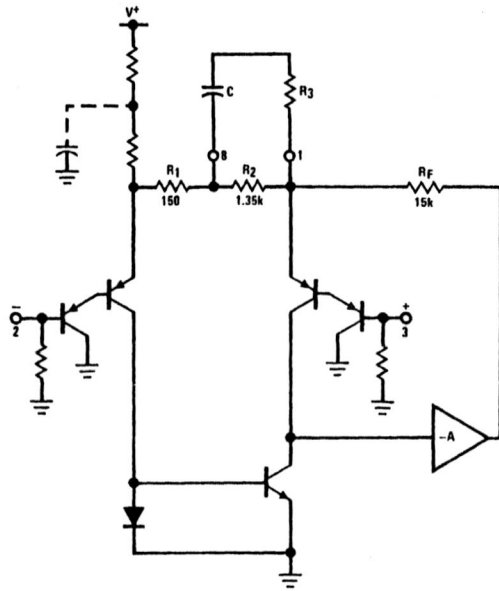

FIGURE 4.7.6 LM386 AC Equivalent Circuit

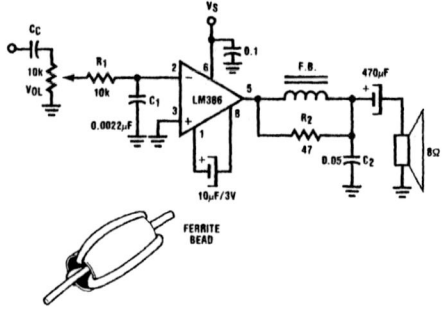

FIGURE 4.7.7 AM Radio Application

gain at frequencies well beyond the audio bandwidth. Noise voltages at these frequencies are amplified and delivered to the load where they can be radiated back to the AM radio ferrite antenna.

Any p.c. board should be layed out to locate the power amplifier as far as possible from the antenna circuit. Extremely tight twisting of the speaker and power supply leads is a must if optimum sensitivity for the radio is to be obtained.

If r.f. radiation still causes a reduction in sensitivity the circuit can be modified as shown in Figure 4.7.7. A typical radio application will use fairly high gain (200V/V) so the device gain is increased by connecting a 10μF capacitor between Pins 1 and 8. To band limit the input signal to 5-10kHz, a two pole filter configuration is used. The first pole is determined by the radio detector circuit and a second pole is added by the R_1C_1 network at the input to the LM386. Any r.f. noise is substantially reduced by placing a ferrite bead (F.B) at the output. A Ferroxcube K5-001-001/3B with 3 turns taken through the bead is suitable for this application. The R_2C_2 network is necessary to stabilize the output stage (Section 4.5.5) but R_2 will also load the ferrite bead, reducing the level of r.f. attenuation. In this instance, a 47Ω resistor is optimum — a smaller value will simply degrade AM sensitivity and a larger value will not ensure stability for all parts. If other ferrite beads are used, a new value for R_2 that will guarantee stability and minimize degradation of AM sensitivity can be found by a few trials.

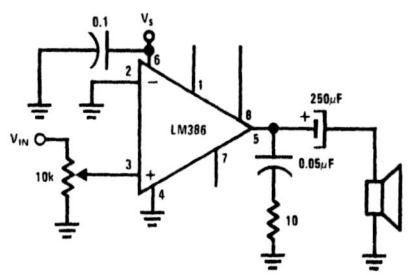

FIGURE 4.7.8 Amplifier with Gain = 20V/V (26dB) Minimum Parts

4.7.7 Typical Applications

Three possible variations of the LM386 as a standard audio power amplifier appear as Figures 4.7.8-4.7.10. Possible gains of 20, 50 and 20V/V are shown as examples of various gain control methods. The addition of the 0.05μF capacitor and 10Ω resistor is for suppression of the "bottom side fuzzies" (i.e., bottom side oscillation occurring during the negative swing into a load drawing high current — see Section 4.5.5).

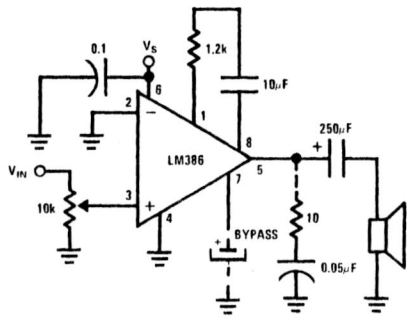

FIGURE 4.7.9 Amplifier with Gain = 50V/V (34dB)

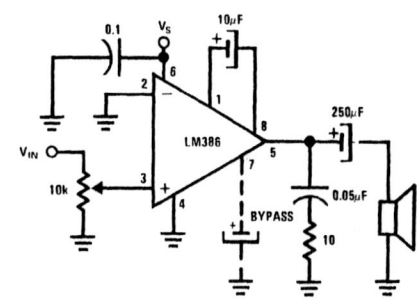

FIGURE 4.7.10 Amplifier with Gain = 200V/V (46dB)

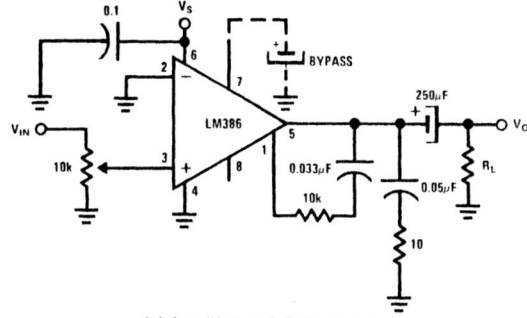

(a) Amplifier with Bass Boost

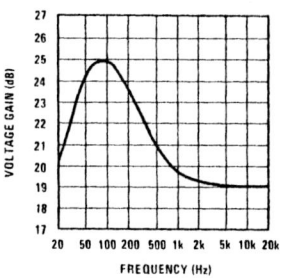

(b) Frequency Response with Bass Boost

FIGURE 4.7.11 LM386 with Bass Boost

Additional external components can be placed in parallel with the internal feedback resistors (Figure 4.7.11) to tailor the gain and frequency response for individual applications. For example, we can compensate poor speaker bass response by frequency shaping the feedback path. This is done with a series RC from pin 1 to 5 (paralleling the internal 15kΩ resistor). For 6dB effective bass boost: R ≈ 15kΩ, the lowest value for good stable operation is R = 10kΩ if pin 8 is open. If pins 1 and 8 are bypassed then R as low as 2kΩ can be used. This restriction is because the amplifier is compensated only for closed-loop gains greater than 9.

4.7.8 Square Wave Oscillator

A square wave oscillator capable of driving an 8Ω speaker with 0.5W from a 9V supply appears as Figure 4.7.12. Altering either R_1 or C_1 will change the frequency of oscillation per the equation given in the figure. A reference voltage determined by the ratio of R_3 to R_2 is applied to the positive input from the LM386 output. Capacitor C_1 alternately charges and discharges about this reference value, causing the output to switch states. A triangle output may be taken from pin 2 if desired. Since DC offset voltages are not relevant to the circuit operation, the gain is increased to 200V/V by a short circuit betwen the pins 1 and 8, thus saving one capacitor.

4.7.9 Power Wien Bridge Oscillator

The LM386 makes a low cost, low distortion audio frequency oscillator when wired into a Wien brige configuration (Figure 4.7.13). Capacitor C_2 raises the "open loop" gain to 200V/V. Closed-loop gain is fixed at approximately ten by the ratio of R_1 to R_2. A gain of ten is necessary to guard against spurious oscillations which may occur at lower gains since the LM386 is not stable below 9V/V. The frequency of oscillation is given by the equation in the figure and may be changed easily by altering capacitors C_1.

Resistor R_3 provides amplitude stabilizing negative feedback in conjunction with lamp L_1. Almost any 3V, 15mA lamp will work.

4.7.10 Ceramic and Crystal Cartridge Phonographs

A large number of inexpensive phonographs are manufactured using crystal or ceramic cartridges. The high output level available from these cartridges enables them to be used without pre-amplifiers in low power phonographs. Because the power amplifier is the only active gain element in such systems, the amplifier design should take into account the unique characteristics of piezo-electric cartridges.

Crystal cartridges are typically made from a single crystal material known as Rochelle Salt (Sodium Potassium Tartrate) which, like quartz, exhibits a natural piezo-electric action – when the crystal is bent or twisted an E.M.F. is developed. Despite a limited operating temperature range and a susceptibility to high relative humidity, the high sensitivity of Rochelle Salt has ensured its continued use. The development of modern ceramic titanates has solved many temperature and humidity problems but the ceramic material is not naturally piezo-electric. To obtain piezo-electric behavior, the ceramics are "poled" at high voltage and temperature. This produces a permanent deformation of the material but the piezo-electric action after "poling" is much lower than that obtainable from Rochelle Salt. Table 4.7.1 summarizes the characteristics of typical crystal and ceramic cartridges.

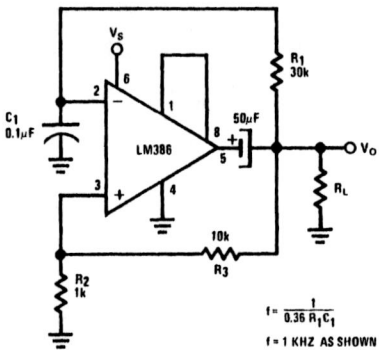

FIGURE 4.7.12 Square Wave Oscillator

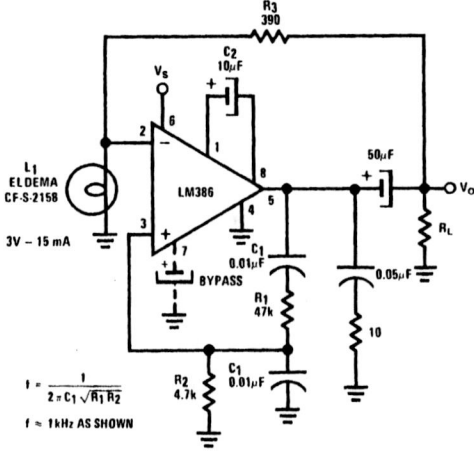

FIGURE 4.7.13 Low Distortion Power Wien Bridge Oscillator

TABLE 4.7.1

CARTRIDGE TYPE	CAPACITANCE	OUTPUT AT 5cm/sec (f = 1kHz)
CERAMIC	800pF	500mV
	2000pF	300mV (Stereo)*
CRYSTAL	800pF	2V (Stereo)*
		3V (Mono)

*Output at 3.5cm/sec

Piezo-electric cartridges (or pick-ups) are operated in the non-resonant mode over a relatively large frequency range and may be represented by the equivalent circuit of 4.7.14 where C_C is the capacitance of the piezo-electric element, R_C the shunt leakage resistance and C_L, R_L are the load capacitance and resistance. R_C is usually several hundred megohms and can be ignored, while typical values for C_C are given in Table 4.7.1.

The E.M.F. generated by any piezo-electric cartridge depends on the amplitude of the movement of the stylus. If discs were recorded with a constant amplitude characteristic, above the cut-off frequency determined by the cartridge capacitance and the load resistance, the response would be essentially flat with frequency, Figure 4.7.15. Note that any load capacitance reduces the output at *all* frequencies above cut-off and that the cut-off frequency moves lower since

$$f_C = \frac{1}{2\pi C_T R_L}$$

where C_T is the paralleled capacitance of the cartridge and the load capacitance.

Since discs are not cut with a constant amplitude versus frequency characteristic (See Section 2.11), when an ideal piezo cartridge plays back a R.I.A.A. recorded disc, there will be a 12.5dB drop in response between 500Hz and 2.1kHz. Before an amplifier response is designed to accomodate this, the designer should realize that crystal cartridges have mechanical compensation to provide relatively flat response through this region, so that a flat amplifier response is all that is required. Ceramic cartridges however, may or may not have mechanical compensation and the decision to compensate electronically will probably depend on the cost objectives (See Section 4.8.7).

4.7.11 LM386 Crystal Cartridge Amplifiers

Where a crystal cartridge is used, the most economical design with the LM386 is shown in Figure 4.7.17. The input stage configuration is the result of a trade-off between cartridge load R_L (which together with the cartridge capacitance will set the low −3dB frequency) and the need to mask variations of input impedance presented by the LM386.

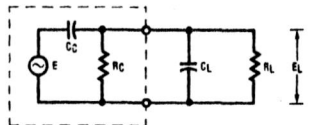

FIGURE 4.7.14 Cartridge Equivalent Circuit

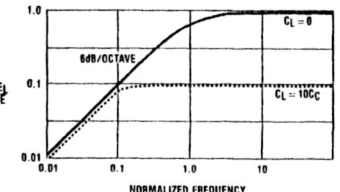

FIGURE 4.7.15 Cartridge Frequency Response

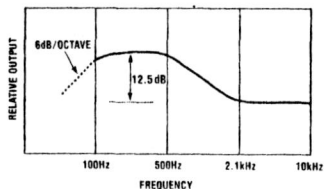

FIGURE 4.7.16 Cartridge Response to RIAA Recorded Disc

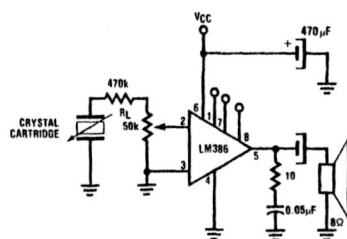

FIGURE 4.7.17 Low Cost Phono Amplifier

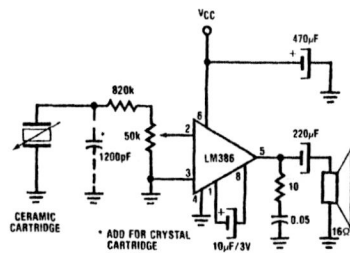

FIGURE 4.7.18 Ceramic Cartridge Amplifier

The resistor R_L is large enough to define the cartridge load for all settings of the volume control, but a signal attenuator is also formed by this resistor and the input resistance of the LM386 (50k) in parallel with the volume potentiometer. With a large valued potentiometer, the amount of signal attenuation will depend of the input resistance of the LM386 which can change by −30% to +100% from device to device. A 50k volume control will mask this variation to less than 4dB for worst case device input resistance change. A smaller volume control will give even less possible variation in output level but the signal become correspondingly more attenuated. Decreasing R_L to restore more signal input to the LM386 will cause further degradation in the cartridge bass response.

4.7.12 Ceramic Cartridge Amplifiers

While the circuit of 4.7.17 can provide a reasonable compromise of output power and bass frequency response with crystal cartridges, the lower output level of ceramic cartridges will require some changes.

In the circuit of 4.7.18 the gain of the LM386 has been raised to 200V/V by connecting a capacitor between pin 1 and pin 8. This will also allow R_L to be increased to 820kΩ, which for a 2000pF capacitance cartridge will give a bass cut-off frequency of under 100Hz. This circuit can be used to accomodate the higher output crystal cartridges without overload simply by adding a 1200pF capacitor across the cartridge terminals. This reduces the crystal cartridge output

by $\frac{800}{800+1200} = 0.4$ or 8dB

and extends the bass response down to 100Hz (compared to the usual bass cut-off of 200Hz). However, for either ceramic or crystal cartridge, extended bass response should be approached with caution, since problems can result from low frequency mechanical feedback between the speaker and the tone arm in complete phonograph units.

This is no problem for stereo units with separated speakers, but for more compact monaural phonographs the circuit of Figure 4.7.18 may cause a low frequency resonance at higher volume settings. It is possible to reconfigure the cartridge loading to prevent this (Figure 4.7.19), by connecting a large valued potentiometer across the cartridge. For ceramic cartridges 500kΩ is suitable and for crystal cartridges 1mΩ is recommended. At low volume setting the cartridge response is dictated by the size of the potentiometer. At higher volume settings where mechanical feedback could occur, the potentiometer becomes shunted by the series resistance (R) and the input resistance of the LM386. Proper choice of R (dependent on the particular phonograph tone arm and speaker arrangement) prevents resonance and will give the impression of a loudness control. The 5kΩ resistor is used to swamp the input resistance of the LM386 and to attenuate the cartridge signal to a level suitable for 16Ω speakers. For 8Ω speakers, this resistor should be increased to 10kΩ.

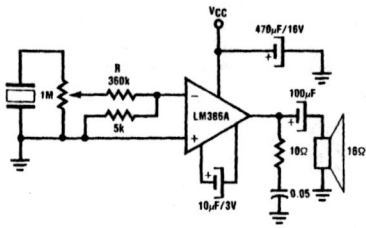

FIGURE 4.7.19 Circuit to Reduce Tone Arm/Speaker Resonance

4.7.13 Phonograph Power Supplies

Most inexpensive phonographs drive the power supply for the electronics from an overwinding on the phonograph motor, and have a no-load voltage from around 12V to 16V. Inspection of the power dissipation curves for an LM386 driving an 8Ω load with this supply, 12V, would indicate that the LM386 is going to be badly over power dissipation limits, even for small output power levels. Fortunately this is not the case since this type of phonograph power supply sags as the power output goes up.

A typical plot of supply voltage versus output current for a half wave rectified, capacitive input filter power supply is given in Figure 4.7.20. The equivalent internal resistance of the supply (contributed mainly by the winding) is approximately 26Ω. Using this supply regulation curve to plot the intenal power dissipation of the LM386 as the load current increases (Figure 4.7.21) shows that at no time does the power dissipation exceed 600mW. Nevertheless, it is important to check that the peak supply voltage under no-load conditions does not exceed the maximum supply voltage rating for the device.

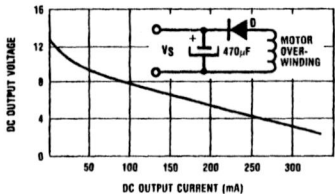

FIGURE 4.7.20 Power Supply Regulation Curve

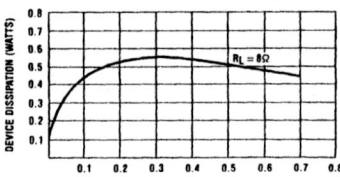

FIGURE 4.7.21 LM386 Power Dissipation on Unregulated Power Supply

4.8 LM389 LOW VOLTAGE AUDIO POWER AMPLIFIER WITH NPN TRANSISTOR ARRAY

4.8.1 Introduction

The LM389 is an array of three NPN transistors on the same substrate with an audio power amplifier similar to the LM386 (Figure 4.8.1).

The amplifier inputs are ground referenced while the output is automatically biased to one half the supply voltage. The gain is internally set at 20 to minimize external parts, but the addition of an external resistor and capacitor between pins 4 and 12 will increase the gain to any value up to 200. Gain control is identical to the LM386 (see Section 4.7.4).

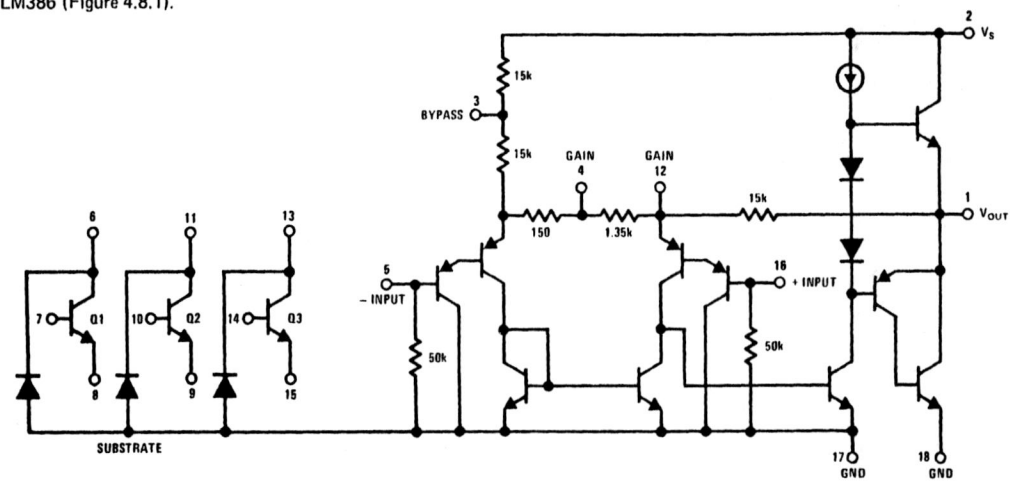

The three transistors have high gain and excellent matching characteristics. They are well suited to a wide variety of applications in DC through VHF systems.

4.8.2 Supplies and Grounds

The LM389 has excellent supply rejection and does not require a well regulated supply. However, to eliminate possible high frequency stability problems, the supply should be decoupled to ground with a $0.1\mu F$ capacitor. The high current ground of the output transistor, pin 18, is brought out separately from small signal ground, pin 17. If the two ground leads are returned separately to supply, the parasitic resistance in the power ground lead will not cause stability problems. The parasitic resistance in the signal ground can cause stability problems and it should be minimized. Care should also be taken to insure that the power dissipation does not exceed the maximum dissipation (825mW) of the package for a given temperature.

4.8.3 Muting

Muting is accomplished in the same manner as for the LM386 (Section 4.7.5), with the exception of applying to different pin numbers.

4.8.4 Transistors

The three transistors on the LM389 are general purpose devices that can be used the same as other small signal transistors. As long as the currents and voltages are kept within the absolute maximum limitations, and the collectors are never at a negative potential with respect to pin 17, there is no limit on the way they can be used.

For example, the emitter-base breakdown voltage of 7.1V can be used as a zener diode at currents from $1\mu A$ to 5mA. These transistors make good LED driver devices; V_{SAT} is only 150mV when sinking 10mA.

In the linear region, these transistors have been used in AM and FM radios, tape recorders, phonographs, and many other applications. Using the characteristic curves on noise voltage and noise current, the level of the collector current can be set to optimize noise performance for a given source impedance (Figures 4.8.2-4.8.4).

4.8.5 Typical Applications

The possible applications of three NPN transistors and a 0.5W power amplifier seem limited only by the designer's imagination. Many existing designs consist of three transistors plus a small discrete power amplifier; redesign with the LM389 is an attractive alternative — typical of these are battery powered AM radios. The LM389 makes a cost-saving single IC AM radio possible as shown in Figure 4.8.5.

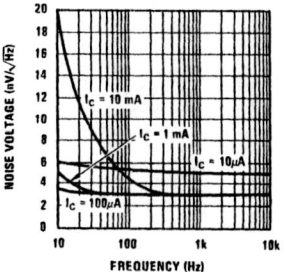

FIGURE 4.8.2 Noise Voltage vs. Frequency

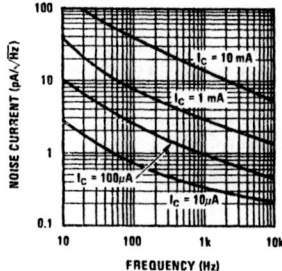

FIGURE 4.8.3 Noise Current vs. Frequency

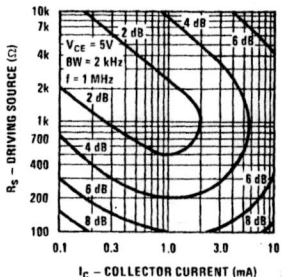

FIGURE 4.8.4 Contours of Constant Noise Figure

4.8.6 Tape Recorder

A complete record/playback cassette tape machine amplifier appears as Figure 4.8.6. Two of the transistors act as signal amplifiers, with the third used for automatic level control during the "record" mode. The complete circuit consists of only the LM389 plus one diode and the passive components.

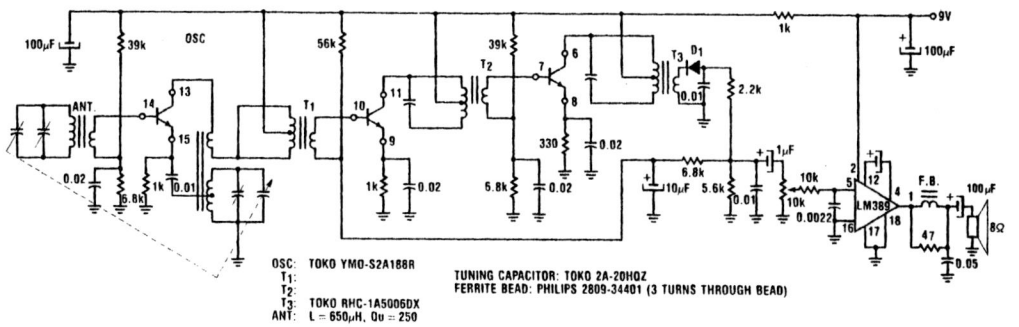

FIGURE 4.8.5 AM Radio

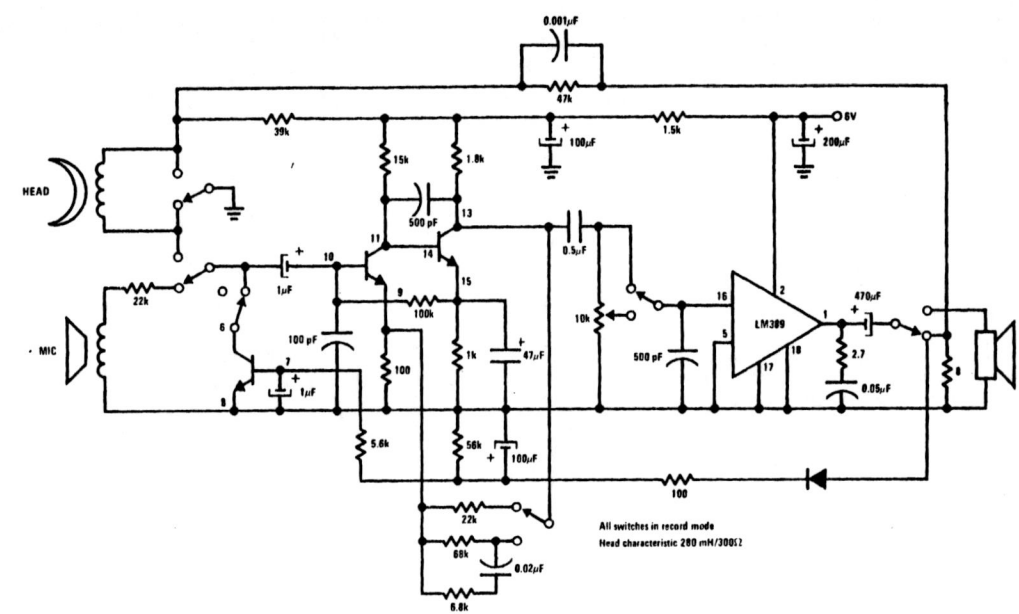

FIGURE 4.8.6 Tape Recorder

4.8.7 Ceramic Phono Amplifier with Compensation for R.I.A.A. Recording Characteristic

All the phonograph amplifiers described up to this point, have been designed on the assumption that the cartridge has mechanical compensation (true for crystal cartridges) or that the 12.5dB fall in response when playing a disc with the R.I.A.A. recording characteristic indicated by Figure 4.7.16, is acceptable. The existence of uncompensated ceramic cartridge implies a need for elecronic compensation — that is an amplifier response that will give 12.5dB boost between 500Hz and 2.1kHz. To achieve this, we can take advantage of the characteristics of piezo-electric cartridges described earlier in Section 4.7.10. Consider the inverting amplifier circuit of Figure 4.8.7. If $R_1C_1 = R_2C_2$ then the frequency response would be flat. Further, if C_1 is the cartridge capacitance, it should be possible to select R_1R_2 and C_2 to compensate for the low frequency roll-off of the cartridge and give a rising reponse between 500Hz and 2.1kHz.

For a cartridge capacitance of 2000pF, R_1 is selected for a break frequency of 2.1kHz. R_2 and C_2 are chosen to give a break frequency of 500Hz. The amplifier response with each of these networks and the combined response is shown in Figure 4.8.8.

It would be difficult to implement this type of equalization with LM386 amplifiers because of the variation of input resistance and the need for a volume control. Instead a single transistor cartridge-compensation stage can be built to precede the power amplifier, Figure 4.8.9. For the 2000pF cartridge, R_1 is 39kΩ. R_2 is chosen to give slightly more than unity gain so that the output at medium to high frequencies is the same as the cartridge rating (measured at 1kHz where the response is −6dB for an uncompensated cartridge on a R.I.A.A. recording). With $R_2 = 62$kΩ, a 0.005μF capacitor gives the 500Hz break frequency.

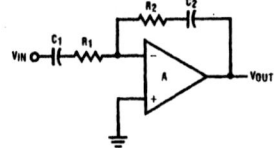

FIGURE 4.8.7 Virtual Ground Inverting Amplifier

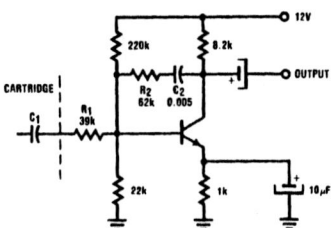

FIGURE 4.8.9 Cartridge-Compensation Stage

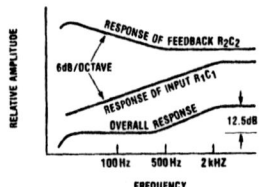

FIGURE 4.8.8 Response of Inverting Amplifier

With the LM389 audio amplifier, the transistor is included within the I/C package so that a complete design, with active tone controls, for good quality uncompensated cartridges appears as shown in Figure 4.8.10. See Section 2.14.7 for the design of the the tone control circuit.

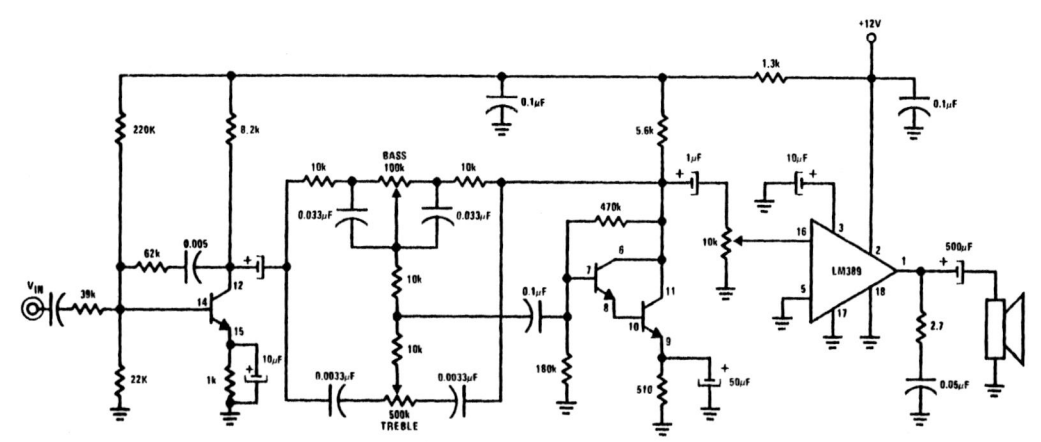

FIGURE 4.8.10 Ceramic Phono Amplifier with Tone Controls

4.8.8 Siren

The siren circuit of Figure 4.8.11 uses one of the LM389 transistors to gate the power amplifier on and off by applying one of the muting techniques discussed in Section 4.8.3. The other transistors form a cross-coupled multi- vibrator circuit that controls the rate of the square wave oscillator. The power amplifier is used as the square wave oscillator with individual frequency adjust provided by potentiometer R_{2B}.

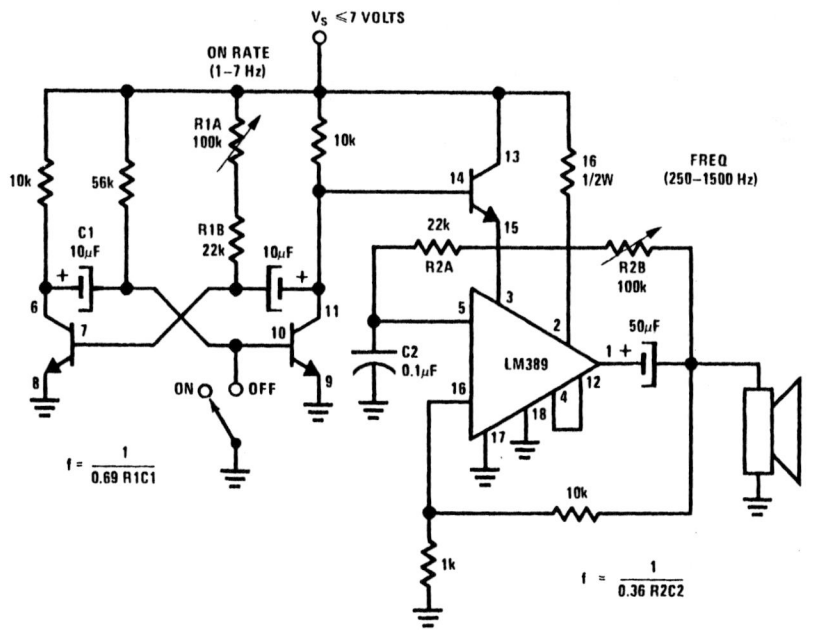

FIGURE 4.8.11 Siren

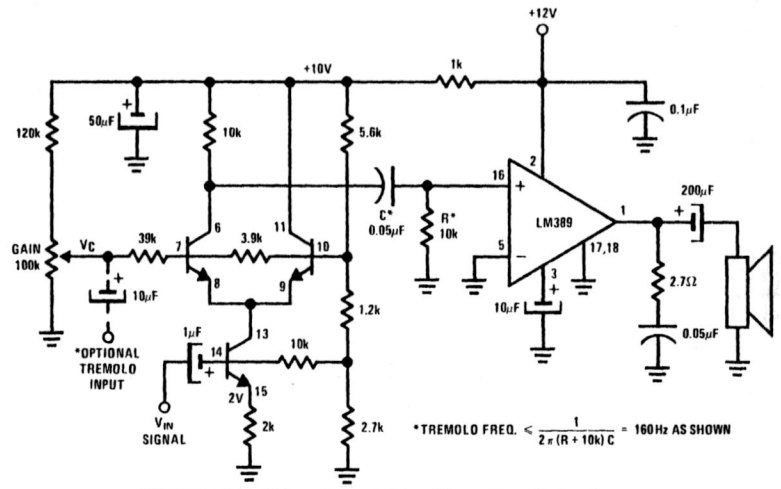

FIGURE 4.8.12 Voltage-Controlled Amplifier or Tremolo Circuit

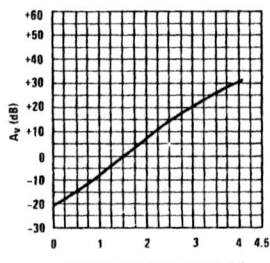

FIGURE 4.8.13 VCA Gain vs. Control Voltage

4.8.9 Voltage-Controlled Amplifier or Tremolo Circuit

A voltage-controlled amplifier constructed from the LM389 appears as Figure 4.8.12. Here the transistors form a differential pair with an active current-source tail. This configuration, known technically as a variable-transconductance multiplier, has an output proportional to the product of the two input signals. Multiplication occurs due to the dependence of the transistor transconductance on the emitter current bias. As shown, the emitter current is set up to a quiescent value of 1 mA by the resistive string. Gain control voltage, V_C, varies from 0 V (minimum gain = -20 dB) to 4.5 V (maximum gain = $+30$ dB), giving a total dynamic range of 50 dB (Figure 4.8.13). V_{IN} signal levels should be restricted to less than 100 mV for good distortion performance. The output of the differential gain stage is capacitively fed to the power amplifier via the R-C network shown, where it is used to drive the speaker.

Tremolo (amplitude modulation of an audio frequency by a sub-audio oscillator — normally 5-15 Hz) applications require feeding the low frequency oscillator signal into the optional input shown. The gain control pot may be set for optimum "depth." Note that the interstage R-C network forms a high pass filter (160 Hz as shown), thus requiring the tremolo frequency to be less than this time constant for proper operation.

4.8.10 Noise Generator

By applying reverse voltage to the emitter of a grounded base transistor, the emitter-base junction will break down in an avalanche mode to form a handy zener diode. The reverse voltage characteristic is typically 7.1 V and may be used as a voltage reference, or a noise source as shown in Figure 4.8.14. The noise voltage is amplified by the second transistor and delivered to the power amplifier stage where further amplification takes place before being used to drive the speaker. The third transistor (not shown) may be used to gate the noise generator similar to Section 4.8.8 if required.

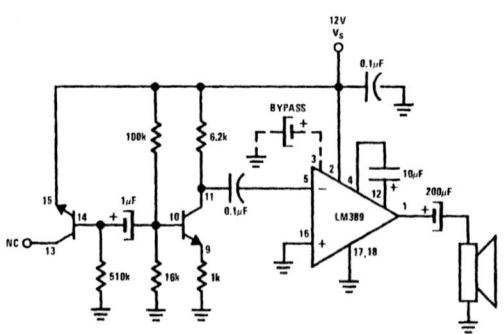

FIGURE 4.8.14 Noise Generator Using Zener Diode

4.8.11 Logic Controlled Mute

Various logic functions are possible with the three NPN transistors, making logic control of the mute function possible. Figures 4.8.15–4.8.17 show standard AND, OR and Exclusive-OR circuits for controlling the muting transistor. Using the optional mute scheme of shorting pin 12 to ground gives NAND, NOR and Exclusive NOR

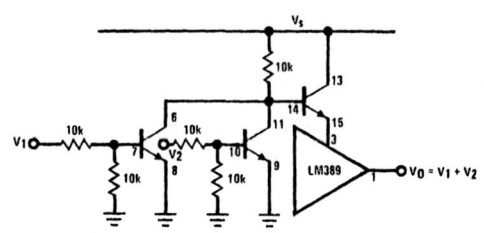

FIGURE 4.8.16 OR Muting

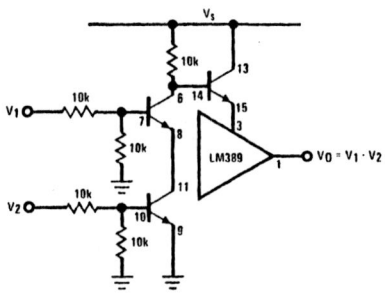

FIGURE 4.8.15 AND Muting

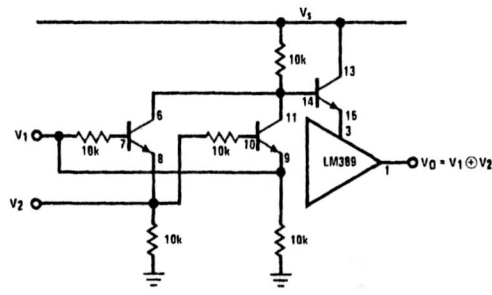

FIGURE 4.8.17 Exclusive-OR Muting

4.9 LM388 BOOTSTRAPPED AUDIO POWER AMPLIFIER

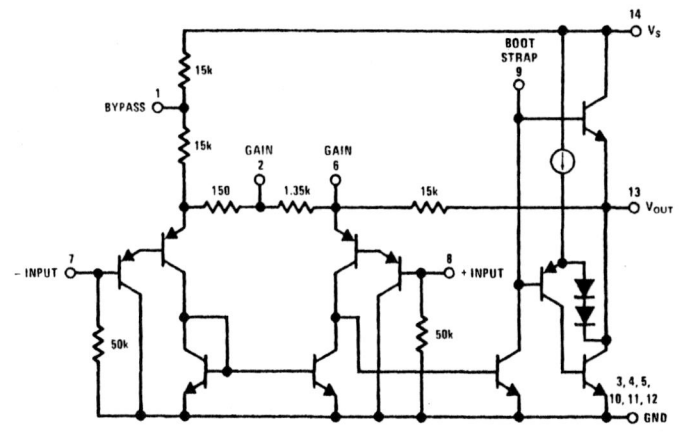

FIGURE 4.9.1 LM388 Simplified Schematic

4.9.1 Introduction

The LM388 audio power amplifier, designed for low voltage, medium power consumer applications, extends the LM386 design concept one step further by incorporating a bootstrapped output stage (Figure 4.9.1). Bootstrapping allows power levels in excess of 1W to be obtained from battery powered products (Figures 4.9.2-4.9.4). Packaging the LM388 into National's 14-pin copper lead-frame (same as LM380) extends maximum package dissipation to values where heatsinking is eliminated for most designs.

4.9.2 General Operating Characteristics

The gain, internally set to 20V/V, is externally controlled in the same manner as the LM386. Consult Section 4.7.4 for details. Input biasing follows LM386 procedures outlined in Section 4.7.3; likewise, muting is the same as Section 4.7.5.

4-41

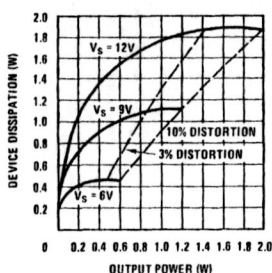

FIGURE 4.9.2 Device Dissipation vs. Output Power — 4Ω Load

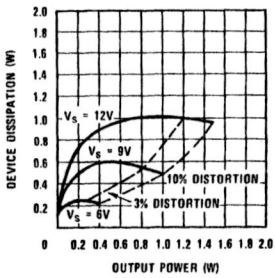

FIGURE 4.9.3 Device Dissipation vs. Output Power — 8Ω Load

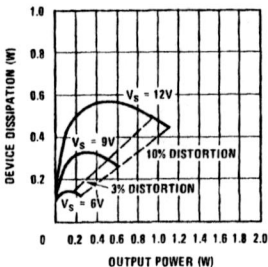

FIGURE 4.9.4 Device Dissipation vs. Output Power — 16Ω Load

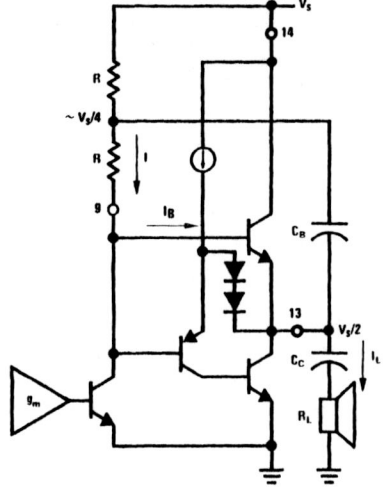

FIGURE 4.9.5 LM388 Output Stage

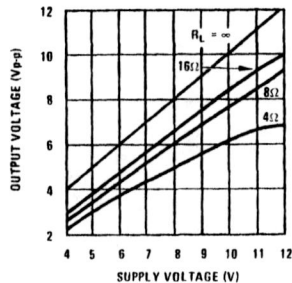

FIGURE 4.9.6 Peak-to-Peak Output Voltage Swing vs. Supply Voltage

4.9.3 Bootstrapping (See also section 4.1.5.)

The base of the top side output transistor is brought out to pin 9 for bootstrapping. The term "bootstrapping" (derived from the expression, ". . . pull oneself up by one's bootstraps") aptly describes the effect. Figure 4.9.5 shows the output stage with the external parts necessary for standard bootstrapping operation. Capacitor C_B charges to approximately $V_S/4$ during the quiescent state of the amplifier and then acts to pull the base of the top transistor up ("by the bootstraps") as the output stage goes through its positive swing — actually raising pin 9 to a *higher* potential than the supply at the top of the swing. This occurs since the voltage on a capacitor cannot change instantaneously, but must decay at a rate fixed by the resistive discharge path.

The stored charge converts to a current with time and supplies the necessary base drive to keep the top transistor saturated during the critical peak period. The net effect allows higher positive voltage swings than can be achieved without bootstrapping. (See Figure 4.9.6.)

For design purposes, resistors (R) and bootstrap capacitor (C_B) can be determined from the following:

$$I_B = \frac{I_L}{\beta} = \frac{V_S/2 - V_{BE}}{2R} \approx \frac{V_S}{4R}$$

$$\therefore I_L = \frac{\beta V_S}{4R}$$

also, $I_{L(max)} = \frac{V_S/2}{R_L}$

so, $\frac{\beta V_S}{4R} = \frac{V_S}{2R_L}$

or, $R = \frac{\beta R_L}{2}$ \hfill (4.9.1)

To preserve low frequency performance the pole due to C_B and $R/2$ (parallel result of R-R) is set equal to the pole due to C_C and R_L:

$$\frac{R}{2} C_B = R_L C_C \qquad (4.9.2)$$

Substituting Equation (4.9.1) into (4.9.2) yields:

$$C_B = \frac{4 C_C}{\beta} \qquad (4.9.3)$$

Letting $\beta = 100$ (nominal) gives:

$$R = 50 R_L \qquad (4.9.4)$$

$$C_B = \frac{C_C}{25} \qquad (4.9.5)$$

For reduced component count the load can replace the upper resistor, R (Figure 4.9.7). The value of bootstrap resistors R+R must remain the same, so the lower R is increased to 2R (assuming speaker resistance to be negligible). Output capacitor (C_C) now serves the dual function of bootstrapping and coupling. It is sized about 5% larger since it now supplies base drive to the upper transistor.

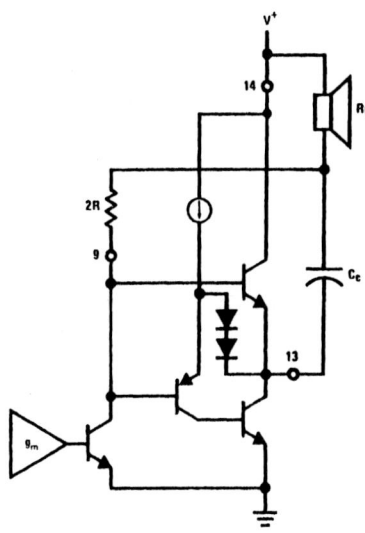

FIGURE 4.9.7 Bootstrapping with Load to Supply

Examples of both bootstrapping methods appear as Figures 4.9.8 and 4.9.9. Note that the resistor values are slightly larger than Equation (4.9.4) would dictate. This recognizes that $I_{L(max)}$ is, in fact, always less than $[V_s/2]/R_L$ due to saturation and V_{BE} losses.

A third bootstrapping method appears as Figure 4.9.10, where the upper resistor is replaced by a diode (with a subsequent increase in the resistance value of the lower resistor). Addition of the diode allows capacitor C_B to be decreased by about a factor of four, since no stored charge is allowed to discharge back into the supply line.

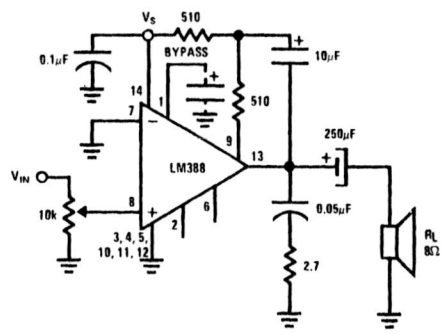

FIGURE 4.9.8 Load Returned to Ground (Amplifier with Gain = 20)

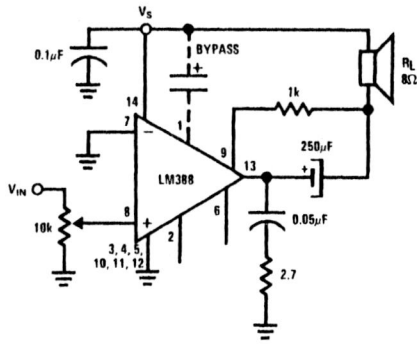

FIGURE 4.9.9 Load Returned to V_S (Amplifier with Gain = 20)

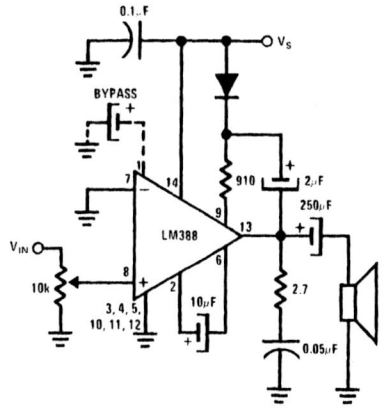

FIGURE 4.9.10 Amplifier with Gain = 200 and Minimum C_B

4.9.4 Bridge Amplifier

For low voltage applications requiring high power outputs, the bridge connected circuit of Figure 4.9.11 can be used. Output power levels of 1.0W into 4Ω from 6V and 3.5W into 8Ω from 12V are typical. Coupling capacitors are not necessary since the output DC levels will be within a few tenths of a volt of each other. Where critical matching is required the 500k potentiometer is added and adjusted for zero DC current flow through the load.

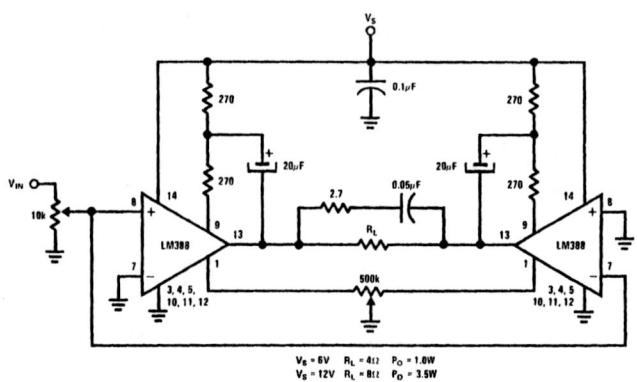

FIGURE 4.9.11 Bridge Amp

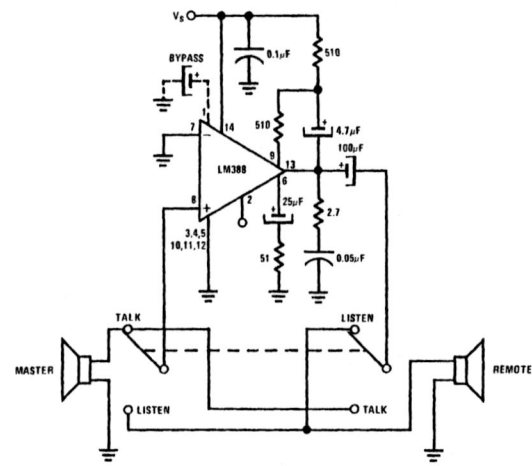

FIGURE 4.9.12 Intercom

4.9.5 Intercom

A minimum parts count intercom circuit (Figure 4.9.12) is made possible by the high gain of the LM388. Using the gain control pin to set the AC gain to approximately 300 V/V ($A_v \approx 15k/51\,\Omega$) allows elimination of the step-up transformer normally used in intercom designs (e.g., Figure 4.5.22). The $2.7\,\Omega$-0.05μF R-C network suppresses spurious oscillations as described for the LM380 (Section 4.5.5).

4.9.6 FM Scanners and Two Way Walkie Talkies

Designed for the high volume consumer market, the LM388 ideally suits applications in FM scanners and two way walkie talkie radios. Requirements for this market generally fall into three areas:

1. Low cost FM scanners; $V_S = 6V$, $P_O = 0.25W$
2. Consumer walkie talkie (including CB); $V_S = 12V$, $P_O = 0.5W$
3. High quality hand-held portables; $V_S = 7.5V$, $P_O = 0.5W$

Since all equipment is battery operated, current consumption is important; also, the amplifier must be squelchable, i.e., turned off with a control signal. The LM388 meets both of these requirements. When squelched, the LM388 draws only 0.8mA from a 7.5V power supply.

A typical high quality hand held portable application with noise squelch appears as Figure 4.9.13. Diodes D_1 and D_2 rectify noise from the limiter or the discriminator of the receiver, producing a DC current to turn on Q_1, which clamps the LM388 in an off condition.

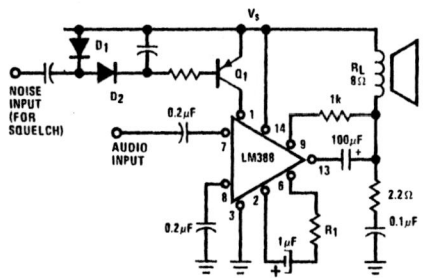

FIGURE 4.9.13 LM388 Squelch Circuit for FM Scanners and Walkie Talkies

As shown, the following performance is obtained:

- Voltage gain equals 20 to 200 (selectable with R_1).
- Noise (output squelched) equals $20\mu V$.
- $P_O = 0.53W$ ($V_S = 7.5V$, $R_L = 8\Omega$, THD = 5%)
- $P_O = 0.19W$ ($V_S = 4.5V$, $R_L = 8\Omega$, THD = 5%)
- Current consumption ($V_S = 7.5V$):
  squelched — 0.8mA
  $P_O = 0.5W$ — 110mA

4.10 LM390 1 WATT BATTERY OPERATED AUDIO POWER AMPLIFIER

Battery operated consumer products often employ 4Ω speaker loads for increased power output. The LM390 meets the stringent output voltage swings and higher currents demanded by low impedance loads. Bootstrapping of the upper output stage (Figure 4.10.1) maximizes positive swing, while a unique biasing scheme (Figure 4.10.2) used on the lower half allows negative swings down to within one saturation drop above ground. Special processing techniques are employed to reduce saturation voltages to a minimum. The result is a monolithic solution to the difficulties of obtaining higher power levels from low voltage supplies. The LM390 delivers 1W into 4Ω (6V) at a lower cost than any competing approach, discrete or IC Figure 4.10.3.

In all other respects (including pin-out) the LM390 is identical to the LM388 (Section 4.9). Gain control, input biasing, muting, and bootstrapping are all as explained previously for the LM386 and LM388.

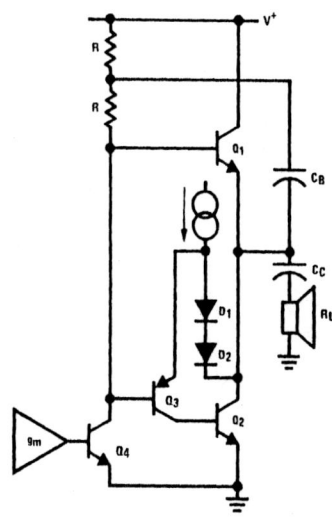

FIGURE 4.10.2 LM390 Output Stage

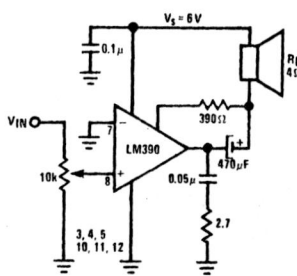

FIGURE 4.10.3 1 Watt Power Amplifier for 6 Volt Systems

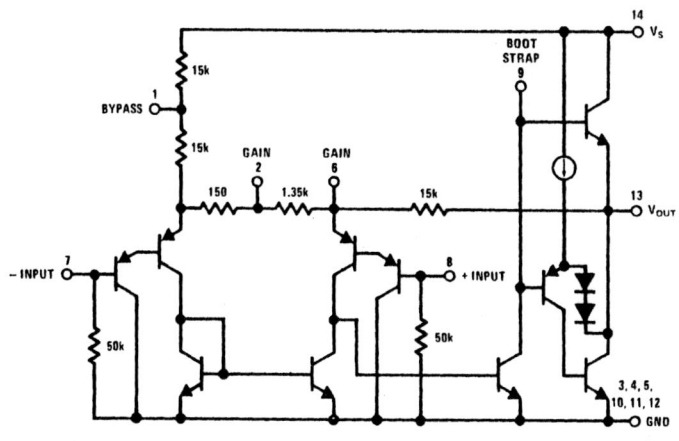

FIGURE 4.10.1 LM390 Simplified Schematic

4.11 LM383 8 WATT AUDIO POWER AMPLIFIER

4.11.1 Introduction

The limited supply voltage available in automotive applications requires amplifiers with an extremely high current output capability to drive low impedance loads if high power outputs are to be obtained. The LM383 is a cost effective, high power amplifer able to continuously deliver 3.5A. Typical output power levels are 5.5 Watts in 4Ω, 8.6 Watts in 2Ω and 9.3 Watts in 1.6Ω — all from 14.4V supplies. In Bridge amplifier circuits as much as 16 Watts into 4Ω can be obtained! Another unique feature of the LM383 is the package style — a five lead TO-220 that permits easy and effective heatsinking. The LM383 output stages are protected with both short circuit current limiting and thermal shut-down circuitry.

4.11.2 Circuit Description

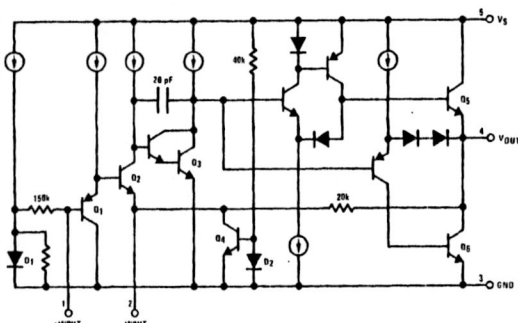

FIGURE 4.11.1 Equivalent Schematic of LM383

An equivalent schematic of the LM383 is given in Figure 4.11.1. The input stage of Q_1 and Q_2 drives the transconductance stage of Q_3. This stage is internally compensated with a fairly large pole splitting capacitor to give a unity gain crossover frequency of around 3mHz. This means that the amplifier is unconditionally stable for all values of closed loop gain, and the restricted bandwidth limits the possibility of r.f. radiation from the output that could cause interference in AM radio applications (see Section 2.3.10). The bandwidth for a closed loop gain of 40dB is still 30kHz (Figure 4.11.9), with careful design of the output stages keeping the *open loop* THD at 1%. The available pin-outs prevent boot-strapping the upper output stage, but the AB bias scheme (see Section 4.1.5) allows a negative swing to within a saturation voltage of ground (Figure 4.11.8). The LM383 uses an interesting dc bias scheme, shown in the simplified schematic of Figure 4.11.2 which has two main advantages. First, the dc gain is set internally unity by the $20k\Omega$ feedback resistor. This will minimize input offset voltages causing shifts in the quiescent output voltage. Secondly, the output voltage is automatically established at one half the supply voltage and will track with supply voltage to maximize the output swing capability. This is accomplished by biasing Q_4 from the $40k\Omega$ resistor and D_2. Since D_2 and Q_4 base-emitter junction have the same voltage across them, then (neglecting base currents and assuming matching geometries) the current flowing in D_2 will be "mirrored" in Q_4. The collector of Q_4 will sink the same current as that flowing in the $40k\Omega$ resistor connected to V^+. The collector current for Q_4 is sourced from the amplifier output stage through the $20k\Omega$ resistor. Since V^+ appears across the $40k\Omega$ resistor, $V^+/2$ is forced across the $20k\Omega$ resistor and the output will track at one half supply voltage.

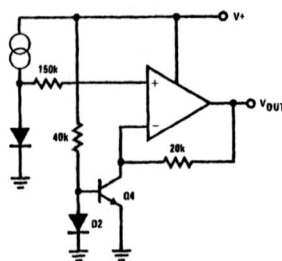

FIGURE 4.11.2 LM383 DC Bias Circuit

4.11.3 General Operating Characteristics

The closed loop gain of the LM383 is set by external components, Figure 4.11.3 showing a typical non-inverting amplifier circuit with A_V set by the ratio of R_1 and R_2. In practical terms the input dynamic range ($\pm 0.5 V_{MAX}$) will determine the lowest useable gain for a given output power and load. The circuit of Figure 4.11.3 is set up for $A_V = 1 + R_1/R_2 = 101$, and it is worth noting the unusually low values of the feedback resistors. This can be attributed to the need for supply ripple rejection. Refering back to Figure 4.11.2, any supply voltage ripple will cause a change in the current in the $40k\Omega$ resistor. This change is "mirrored" in Q_4 and without any external ac feedback the ripple voltage would appear attenuated by only $-6dB$ at the output. However, if C_1 is large enough, the feedback network works to prevent any ac voltage change at the amplifier inverting input, so that the ripple appears as a current in the feedack resistor R_1. To a first order approximation therefore, the ripple at the output is given by the ratio of R_1 to the internal $40k\Omega$ resistor.

By using a 220Ω feedback resistor, the ripple rejection ratio obtained is better than 40dB. Although low resistor values mean that more power will be dissipated in the feedback network, the dc voltage across R_2 is 0.7V, and that across R_1 is typically 6.5V, giving the power ratings shown.

The non-inverting input has a relatively high input impedance, but a large input coupling capacitor is recommended. Before turn-on, both the input capacitor and the feedback capacitor are at ground potential. At turn-on, the feedback capacitor can charge up more quickly than the internal bias resistor can charge the input coupling capacitor. This prevents the output from rapidly going up to the positive supply rail and producing a "pop" in the speaker.

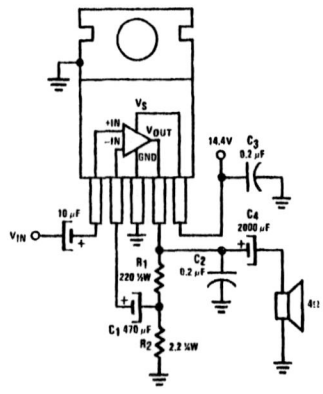

FIGURE 4.11.3 Non Inverting Amplifier ($A_V \approx 40dB$)

4.11.4 Layout, Ground Loops and Supply Bypassing

The very high output current capability of the LM383 means that careful attention should be paid the p.c.b. layout. A suitable component layout is shown in Figure 4.11.4. Parts worth noting are:

1) Supply decoupling capacitor (C_3) is located right at the supply pin.
2) A $0.2\mu F$ capacitor (C_2) is located at the output pin to prevent negative swing parasitic oscillation — *there is no damping resistor in series with this capacitor.*
3) The input ground is returned to the center pin of the I/C — the output or power ground is through the tab via the heatsink.

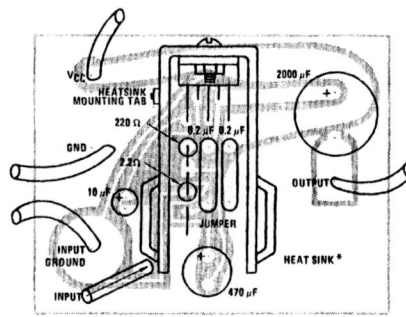

FIGURE 4.11.4 LM383 Board Layout

4.11.5 Output Power and Heat Sinking.

Device power dissipation vs. power output is indicated in Figure 4.11.6 for 4Ω loads and 4.11.7 for 2Ω loads. The ability of the LM383 to sustain these dissipation levels is given by Figure 4.11.5. For example, when driving a 4Ω load, the circuit of Figure 4.11.3 will have a maximum device dissipation of 3.5 Watts. This can be comfortably handled by a $13°C/W$ heatsink, such as the Staver V-5. If the load resistance is 2Ω, considerably more heatsink capability is required since the maximum device dissipation is now over 6 watts. In this instance a heatsink equivalent to the Staver V3-3-2 would be suitable.

For most applications, since the tab of the LM383 package is grounded, the device can be bolted to the chassis to provide adequate heat sinking.

4.11.6 High Voltage Operation

The LM383 has a maximum supply voltage rating of 20V, above which the amplifier will shut down. The LM383A selection will withstand momentary peak supply voltages of 40V caused by supply-line transients. In an automotive application, a worst case transient is usually caused by alternator "load-dump" or loss of the battery charging load. When a 50 amp alternator loses the load, a peak output voltage of about 120V is generated and the transient on the supply line lasts for many milliseconds. Fortunately for the radio, this transient is clamped by an "A" line L-C filter to between 35-40 volts. The LM383A is rated to withstand 40 volts for 50 mSecs.

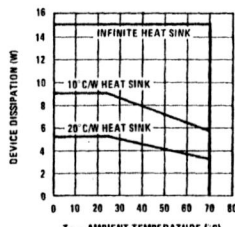

FIGURE 4.11.5 Device Dissipation vs. Ambient Temperature

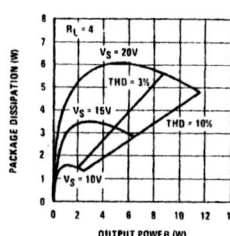

FIGURE 4.11.6 Power Dissipation vs. Output Power

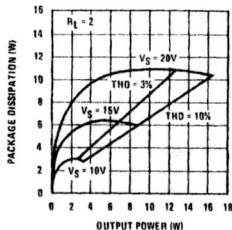

FIGURE 4.11.7 Power Dissipation vs. Output Power

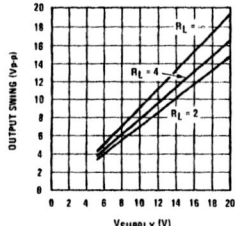

FIGURE 4.11.8 Output Swing vs. Supply Voltage

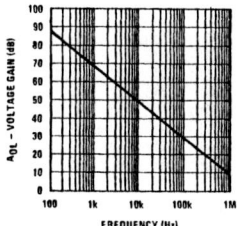

FIGURE 4.11.9 Open Loop Gain vs. Frequency

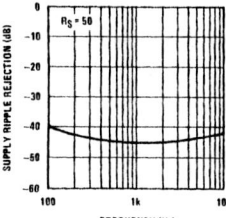

FIGURE 4.11.10 Supply Ripple Rejection vs. Frequency

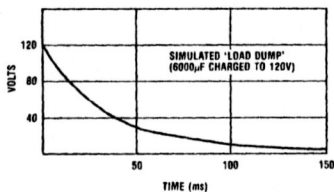
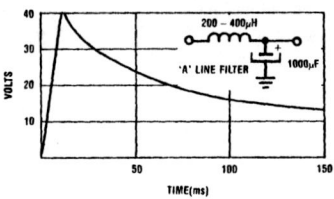

FIGURE 4.11.11 Effect of "A" Line Filter on Automotive Transient

Where more power into a given load is required, two LM383's can be used in a bridge configuration (Figure 4.11.12). The power output and device dissipation per amplifier may be estimated by assuming each amplifier is driving a load of $R_L/2$, in this case 2Ω. On a 14.4V supply, each amplifier can deliver 8W at the 10% THD level for a total output power of 16W (Figure 4.11.7). Each amplifier is dissipating a maximum of 6 watts requiring a *total* heat-sink capability of $5°C/W$ for the complete system (Figure 4.11.5).

A 100Ω potentiometer is used to trim out the differences in individual LM383 dc output levels since, with a direct connected load, substantial dc power consumption can result if the quiescent output levels are not matched. Although the LM383 is designed to be proof against ac short circuits, this will not be the case for the circuit of Figure 4.11.12. If there is a chance that either side of the load could be shorted to ground, coupling capacitors should be included in each output.

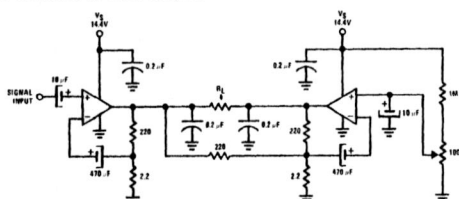

FIGURE 4.11.12 16 Watt Bridge Amplifier

4.12 POWER DISSIPATION

Power dissipation within the integrated circuit package is a very important parameter requiring a thorough understanding if optimum power output is to be obtained. An incorrect power dissipation (P_D) calculation may result in inadequate heatsinking, causing thermal shutdown to operate and limit the output power. All of National's line of audio power amplifiers use class B output stages. Analysis of a typical (ideal) output circuit results in a simple and accurate formula for use in calculating package power dissipation.

4.12.1 Class B Power Considerations

Begin by considering the simplest audio circuit as in Figure 4.12.1, where the power delivered to the load is:

$$P_o = \frac{V_O^2}{R_L} = I_O^2 R_L \quad (4.12.1)$$

where: P_o = power output

V_O = RMS output voltage

I_O = RMS output current

Transforming Equation (4.12.1) into peak-to-peak quantities gives:

$$P_o = \frac{V_{OPP}^2}{8 R_L} = \frac{R_L I_{OPP}^2}{8} \quad (4.12.2)$$

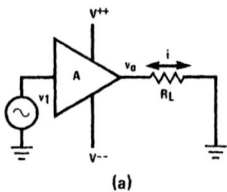

(a)

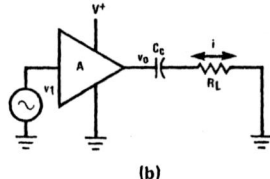

(b)

FIGURE 4.12.1 Simple Audio Circuits

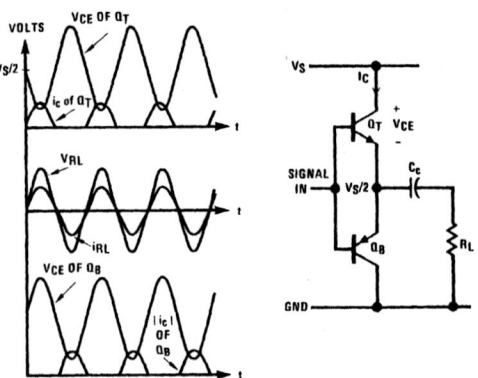

FIGURE 4.12.2 Class B Waveforms

Figure 4.12.2 illustrates current and voltage waveforms in a typical class B output. Dissipation in the top transistor Q_T is the product of collector-emitter voltage and current, as shown on the top axis. Certainly Q_T dissipates zero power when the output voltage is not swinging, since the collector current is zero. On the other hand, if the output waveform is overdriven to a square wave (delivering maximum power to the load, R_L) Q_T delivers large currents, but the voltage across it is zero — again resulting in zero power. In the range of output powers between these extremes, Q_T goes through a point of maximum dissipation. This point always occurs when the peak-to-peak output voltage is 0.637 times the power supply. At that level, assuming all class B power is dissipated in the two output transistors, the chip dissipation is:

$$\max P_D = \frac{V_s^2}{2\pi^2 R_L} \approx \frac{V_s^2}{20 R_L} \quad (4.12.3)$$

Inserting the applicable supply voltage and load impedance into Equation (4.12.3) gives the information needed to size the heat sink for worst case conditions.

4.12.2 Derivation of Max P_D

The derivation of Equation (4.12.3) for maximum power dissipation follows from examination of Figure 4.12.2 and application of standard power formulas:

Neglect X_{C_C} and let V_L' = voltage across the load (resistive)

then

$$V_L' = V_L \sin \omega t$$

$$V_{CE} = V_s - \left(\frac{V_s}{2} + V_L \sin \omega t\right) = \frac{V_s}{2} - V_L \sin \omega t$$

$$I_C = \frac{V_L \sin \omega t}{R_L}$$

since

$$\overline{P_D} = \frac{1}{2\pi} (2) \int_0^\pi p_d \, d(\omega t)$$

— two transistors operated Class B (since both transistors are in the same IC package)

where: $\overline{P_D}$ = average power

p_d = instantaneous power

then

$$\overline{P_D} = \frac{1}{\pi} \int_0^\pi \left(\frac{V_s}{2} - V_L \sin \omega t\right)\left(\frac{V_L \sin \omega t}{R_L}\right) d(\omega t)$$

$$= \frac{V_s V_L}{2\pi R_L} \int_0^\pi \sin \omega t \, d(\omega t) - \frac{V_L^2}{2\pi R_L} \int_0^\pi (1 - \cos 2\omega t) \, d(\omega t)$$

$$= \frac{V_s V_L}{2\pi R_L} (2) - \frac{V_L^2}{2\pi R_L} (\pi)$$

$$= \frac{V_s V_L}{\pi R_L} - \frac{V_L^2}{2 R_L} \quad (4.12.4)$$

Equation (4.12.4) is the average power dissipated; the *maximum* average power dissipated will occur for the value of V_L that makes the first derivative of Equation (4.12.4) equal to zero:

$$\frac{d(\overline{P_D})}{d(V_L)} = \frac{V_s}{\pi R_L} - \frac{V_L}{R_L} = 0 \text{ at maximum}$$

$$\therefore V_{Lp} = \frac{V_s}{\pi} \quad (4.12.5)$$

Equation (4.12.5) is the peak value of V_L that results in max P_D; multiplying by two yields the peak-to-peak value for max P_D:

$$V_{Lp-p} = \frac{2 V_s}{\pi} = 0.637 V_s \quad (4.12.6)$$

Substitution of Equation (4.12.5) into Equation (4.12.4) gives the final value for max P_D:

$$\max \overline{P_D} = \frac{V_s^2}{2\pi^2 R_L} \approx \frac{V_s^2}{20 R_L} \quad (4.12.7)$$

Another useful form of Equation (4.12.7) is obtained by substitution of Equation (4.12.2):

$$\max \overline{P_D} = \frac{4}{\pi^2} P_O(\max) \quad (4.12.8)$$

4.12.3 Application of Max P_D

Max P_D determines the necessity and degree of external heatsinking, as will be discussed in Section 4.14.

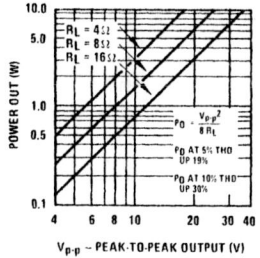

FIGURE 4.12.3 Power Out

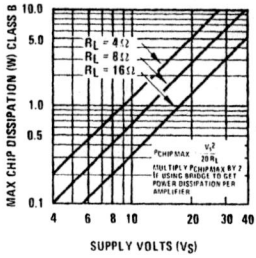

FIGURE 4.12.4 Max Chip Dissipation

The nomographs of Figures 4.12.3 and 4.12.4 make it easy to determine package power dissipation as well as output VI characteristics for popular conditions. Since part of the audio amplifier specmanship game involves juggling output power ratings given at differing distortion levels, it is useful to know that:

P_o increases by 19% at 5% THD

P_o increases by 30% at 10% THD

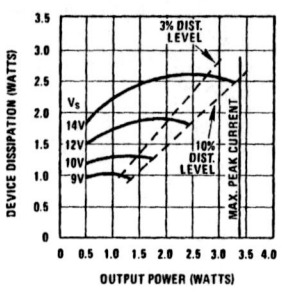
Device Dissipation vs. Output Power — 4Ω Load

Device Dissipation vs. Output Power — 8Ω Load

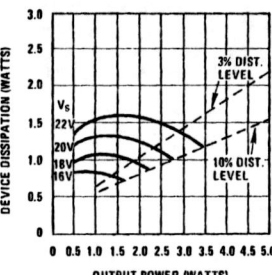
Device Dissipation vs. Output Power — 16Ω Load

FIGURE 4.12.5 Data Power Curves as Shown on Many Data Sheets

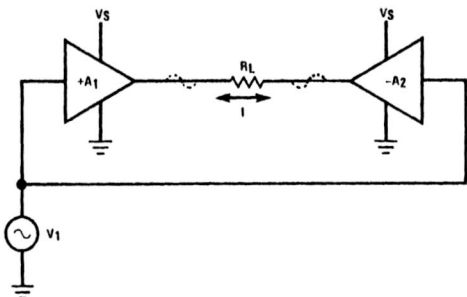

FIGURE 4.12.6 Bridge Audio

Equation (4.12.6) raises an intriguing question: If max P_D occurs at peak-to-peak output voltages equal to 0.637 times the power supply, will P_D go *down* if the output swing is *increased*? The answer is yes — indeed if an amplifier runs at $0.637 V_s$ to the load, and then is driven harder, say to $0.8 V_s$, it will cool off, a phenomenon implied in the power curves given on many audio amplifier data sheets (Figure 4.12.5).

4.12.4 Max P_D of Bridge Amplifiers

Bridge connecting two amplifiers as in Figure 4.12.6 results in a large increase of output power. In this configuration the amplifiers are driven antiphase so that when A_1's output voltage is at V_s, A_2's output is at ground. Thus the peak-to-peak voltage is ideally twice the supply voltage. Since output power is the square of voltage, four times more power can be obtained than from one of these same amplifiers run alone. Note, however, that since the peak voltage across the bridged load is twice that run as a single, the amplifiers must be capable of twice the peak currents. This, along with the fact that no real power amplifier can swing its output completely to V_s and ground, explains why actual bridge circuits never fully realize four times their single circuit output power.

Power dissipation in a bridge is calculated by noting that the voltage at the center of the load does not move. Thus, Equation (4.12.3) can be applied to half the load resistor:

$$P_{A1 \text{ or } A2} = \frac{V_s^2}{\pi^2 R_L} = \frac{V_s^2}{10 R_L} \qquad (4.12.9)$$

4.13 BOOSTED POWER AMPS

4.13.1 Introduction

When output power requirements exceed the limits of available monolithic devices, boosting of the output with two external transistors may be done to obtain higher power levels. The simplest approach involves adding a complementary emitter follower output stage within the feedback loop. The limiting factor is the limitation upon output voltage swing imposed by the B-E drop from the driver's output. Such designs cannot swing closer to the rail voltages than about one volt less than the IC's swing.

4.13.2 Output Boost with Emitter Followers

The simple booster circuit of Figure 4.11.1 allows power output of 10W/channel when driven from the LM378. The circuit is exceptionally simple, and the output exhibits lower levels of crossover distortion than does the LM378 alone. This is due to the inclusion of the booster transistors within the feedback loop. At signal levels below 20mW, the LM378 supplies the load directly through the 5Ω resistor to about 100mA peak current. Above this level, the booster transistors are biased ON by the load current through the same 5Ω resistor.

The response of the 10W boosted amplifier is indicated in Figure 4.13.2 for power levels below clipping. Distortion is below 2% from about 50Hz to 30kHz. 15W RMS power is available at 10% distortion; however, this represents extreme clipping. Although the LM378 delivers little power, its heat sink must be adequate for about 3W package dissipation. The output transistors must also have an adequate heat sink.

The circuit of Figure 4.13.3 achieves about 12W/channel output prior to clipping. Power output is increased because there is no power loss due to effective series resistance and capacitive reactance of the output coupling capacitor required in the single supply circuit. At power up to 10W/channel, the output is extremely clean, containing less than 0.2% THD midband at 10W. The bandwidth is also improved due to absence of the output coupling capacitor. The frequency response and distortion are plotted in Figures 4.13.4 and 4.13.5 for low and high power levels. Note that the input coupling capacitor is still required, even though the input may be ground referenced, in order to isolate and balance the DC input offset due to input bias current. The feedback coupling capacitor, C_1, maintains DC loop gain at unity to insure zero DC output voltage and zero DC load current. Capacitors C_1 and C_2 both contribute to decreasing gain at low frequencies. Either or both may be increased for better low frequency bandwidth. C_3 and the 27k resistor provide increased high frequency feedback for improved high frequency distortion characteristics. C_4 and C_5 are low inductance mylar capacitors connected within 2 inches of the IC terminals to ensure high frequency stability. R_1 and R_f are made equal to maintain $V_{OUT DC} = 0$. The output should be within 10 to 20 mV of zero volts DC. The internal bias is unused; pin 1 should be open circuit. When experimenting with this circuit, use the amplifier connected to terminals 8, 9 and 13. If using only the amplifier on terminals 6, 7 and 2, connect terminals 8 and 9 to ground (split supply) to cause the internal bias circuits to disconnect.

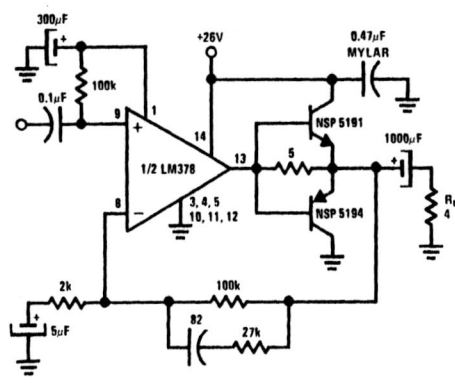

FIGURE 4.13.1 10-Watt Power Amplifier

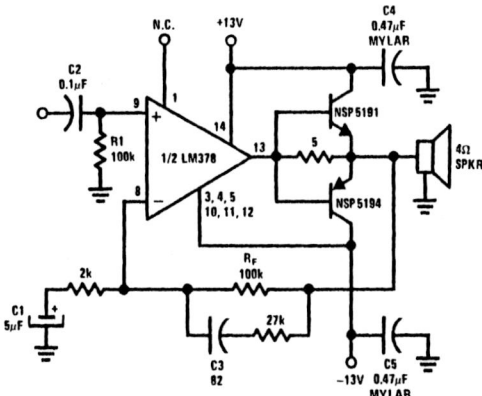

FIGURE 4.13.3 12-Watt Low-Distortion Power Amplifier

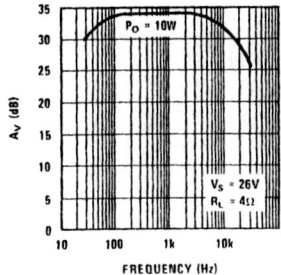

FIGURE 4.13.2 10-Watt Boosted Amplifier, Frequency Response

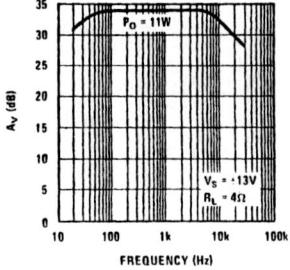

FIGURE 4.13.4 Response for Amplifier of Figure 4.13.3

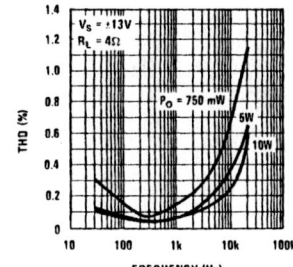

FIGURE 4.13.5 Distortion for Amplifier of Figure 4.13.3

4.13.3 Power Drivers

Using external transistors to boost the output of monolithic power amplifiers does increase the output power available, but with the limitation that the supply voltage cannot be increased beyond the maximum rating of the I/C. Also the output swing on that supply voltage is always less than the I/C output swing. To accomodate higher voltages and larger output swings, National has developed the Power Driver I/C's, the LM391 and LM2000 series. The LM391 is designed to drive power transistors in 10 watt to 50 watt power amplifiers operating from split supplies as high as ±50 volts. The LM2000 and LM2001 are designed for battery operated use to obtain 4 watts in 4Ω on a 12 volt supply and 2 watts in 2Ω on 6 volt supplies.

4.13.4 LM391 Circuit Description

An equivalent schematic for the LM391 is shown in Figure 4.13.6. A PNP differential input stage is used with emitter degeneration provided by 5kΩ resistors to give a good slew rate and a large linear input voltage range (see Section 4.1.2).

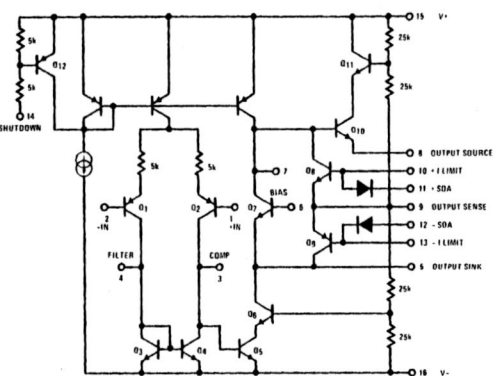

FIGURE 4.13.6 LM391 Equivalent Circuit

The amplifier compensation capacitor is external and connected between pins 3 and 5. This capacitor will normally be selected to be smaller than the value required for unity gain stability to ensure that there is adequate loop gain at the higher audio frequencies to reduce distortion. For stable designs using amplifier closed loop gains of 20V/V or more, the high frequency pole set by the compensation capacitor C_C should be below 500kHz.

Since $f_H A_V = \dfrac{g_m}{2\pi C_C}$ (4.13.1)

g_m = transconductance of input stage $\approx \dfrac{1}{5.5 \times 10^3}$

$\therefore C_C \geq \dfrac{1}{2\pi 5.5 \times 10^3 \times 20 \times 500 \times 10^3}$ ($A_V = 20V/V$)

$\geq 3pF$

The size of C_C will also determine the maximum possible slew rate. Since the largest current available to charge C_C is 100μA (input stage fully switched),

Slew Rate = $\dfrac{I}{C_C} \leq \dfrac{100 \times 10^{-6}}{3 \times 10^{-12}}$ (4.13.2)

i.e. Slew Rate ≤ 33V/μS

To improve the negative supply ripple rejection, a capacitor equal in value to C_C should be connected from pin 4 to ground (Figure 4.13.9).

In order to accomodate different output stage configurations, the AB Bias circuit of the LM391 can be programmed externally. This type of bias circuit is known as a "V_{BE} Multiplier" and is set up as shown if Figure 4.13.10.

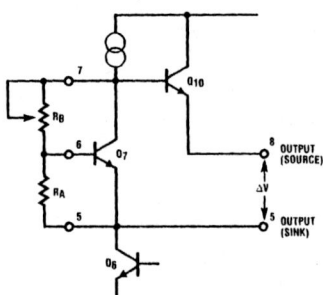

FIGURE 4.13.10 AB Bias Current Circuit

The voltage across the lower resistor R_A must always be equal to the base-emitter voltage of Q_7 ... V_{BE} (Q_7). If the base current of Q_7 is assumed to be negligible, the current producing this voltage across R_A must also be flowing through the upper resistor R_B. Q_7 collector will absorb the additional current from the upper current source so that Q_7 collector base voltage is defined by the current through R_B, and will be a multiple of V_{BE} determined by the ratio of R_A and R_B.

$V_{CB}(Q_7) = V_{BE}(Q_7) \times \dfrac{R_B}{R_A}$ (4.13.3)

The total differential voltage produced between the output pin 5 and 8 is

$V_{BIAS} = V_{BE}(Q_7) + V_{BE}(Q_7)\dfrac{R_B}{R_A} - V_{BE}(Q_{10})$

If $V_{BE}(Q_7) = V_{BE}(Q_{10})$

$V_{BIAS} = V_{BE}(Q_7)\dfrac{R_B}{R_A}$ (4.13.4)

By making R_B 3.9kΩ and R_A a 10kΩ potentiometer, a wide range of AB bias voltages can be obtained.

Output sink and source currents are guarranteed to be 5mA minimum, and the protection devices Q_8 and Q_9 can be connected to reduce these drive currents automatically when damage to the external output transistors could occur. Also for protection a thermal shut-down pin (Pin 14) is provided that will remove all output current capability when it is pulled low. Turn-on delay to prevent preamplifier pops from reaching the speakers is another use of this pin.

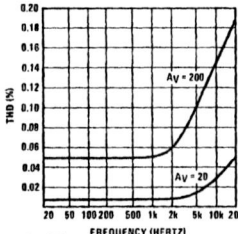

FIGURE 4.13.7 Total Harmonic Distortion $R_L = 8$

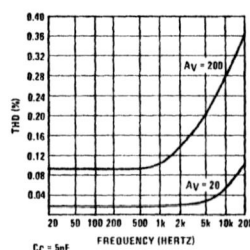

FIGURE 4.13.8 Total Harmonic Distortion $R_L = 4$

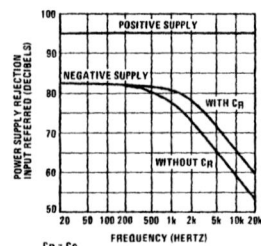

FIGURE 4.13.9 Input Referred Power Supply Rejection

4.13.5 Non-Inverting Amplifier Application

A typical amplifier set-up, without using the protection circuitry, is shown in Figure 4.13.11. R_{IN} provides a dc bias path for the input stage and sets the amplifier input resistance. A good value is 100kΩ. If very high resistor values are used for R_{IN}, layout induced oscillations will become probable and there will be larger dc offset voltages produced at the output. To minimize input bias currents producing output offsets, the feedback resistor R_{f1}, should be made equal to R_{IN}. R_{f1}, together with R_{f2} and C_F will set the amplifier mid-band gain, A_V,

$$A_V = 1 + \frac{R_{f1}}{R_{f2}} \quad (4.13.5)$$

C_F reduces the amplifier gain to unity at dc for minimum offset voltage and gives a low frequency pole with R_{f2},

$$f_L = \frac{1}{2\pi R_{f2} C_F} \quad (4.13.6)$$

For amplifier gains of 20V/V and above, the recommended value for the compensation capacitor C_C is 5pF which, by rearrangement of Equation 4.13.1, will give an amplifier closed loop bandwidth of 320kHz. If C_R is used for improved ripple rejection, this also should be 5pF.

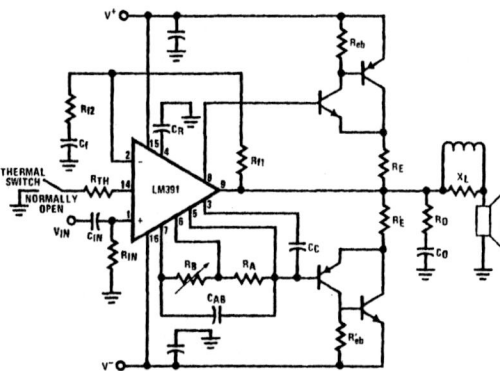

FIGURE 4.13.11 LM391 With External Components-Protection Circuitry not Shown

Connecting a 0.1μF capacitor across the AB Bias network will improve the transient response of the amplifier and reduce distortion at high frequencies. The AB bias current in the output stages should be set by R_B to above 20mA to ensure low THD (Figure 4.13.12).

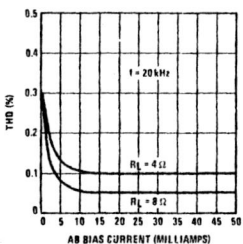

FIGURE 4.13.12 THD vs AB Bias Current

The resistor R_{TH} in series with pin 14 and a thermal shutdown switch determines the amount of current pulled from pin 14 during shut-down. Since this current shold not exceed 1mA, the value of R_{TH} is given by,

$$R_{TH} \geq \frac{V^+}{1 \times 10^{-3}} \quad (4.13.7)$$

4.13.6 Output Stage Stability (External Components)

The output stage of Figure 4.13.11 is a composite NPN/PNP arrangement. A resistor-capacitor network $R_O C_O$ is used to compensate the output power devices and a resistor R_{eb} is included to "bleed" off stored charge in these output devices caused by their large input diffusion and depletion capacitance. Between the output devices and the speaker an inductor X_L is placed to protect the amplifier from instabilities caused by driving capacitance loads.

Both output devices have a low valued resistor R_E to help maintain thermal stability of the AB bias current. When a power transistor goes through a power cycle, the chip temperature can change by many degrees with a corresponding change in the base-emitter voltage (the temperature coefficient k of the base-emitter voltage is typically $-2mV/°C$). It is unlikely that the V_{BE} multiplier providing the output bias voltage will be able to track the temperature change of the output device and the drop in output transistor V_{BE} will cause an increase in the AB Bias current and an increase in power dissipation. At very low frequencies these changes can occur during a single cycle of the output swing contributing to increased distortion levels. A more serious problem is that the AB Bias transistor junction (Q7) can cool more rapidly than the output transistor junction following a period of sustained power output. Thermal instability can result if the then increased bias current causes a higher power dissipation level than that previously being sustained. Using R_E helps to maintain thermal stability since an increased bias current in the output stage will cause an additional voltage drop across R_E to compensate for the decrease in V_{BE}.

For an output stage of the type shown in Figure 4.13.11, a simple expression can be developed for the least value of R_E that will ensure thermal stability of the output stage,

$$R_E \geq \frac{\theta_{JA} V_{MAX} K}{(\beta + 1)} \quad (4.13.8)$$

where: θ_{JA} = Thermal Resistance of Driver Transistor (Junction to Ambient)

β = Minimum beta of the *output* device

V_{MAX} = Maximum supply voltage (V^+ or V^-)

K = 2mV/°C

4.13.7 Output Device S.O.A. Protection

For higher power output amplifiers, the Safe Operating Area (S.O.A.) of the output devices becomes important. Operating within the power-temperature ratings and avoiding thermal run away will not guarantee circuit reliability. To avoid failure the operating load line must be maintained within safe voltage-current limits and, particularly in the case of second breakdown which is energy dependent, time limits must be observed.

For a given device, reliable areas of operation are specified by an S.O.A. chart, of which Figure 4.13.13 is a typical example. The S.O.A. boundary can be defined by a few specific limits, current, power, second breakdown and voltage. Note that for high voltage and medium currents the available power is limited primarily by second breakdown considerations. A second important point to note is that the DC operating curve is based on a 25°C case temperature. Therefore the dc or low frequency operation is usually thermally limited to limits less than shown by this chart.

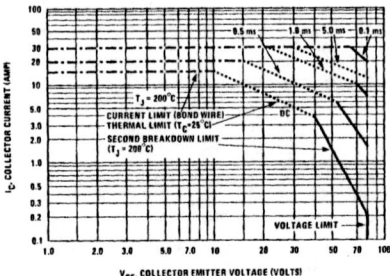

FIGURE 4.13.13 Active Region Safe Operating Area

Clearly the choice of output device as the power amplifier rating is increased in not a simple matter. Also, to keep the output stage compensation straight forward, the driver device is usually chosen to have a good frequency response, yet as the output power goes up the S.O.A. of these devices will come into play.

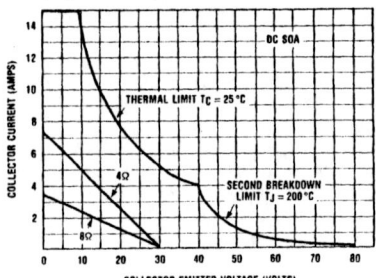

FIGURE 4.13.14 Amplifier Load Lines On S.O.A. Chart

If we add 8Ω and 4Ω load lines to the S.O.A. chart (Figure 4.13.14 -- note the change in axes scale) which would represent *practical* 40 watt and 60 watt amplifiers respectively, we can see that the load lines are all safely within the S.O.A. of the device. So where is the problem, since a number of devices with a lot less S.O.A. than the example (2N5884/86) given here would do the job? The problem is that real world speakers are anything but resistive and this will cause a substantial increase in the heatsink and S.O.A. requirements than is indicated by a simple resistive load.

4.13.8 Effect of Speaker Loads

Figure 4.13.15 shows an impedance curve for a typical dynamic loudspeaker. As can be seen, there is a wide variation in impedance between 20 Hz and 20 kHz. The impedance at the resonant frequency can commonly measure five times or more the rated impedance. Indeed, many speakers will only display their rated impedance at one frequency (typically 400 Hz). The actual impedance is a complex value of dc resistance, inductive reactance of the voice coil, coupling capacitor reactance, crossover network impedance and frequency. In general, though, loudspeakers appear inductive with a worst case phase angle of 60°. This means that the voltage across the speaker leads the current by 60°.

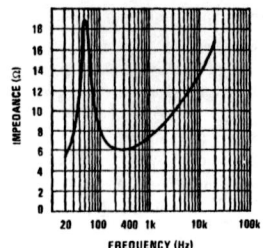

FIGURE 4.13.15 Impedance Curve for a Typical Dynamic Loudspeaker

An intuitive approach to what phase angle does to maximum average power dissipation produces the realization that the worst case load for power dissipation is purely reactive, i.e., 90° phase angle. This becomes clear by considering the resistive case of zero phase angle depicted in Figure 4.13.16(a) where the maximum voltage across the load, V_L, results in maximum current, I_L; but since they are in phase there exists zero volts across the device and no package disipation results. Now, holding everything constant while introducing a phase angle causes the voltage waveforms to shift position in time, while the current stays the same. The voltage across the load becomes smaller and the voltage across the package becomes larger, so with the same current flowing package dissipation increases. At the limit of 90° phase difference Figure 4.13.16(b) results, where there exists zero volts across the load, maximum voltage across the package, and maximum current flowing through both, producing maximum package dissipation.

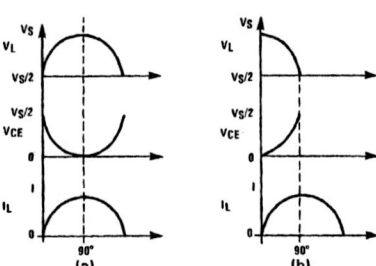

FIGURE 4.13.16 Phase Angle Relationship Between Voltage and Current (a) 0° and (b) 90°

If we consider the effect of load phase angle Φ, the formula obtained in Section 4.12.2 must be modified. Equation (4.12.8) becomes:

$$\max \overline{P_D} = \frac{4P_{o(MAX)}}{\pi^2 \cos\Phi} \qquad (4.13.9)$$

and the maximum peak instantaneous power dissipation,

$$P_{d(MAX)} = \frac{V_s^2}{4R_L}[\sin(wt-\Phi) - \sin wt \sin(wt-\Phi)]$$

or,

$$P_{d(MAX)} = 2P_{o(MAX)}[\sin(wt-\Phi) - \sin wt \sin(wt-\Phi)] \qquad (4.13.10)$$

Equation (4.13.9) can be used to plot the curve shown in Figure 4.13.17 which gives the output stage maximum average power dissipation for reactive loads up to 60° phase angle. The importance of Figure 4.13.17 is seen by comparing the power ratio at zero degrees (0.405) with that at 60° (0.812) ... double! This means that the maximum Class B output stage dissipation can be *twice as much* for a speaker load as for a resistive load.

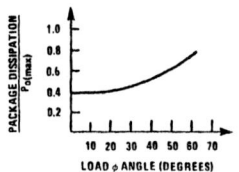

FIGURE 4.13.17 Class B Package Dissipation for Reactive Loads

The maximum power dissipation in *each* output device will be half that given by Equation (4.13.9)

$$\overline{P}_{D(MAX)} = \frac{2P_{o(MAX)}}{\pi^2 \cos\Phi} \qquad (4.13.11)$$

and the maximum power dissipation for *each* driver transistor in a composite output stage is given by,

$$\overline{P}_{D(Driver)} = \frac{\overline{P}_{o(MAX)}}{\beta_{MIN}} \qquad (4.13.12)$$

($\beta_{(MIN)}$ is that of the *output* device)

Equations (4.3.11) and (4.3.12) can be used to determine the heatsinking requirements for *each* device in the output stage.

$$\theta_{JA} \leq \frac{T_{JMAX} - T_{AMAX}}{P_{DMAX}} \qquad (4.13.13)$$

The heatsink thermal resistance is given by,

$$\theta_{SA} \leq \theta_{JA} - \theta_{JC} - \theta_{CS} \qquad (4.13.14)$$

where: T_{JMAX} is maximum transistor junction temperature

T_{AMAX} is maximum ambient temperature

θ_{JA} is thermal resistance junction to ambient

θ_{SA} is thermal resistance sink to ambient

θ_{JC} is thermal resistance junction to case

θ_{CS} is thermal resistance case to sink

For driver devices mounted on a common heatsink, the thermal resistance given by Equation (4.13.14) should be divided by the number of driver devices for the total heatsink requirement. A similar calculation can be made for the output transistors but it is worth noting that the output heatsinks may depend more on the ability or need for the amplifier to withstand a continuously shorted output (see Section 4.13.9).

The effect of a reactive load on the heatsink is easy to calculate from Equations (4.13.11) and (4.13.12). To understand the effect of reactive loads on the output device S.O.A. requirements we need to refer to Figure 4.13.18. This shows normalized curves for several reactive load lines up to 60° phase angle, and the locus of the maximum peak instantaneous power dissipation in the output transistors [given by Equation (4.13.10)]. Now it is apparent that with a 60° phase angle load, the load line will approach very closely the S.O.A. limits. In fact, if the S.O.A. limits for the 60 watt/4Ω amplifier are superimposed on Figure 4.13.18 (dashed line), the dc S.O.A. will actually *be exceeded* if the load angle increases beyond 40°!

The locus of $p_{d(MAX)}$ on Figure 4.13.18 can be used to specify the required device S.O.A. for the worst reactive load for which the amplifier will be designed. For example, with a 60° load angle the S.O.A. should be specified at a voltage of 0.59 ($V^+ + V^-$) with a current of 0.77 I_{MAX}. It is instructive to contrast this with 0.25 ($V^+ + V^-$) and 0.5 I_{MAX} for a purely resistive load!

4.13.9 Protection Circuits

If we assume that an output device can be selected with a sufficient S.O.A. to accomodate the worst load angle condition designed for, the next step is to protect this output device from abnormal or transient conditions that could place operation outside the S.O.A. — overload or short circuited outputs for example.

The simplest form of protection — current limiting — has already been made possible by the inclusion of the resistors R_E in the output stage.

During normal operation, the load current flowing through R_E does not produce a large enough voltage drop to turn on either Q_8 (positive side) or Q_9 (negative side). The limit current is reached when the voltage across R_E exceeds V_{BE} (Q_8 or Q_9). Then the protection transistors turn on and begin to bleed the available base current from the output stages to hold the output current at the limit level. Although the normal "on"

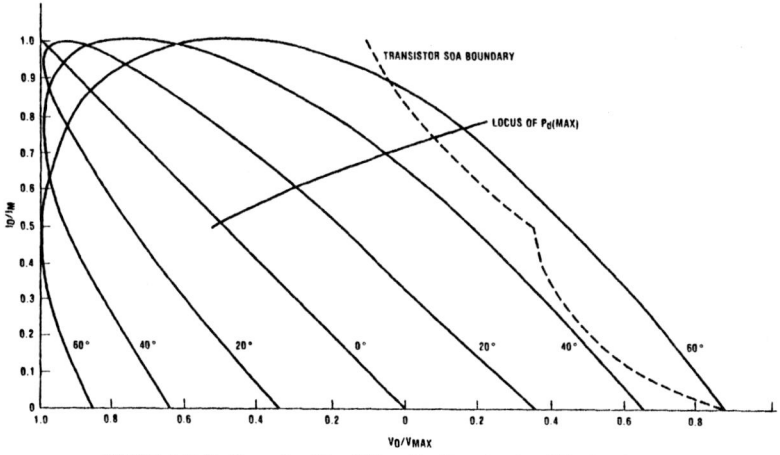

FIGURE 4.13.18 Normalized Load Lines For Reactive Amplifier Loads

voltage for Q_8 is 710mV and for Q_9 is 660mV at 25°C, equal turn-on voltages are assumed of 650mV at 55°. The load current limit is given by,

$$I_L = \frac{R_E}{0.65} \text{ AMPS} \qquad (4.13.15)$$

Note that R_E also provides thermal run away protection for the AB Bias current. When R_E has been chosen for a given current limit a check should be made to see that R_E exceeds the value given by Equation (4.13.8).

Simple current limiting will not necessarily prevent the output devices from failing if a transient causes operation outside the device S.O.A. Also, during shorted conditions the average power being dissipated in each output transistor is given by,

$$\overline{P}_{D(SHORT)} = \tfrac{1}{2} I_{LIMIT} V_{CE}$$

or

$$\overline{P}_{D(SHORT)} = \frac{I_{LIMIT}(V^+ + V^-)}{4} \qquad (4.13.16)$$

This power dissipation is substantially more than that obtained during normal operation and the heatsinks may not be able to handle this long term.

An improvement over current limiting is "single slope" load line protection (Figure 4.13.20). Two more resistors, R_1 and R_2 have been added to each output with a compensation capacitor C connected across R_2. Now the voltage across the output stage as well as the current through it is being monitored.

Now Q_8 and Q_9 operate to reduce the output current from the limit value set by the choice of R_E (obtained when either output device is saturated — i.e. $V_{CE} = 0V$) down to zero current when the voltage across the output device reaches the maximum rated collector-emitter voltage. In order for the current to be zero at $V_{CE} = V_M$,

$$V_{BE}(Q_8) = \frac{R_2}{(R_1 + R_2)} V_M$$

or

$$R_1 = \frac{R_2}{0.65}(V_M - 0.65) \qquad (4.13.17)$$

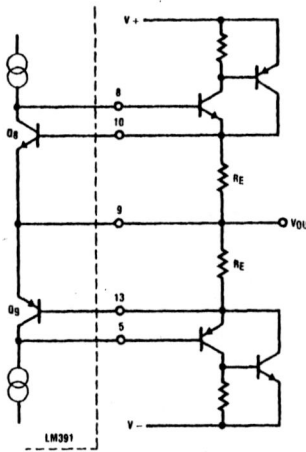
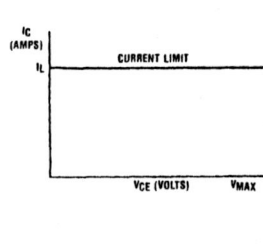

FIGURE 4.13.19 Use of LM391 To Limit Output Current

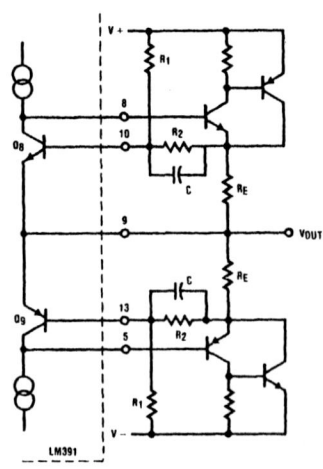
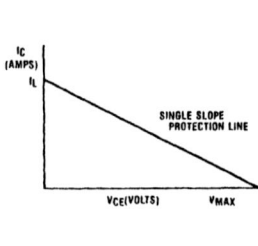

FIGURE 4.13.20 Single Slope Load Line Protection

With a shorted output, the average power dissipation in the output stage will be about half that obtained when only current limiting is used. However, if I_L is set close to the maximum load current for rated output (I_M), inspection of the reactive load lines given in Figure 4.13.18 will show that there is a good chance that the protection circuit will be activated even though the S.O.A. is not being exceeded. I_L can be increased above I_M until the protection line is asymptotic to the nearest point of the S.O.A. boundary (or to the maximum current rating of the device in some instances). This will cause a corresponding increase in the short circuit power dissipation.

R_2 is usually arbitrarily chosen to be 1kΩ leaving R_1 to be defined by Equation (4.13.17). A good choice for C is 1000pF.

To permit operation over most of the output transistor S.O.A. without activating protection circuits, dual slope load line protection is recommended, Figure 4.13.21. The corresponding protection lines superimposed on a typical transistor S.O.A. chart are shown in Figure 4.13.22.

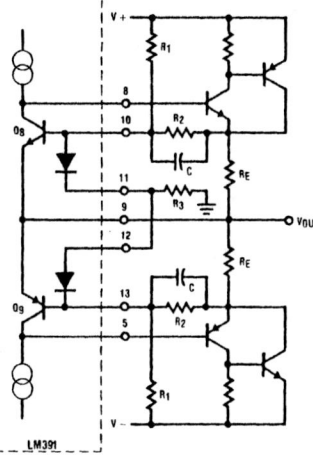

FIGURE 4.13.21 Dual Slope Load Line Protection

The internal diodes on the protection transistor bases are connected through a resistor R_3 to ground. R_E, R_1, R_2 and C are selected as before in the single slope protection circuit and with R_3 connected to ground, the break point where the protection line changes slope will be at the midpoint between the supply rails (equal to V^+ or V^- as far as the output devices are concerned).

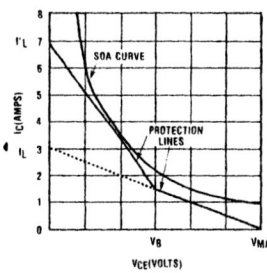

FIGURE 4.13.22 Dual Slope Load Lines on S.O.A. Chart

The design formula can be determined by assuming the output transistor is saturated and delivering the new upper limit current I'_L. The voltage across R_2, R_3 and the corresponding diode will be approximately V^+ (or V^-) and the current through R_2 is given by,

$$i = \frac{I'_L R_E - V_{BE}}{R_2}$$

Therefore,

$$R_3 = R_2 \left(\frac{V^+}{I'_L R_E - 0.65} - 1 \right) \quad (4.13.18)$$

Assuming $V^+ \gg V_{SAT}$
$V^+ \gg V_{BE}$

Again the short circuit power dissipation will be half that obtained with simple current limiting, but the device can now operate over most of the S.O.A.

For convenience, Table 4.13.1 summarizes the necessary formula to determine component values for any degree of protection.

4.13.10 Power Supply Requirements

The power supply voltage and current capability depends on the amplifier rated ouput and load impedance. For a given power output P_O and load R_L,

$$V_{O\,PEAK} = \sqrt{2 R_L P_O} \quad (4.13.19)$$

$$I_{O\,PEAK} = \frac{\sqrt{2 P_O}}{R_L} \quad (4.13.20)$$

To obtain these output swings the power supply voltage will have to be higher to allow for transistor saturation voltage drops. Since, in a large number of cases, an unregulated supply will be used for economic reasons, the unloaded supply voltage will be about 15% higher than when delivering the rated current output. If we allow an additional 10% for high line conditions, the maximum supply voltage is given by,

$$\text{MAX}\,V_{SUPPLY} = \pm (V_{O\,PEAK} + V_{SAT}) \times 1.15 \times 1.1 \quad (4.13.21)$$

The requirement that rated power output be obtained with low line conditions will add another 10% to the number obtained from Equation (4.13.21). Assuming that transistor saturation and protection circuit voltage drops total 5V (per side), Table 4.13.2 lists the voltage and current requirements for 20 watt through 100 watt amplifiers operating on unregulated supplies. The final column of Table 4.13.2 lists the collector breakdown voltage requirement for the output transistors (including the drivers). This column will also define the breakdown voltage required of the LM391.

4.13.11 Amplifier Design

Although the preceding text may imply that power amplifier design is fraught with pitfalls, the following examples should illustrate how an amplifier can be designed using the LM391 in a very straightforward fashion.

Example 4.13.1

Design an amplifier capable of delivering an average power of 20 watts into 8Ω and 30 watts into 4Ω. The input sensitivity should be lower than $1V_{MAX}$ with an input impedance of 100kΩ or more. A 20Hz to 20kHz ±0.25dB bandwidth is required.

TABLE 4.13.1 Protection Circuit Formula

TYPE OF PROTECTION	R_E	R_1	R_2	R_3	C
CURRENT LIMIT	$R_E = \dfrac{0.65}{I_L}$		SHORT		
SINGLE SLOPE PROTECTION LINE	$R_E = \dfrac{0.65}{I_L}$	$R_1 = \dfrac{R_2(V_m - 0.65)}{0.65}$	$1k\Omega$		1000 PF
DUAL SLOPE PROTECTION LINE	$R_E = \dfrac{0.65}{I_L}$	$R_1 = \dfrac{R_2(V_m - 0.65)}{0.65}$	$1k\Omega$	$R_3 = R_2\left[\left(\dfrac{V+}{I'_L R_E - 0.65}\right) - 1\right]$	1000 pF

TABLE 4.13.2 Power Supply Requirements For 20 Watt To 100 Watt Amplifiers

P_o POWER(WATTS)	LOAD (Ω)	IpAMPS ($\sqrt{2P_o/R_L}$)	OUTPUT SWING ($\pm \sqrt{2P_o R_L}$) Volts	TOTAL SUPPLY VOLTAGE REGULATED	TOTAL SUPPLY VOLTAGE UNREGULATED (NO LOAD)	TRANSISTOR V_{CEO} (SUS) (UNREGULATED SUPPLY VOLTAGE)
20	8	2.24	±17.9	47.1	54.1	59.6
	4	3.16	±12.6	36.6	42.1	46.3
30	8	2.74	±21.9	55.1	63.4	69.7
	4	3.87	±15.5	42.3	48.6	53.5
40	8	3.16	±25.3	61.9	71.2	78.3
	4	4.47	±17.9	47.1	54.1	59.6
50	8	3.54	±28.3	67.9	78.1	85.9
	4	5.00	±20.0	51.3	59.0	64.9
60	8	3.87	±31.0	73.1	84.1	92.5
	4	5.48	±21.9	55.1	63.4	69.2
70	8	4.18	±33.5	78.2	89.9	98.9
	4	5.92	±23.7	58.6	67.4	74.2
80	8	4.47	±35.8	82.9	95.3	104.8
	4	6.32	±25.3	61.9	71.2	78.3
90	8	4.74	±37.9	85.9	98.8	108.6
	4	6.71	±26.8	65.0	74.7	82.2
100	8	5.00	±40.0	91.3	100.5	115.5
	4	7.07	±28.3	67.9	78.1	85.9

Solution.

1. From Table 4.13.2:

   Voltage swing for rated power in $8\Omega = \pm 17.9$ volts.

   Peak current for rated power in $4\Omega = 3.87$ amps

   For an unregulated supply, no load voltage

   $V_{SUPPLY} = 54$ volts or ± 27 volts.

2. $A_V \geq \dfrac{17.9/\sqrt{2}}{1} = 12.66$

   If we use an LM391 with a gain of $A_V = 20 V/V$ the resulting sensitivity is 630mVrms which is well within the required specification.

3. Letting $R_{IN} = 100k$ gives the required input impedance, and to ensure low dc offset voltages, $R_{F1} = 100k$. From Equation (4.13.5)

   $A_V = 20 = 1 + \dfrac{R_{f1}}{R_{f2}}$

   $\therefore R_{f2} = 5.26 k\Omega$

   Put $R_{f2} = 5.1 k\Omega$

4. Two octaves below the high frequency pole (f −3dB), the amplitude response will be 0.25dB down,

   i.e. $f_H \geq 20 \times 10^3 \times 4 = 80 kHz$

   Similarly two octaves above the low frequency pole, the response will again be −0.25dB,

   i.e. $f_L \leq \dfrac{20}{4} = 5 Hz$

   From Equation (4.13.6)

   $C_F \geq \dfrac{1}{2\pi f_L R_{f2}} \geq 6.2 \mu F$

   Use $C_F = 10 \mu F$

   For $A_V = 20 V/V$, the recommended value for C_C is 5pF

   From Equation (4.13.1)

   $f_H = \dfrac{1}{2\pi \times 5 \times 10^{-12} \times 20 \times 5 \times 10^3} = 318 kHz$

5. The I/C and output transistor breakdown voltages must be greater than the maximum supply voltage ($V^+ + V^-$). From Table 4.13.2

   $V_{MAX} = 59.6$ volts (unregulated high line)

   Use LM391N-60

6. A suitable power transistor complementary pair is the National BD346 and BD347 (2N6487, 2N6490) with a V_{CEO} (sus) of 60 volts and a minimum beta of 30 at 4 amps. Since the guaranteed minimum drive current from the LM391-60 is 5mA, the driver transistors must have a minimum beta given by,

$$\text{Driver } \beta_{MIN} \geq \frac{3.87}{30 \times 5 \times 10^{-3}} \approx 26 \text{ (@ 130mA)}$$

The National complementary pair BD344, 345 (MJE171, MJE181) are 60V devices with a minimum beta of 40 at 200mA.

7. For each output transistor the maximum average power dissipation is given by Equation (4.13.11)

$$P_{d(MAX)} = \frac{2P_{D(MAX)}}{\pi^2 \cos \phi}$$

Assume $\Phi = 60°$ MAX

$$P_{d(MAX)} = \frac{2 \times 30}{\pi^2 \times 0.05} = 12.2 \text{ watts}$$

8. From Equation (4.13.13)

$$\theta_{JA} \geq \frac{150°C - 55°C}{12.2W} \text{ for } T_A = 55°C$$

$$= 7.8°C/W$$

From Equation (4.13.14)

$\theta_{SA} \leq 7.8 - 2.1 - 1.0 = 4.8°C/W$

If both transistors for one amplifier are mounted on a single heatsink.

$\theta_{SA} \leq 2.4°C/W$

9. The driver maximum average power dissipation is given by Equation (4.13.12)

$$P_{D(Driver)} = \frac{\bar{P}_{DMAX}}{\beta_{MIN}} = \frac{12.2W}{30} = 410 \text{mW}$$

Using Equation (4.13.13) again

$$\theta_{JA} \leq \frac{155°C - 55°C}{0.410} = 244°C/W$$

The free-air thermal resistance of the BD344, 345 is 100°C/W so that no additional heatsinking is required.

10. The least value of R_E to prevent AB bias thermal runaway is obtained from Equation 4.13.8

$$R_E \geq \frac{100(30) \times 2 \times 10^{-3}}{(30 + 1)} = 0.19 \Omega$$

11. Figure 4.13.23 is the S.O.A. chart for the BD346 and BD347 transistors. Also shown are the 4Ω and 8Ω load lines and desired protection lines for dual slope protection. From Figure 4.13.23

$V_m = 60$ volts; $V_B = 23$ volts; $I_L = 3$ amps;

$I'_L = 7$ amps

From Table 4.13.1

$$R_E = \frac{0.65}{3} = 0.22\Omega$$

(This value of R_E also satisfies Equation 4.13.8, see Step 10)

The completed amplifier schematic is shown in Figure 4.13.24. One final point, the heatsink capability calculated in Steps 8 and 9 were for continuous operation into a 4Ω load. If that load is inadvertently shorted, then the average power being dissipated is given by Equation (4.13.16)

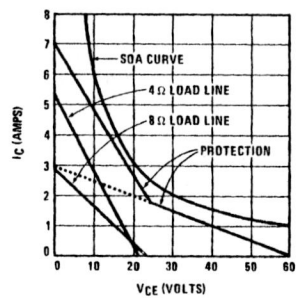

FIGURE 4.13.23 DC-S.O.A. for BD346 and BD347

$$P_{D(SHORT)} = \frac{I_{LIMIT}(V^+ - V^-)}{4}$$

I_{LIMIT} is obtained from Figure 4.13.23 and in this case is 1.8 amps.

$$\bar{P}_{D(SHORT)} = \frac{1.8(56)}{4} = 25.2 \text{ watts}$$

Sustained operation under shorted conditions will require a much larger output device heatsink or a thermal sensor to pull down Pin 14 of the LM391 when the original heatsink temperature exceeds 111°C.

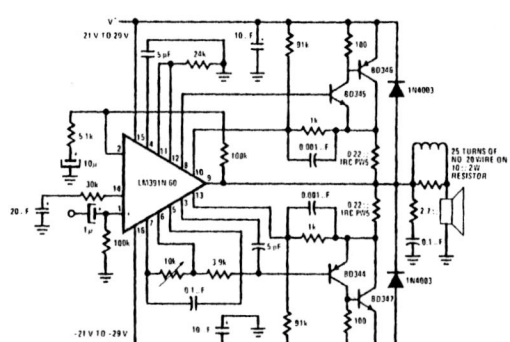

FIGURE 4.13.24 20W-8Ω, 30W-4Ω Amplifier

Example 4.13.2

Design an audio power amplifier with an input sensitivity of 1V$_{rms}$, to drive 8Ω and 4Ω loads to power levels of 40 watts and 60 watts respectively. The maximum load phase angle is 60° and the design should include S.O.A. protection for the output stage.

Solution

Following the same steps as in the previous example:

1. I_{PEAK} = 5.48 amps
   V_{PEAK} = ±25.3 volts
   V_{SUPPLY} = 71.2 volts (No Load)
   = ±31 volts (Full load)

2. Put A_V = 20V/V
   Input sensitivity = 900mV$_{rms}$

4-59

3. $R_{f1} = 100\,k\Omega \quad R_{IN} = 100\,k\Omega$
   $R_{f2} = 5.1\,k\Omega$
4. $C_F = 10\,\mu F \quad C_C = 5\,pF \quad C_R = 5\,pF$
5. Device voltage ≥ 78.3 volts
   $= 80$ volts
   Choose LM391N-80
6. Output devices: BD350, BD351 (2N5880, 2N5882),
   $V_{CEO(SUS)} = 80\,V$, $\beta_{MIN} = 50$ at 5.5 amps.
   Driver devices: BD348, BD349 (MJE172, MJE182),
   $V_{CEO(SUS)} = 80\,V$, $\beta_{MIN} = 50$ at 250 mA.
7. Maximum output device dissipation
   $P_{O(MAX)} = 24.3$ watts
8. $\theta_{JA} \leq \dfrac{200°C - 55°C}{24.3} = 6°C/W$

   $\theta_{SA} \leq 6 - 1.1 - 1.0 = 3.9°C/W$

   For both devices on a common heatsink
   $\theta_{SA} \leq 2.0°C/W$

   It is worth noting and comparing the heatsinking for this amplifier with that of the previous example. Using TO-3 case style transistor with higher junction temperature and lower thermal resistance has kept the heatsink size down more than might be expected. (Example 4.13.1 Specified Case Style TO-220)

9. $\overline{P}_{O(DRIVER)} = \dfrac{24.3}{22.5} = 1.1$ watts

   $\theta_{JA} \leq \dfrac{150°C - 55°C}{1.1} = 86.4°C/W$

   $\therefore \theta_{SA} \leq 86.4 - 6 - 1 = 79.4°C/W$

10. $R_E \geq \dfrac{86.4(40)2 \times 10^{-3}}{22.5 + 1} = 0.29\,\Omega$

11. Using Figure 4.13.25
    $V_m = 80$ volts, $V_B = 47$ volts, $I_L = 3$ amps,
    $I'_L = 11$ amps

    For $I_L = 3$ amps, from Table 4.13.1
    $R_E = \dfrac{0.65}{3} = 0.22\,\Omega$

    This does not simultaneously satisfy Equation 4.13.8 — Step 10. Recalculating driver device heatsink from Equation 4.13.8

    $\theta_{JA} \leq \dfrac{0.22(22.5 + 1)}{40(2 \times 10^{-3})} = 65°C/W$

    $R_2 = 1\,k\Omega$ (arbitrary)

    $R_1 = 1 \times 10^3 \dfrac{(80 - 0.65)}{0.65} \approx 120\,k\Omega$

    Since to obtain the desired protection lines, V_B is not centered between the supply rails R_3 is replaced with a resistive divider ($R_A R_B$) between the positive supply and ground for the lower output device (Pin 12). A similar divider is connected at Pin 11 between the negative supply and ground for protection of the upper output device.

    Now $R_A \parallel R_B = R_3$

    $\therefore R_A \parallel R_B = 10^3 \left\{ \dfrac{47}{11(0.22) - 0.65} \right\} - 1 = 25.55\,k\Omega$

    Since V_B is 17 volts away from the center of the output swing (with supply loaded to ± 31 volts)

    $\dfrac{R_A}{(R_A + R_B)} \cdot 31 = 17$

    $\therefore R_B = 0.82\,R_A$
    a) guess $R_A = 62k$
       $\therefore R_B = 51k$, $R_A \parallel R_B = 28\,k\Omega$
    b) guess $R_A = 56k$
       $\therefore P_B = 45.92$ Put $R_B = 47\,k\Omega$
       $\therefore R_A \parallel R_B = 25.55k$ which is close enough.

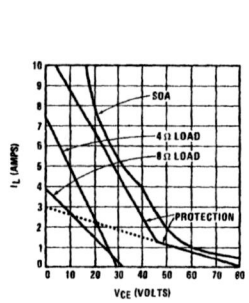

FIGURE 4.13.25 S.O.A. for BD350, and BD351

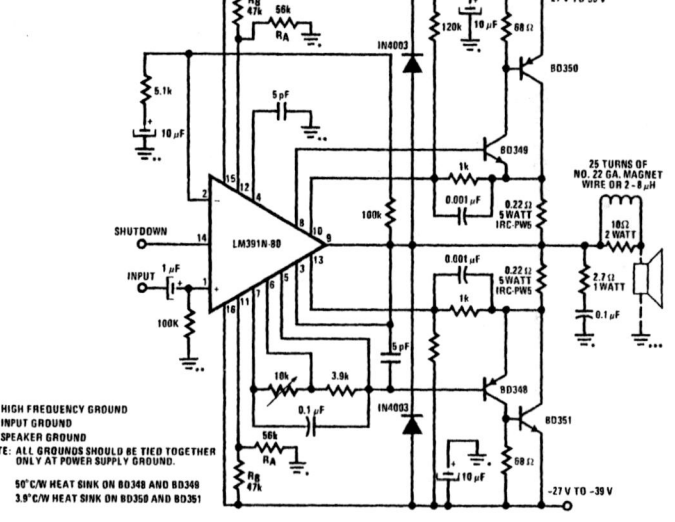

* HIGH FREQUENCY GROUND
** INPUT GROUND
*** SPEAKER GROUND
NOTE: ALL GROUNDS SHOULD BE TIED TOGETHER ONLY AT POWER SUPPLY GROUND.

50°C/W HEAT SINK ON BD348 AND BD349
3.9°C/W HEAT SINK ON BD350 AND BD351

FIGURE 4.13.26 40W-8Ω, 60W-4Ω Amplifier

The complete amplifier is shown in Figure 4.13.26. Two additional diodes are shown clamping Pin 9 to within a diode drop greater than either supply. This is to prevent the output devices being damaged when the output voltage exceeds either supply — which can occur if the protection circuitry is activated while the load appears inductive.

Under shorted output conditions the limit current will be 4.5amps and the short circuit power dissipation is given by,

$$\bar{P}_{D(SHORT)} = \frac{4.5 \times 60}{4} = 67.5 \text{ watts}$$

Again, the heatsink capability caluclated in Steps 8 and 11 will not permit a continuous short on the output. However, for normal music inputs into typical speaker loads, these heatsinks are actually quite conservative and will have the thermal capacity to ride out intermittent shorts of limited duration. Where a designer has a specific knowledge of the load and operating conditions for his amplifier, smaller heatsinks may be used for economic reasons. Even so, especially when smaller output devices are selected by the same reasoning, any amplifier should be thoroughly tested for reliability under actual operating conditions.

4.13.12 Oscillations and Grounding

Most power amplifiers will work the first time they are turned on. They also tend to oscillate, sometimes with catastrophic results, and have excess THD. The majority of oscillation problems are caused by inadequate power supply bypassing and by ground loops (see Section 2.2.2 and 2.2.3). 10μF capacitors on the supply leads, close to the circuit rather than to the power supply, will stop supply related oscillations. If the signal ground is used for these bypass capacitors, the THD will probably be further increased. To avoid this, the signal ground must return to the power supply alone, as should the output load or speaker ground. The bypass capacitor, output R-C and protection grounds can be connected together. Figure 4.13.26 shows the recommended grounding arrangement for power amplifiers using the LM391.

4.13.13 Turn-On Delay

It is often desirable to delay the turn-on of the power amplifier so that turn-on pops generated in the pre-amplifier section do not go to the speakers. This can be achieved with the LM391 simply by using the shutdown pin (Pin 14). A series capacitor-resistor combination is used to set the turn-on delay (Figure 4.13.27). At turn-on, the capacitor is at ground potential, holding Pin 14 low through the resistor and there is no current drive available for the output stage. After approximately two time constants, the capacitor has charged sufficiently that output drive current is enabled and normal amplifier operation can take place. The minimum value for the resistor is given by Equation (4.13.7)

$$R = \frac{V_+ \text{(max)}}{10^{-3}} \tag{4.13.7}$$

For an amplifier with ±30 volts supplies

$$R = \frac{30}{10^{-3}} = 30k$$

Turn-on delay in seconds is given by

$$T = 2RC \tag{4.13.22}$$

If we use a 33kΩ resistor

$$C = \frac{1}{2 \times 33 \times 10^3} = 15\mu F$$

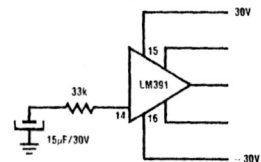

FIGURE 4.13.27 Turn-On Circuit Delay

4.13.14 Transient Distortion

The topic of transient distortion is one that is still subject to a great deal of discussion at this time. Nevertheless several design criteria have been evolved for the avoidance of distortion effects produced by transients in the program material. This section will discuss two of these criteria with respect to LM391 amplifiers.

Slew rate limits have been mentioned several times with respect to the compensation capacitors used in monolithic audio amplifiers (Sections 1.2.1, 4.1.1, etc.). Simple expressions have been developed for the frequency at which slew rate limiting will occur for a given output voltage swing and amplifier slew rate. What has not been emphasized is that at that frequency, distortion of the signal has already begun — of the order of 1% to 3% THD. To minimize distortion, the frequency at which slew rate limiting occurs (with maximum output swing) should be well above the audio bandwidth. Current practice indicates that a slew rate of 0.5V/μS *per output peak volt* is acceptable, with 1V/μS per output peak volt being conservative. The LM391 has a slew rate of 20V/μS with a 5pF compensation capacitor — reference to Table 4.13.2 shows that this is adequate for amplifiers up to 100 watts.

It may not be possible to avoid slew induced distortion simply by being able to slew at frequencies substantially above the audio bandwidth. If an input transient causes the amplifier input stage to overload, the output will be in slew limiting until the feedback loop responds. This can be prevented by using a low pass filter at the input stage. The cut-off frequency of this filter must be above the audio bandwidth, and how far above will depend on the amplifier input stage dynamic range and transconductance, and the open loop pole frequency of the amplifier. A detailed paper by Peter Garde in the May, 1978 *Journal of the Audio Engineering Society* derives the following criterion to prevent input stage overload.

$$\frac{I}{g_m} \geq V_{IN} \frac{(2f_f - f_o)K}{f_c} \tag{4.13.23}$$

where: V_{IN} = maximum peak input voltage to the amplifier

I = input stage maximum current

g_m = input stage transconductance

f_o = open loop pole frequency

f_c = closed loop pole frequency

f_f = input filter pole frequency

K = constant(dependent on ratio of f_c/f_f)

For the LM391; f_o = 1kHz, $g_m = \frac{1}{5.5 \times 10^3}$, $I = 100\mu A$. If the closed loop gain is 20V/V with C_C = 5pF as in the previous design examples, the closed loop pole frequency is 300kHz. In a 40 watt, 8Ω amplifier, the maximum input voltage

$$V_{IN} = \frac{25.3}{20} = 1.3 \text{ volts peak}$$

For an input filter pole frequency of 100kHz, K = 0.6 and Equation 4.13.23 becomes

$$\frac{100 \times 10^{-6}}{0.18 \times 10^{-3}} \geqslant \frac{1.3(2 \times 100 \times 10^{-3} - 10^3) \times 0.6}{300 \times 10^3}$$

or $0.56 \geqslant 0.52$ which is so.

Figure 4.13.28 shows a simple input filter for the LM391 to prevent input stage clipping up to the rated amplifier output.

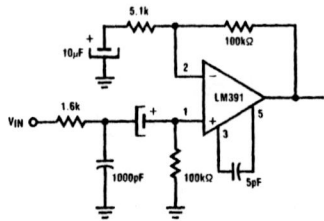

FIGURE 4.13.28 LM391 Input Filter

4.13.15 Low Voltage Power Drivers

High voltage power amplifiers require careful selection of the output transistors in order for them to be able to handle the power requirements. At the other end of the scale with low voltage, battery operated equipment, the concern is to obtain sufficient output swing into the available load impedance. The LM2000 and LM2001 amplifiers are designed to drive low cost external transistors to within a collector saturation voltage drop of the supply rails. For operation from 12 volts down to 2.5 volts the LM2000 is recommended, and for operation below 6 volts down to 1.8 volts the LM2001 is the device to use. Both have similar circuit configurations except for slight differences in the output stage. The output stage gain setting resistors are external for the LM2000 because of the higher levels of power dissipation.

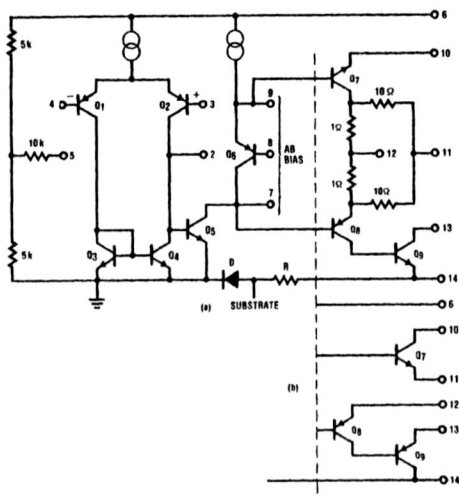

FIGURE 4.13.29 LM2000/2001 Equivalent Schematic
(a) LM2001
(b) LM2000

An equivalent schematic for both amplifiers is shown in Figure 4.13.29 (a) & (b). A fully differential PNP input stage with active load is used and a half supply voltage bias point (Pin 5) is provided to set up the output midway between the positive supply and ground. The external compensation around Q_5 (Pins 2 and 7) and a V_{BE} multiplier Q_6 for the output stage AB Bias (Pins 9, 8 and 7) are similar to those already described for the LM391 (Section 4.13.1). Both amplifiers have output driver stages designed for voltage gain and current drive to the external transistors.

4.13.16 Output Stage Operation — Upper Side

Referring to Figure 4.13.30, the external potentiometer R_B in the collector circuit of Q_6 is adjusted to set the current level in the driver transistor Q_7. The local feedback resistors R_2 and R_1 set the output stage voltage gain at,

$$A_{V(OUTPUT)} = 1 + \frac{R_2}{R_1} \qquad (4.13.24)$$

This voltage gain allows Q_{10} to be driven into saturation without Q_7, Q_6 or the bias current source also saturating. A_V is determined externally for the LM2000 and fixed at 11 internally for the LM2001. Notice that there is no bleed resistor in the Q_{10} base circuit so that the collector current of Q_7 (I_{C7}), set by R_B, is also the base current of the output device Q_{10}. This means that the AB Bias current is $\beta \times I_{C7}$ where β is the dc beta of Q_{10}. Now the AB Bias current thermal stability depends on the T.C. of the output device β rather than its V_{BE}.

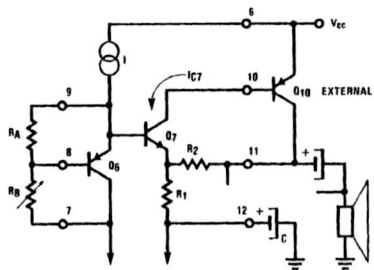

FIGURE 4.13.30 LM2001 Upper Side Driver Stage

4.13.17 Output Stage Operation — Lower Side

To enable the lower external transistor to be driven into saturation, the output load coupling capacitor is used to bootstrap the lower side output driver, which is isolated by the resistor R from the substrate (Diode D clamps the substrate when the output swings above ground). Figure 4.13.31.

4.13.18 Inverting Amplifier Applications

Figure 4.13.32 demonstrates a typical use of the LM2001 driving a 2Ω load through two PNP output transistors. The mid-band voltage gain is set externally by the ratio of Rf_1 and Rf_2 to 101, with C_F reducing the dc gain to unity to minimize output voltage offsets and giving a low frequency pole at $1/2\pi \times C_F Rf_2$. R_{IN} connected to Pin 5 establishes the output dc bias at half supply and sets the input resistance of the amplifier. R_B adjusts the output stage AB Bias current to about 15mA.

The compensation capacitor C_C is selected on the basis of Figure 4.13.33 which has curves of the ±3dB bandwidth for values of C_C and amplifier closed loop gain. Since the bandwidth and C_C are inversely proportional (Equation 4.13.1), for a given closed loop gain, other curves for different bandwidths are easily extrapolated. For example, for a gain of 100, and a bandwidth of 50kHz, C_C will be 20/50 × 120pF, or approximately 50pF.

An additional resistor R_C can be added between Pin 8 and the inverting input to improve the amplifier response when the output stage is recovering from clipping. If the input level is sufficiently high that the output external transistors are driven into clipping, the amplifier will momentarily be open loop. The internal node at Pin 8 will come out of overload before the external devices (since the output driver stage has a gain of 11) and the feedback loop via R_C allows the output to make a much smoother transition from clipping than it would otherwise.

A large valued capacitor is connected at Pin 12 to enable the LM2001 driver transistors to sink large currents at low frequencies. At least 100mA sink current guarantees that over 1amp can be delivered to the load with external transistors having a forced beta of 10 or better.

In the case of the LM2000 a similar circuit hook-up is used with the output driver stage gain set resistors external. Operation to 12V supplies is permissible (see Figure 4.13.34).

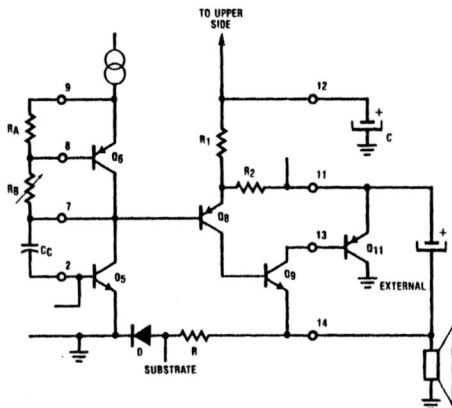

FIGURE 4.13.31 LM2001 Lower Side Driver Stage

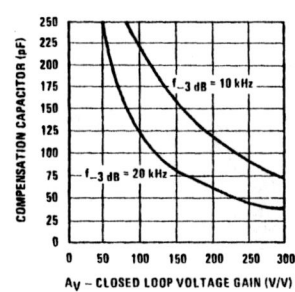

FIGURE 4.13.33 Compensation Capacitance and Closed Loop Bandwidth

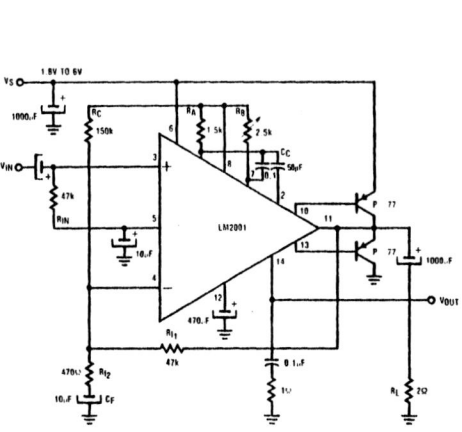

FIGURE 4.13.32 LM2001 Inverting Amplifier

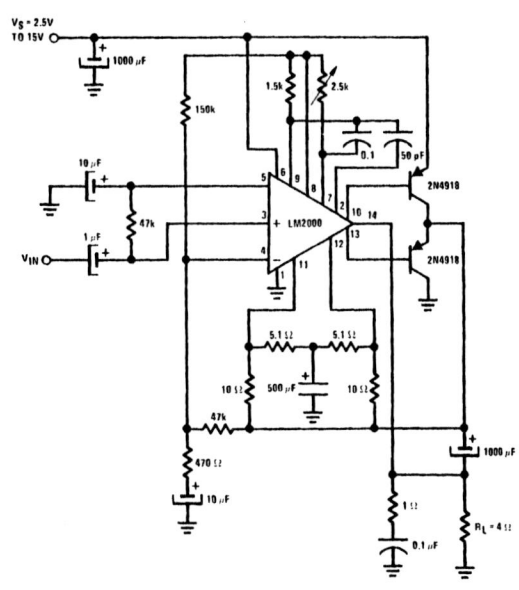

FIGURE 4.13.34 LM2000 Power Amplifier

4-63

Typical performance characteristics for the amplifiers are summarized in Figures 4.13.35 through 4.13.38.

4.13.19 Complementary Output Stages

Both amplifiers can be used to drive complementary external transistors by reconfiguring the lower side output driver stage (Figure 4.13.39). The NPN output device is driven from the emitter of Q_9 with Q_9 collector being connected to the positive supply rail.

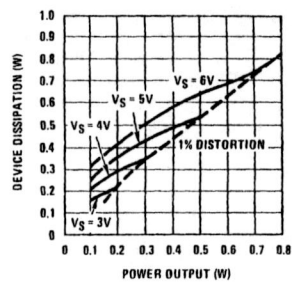

FIGURE 4.13.35 Device Dissipation – 4Ω Load
LM2001 Only

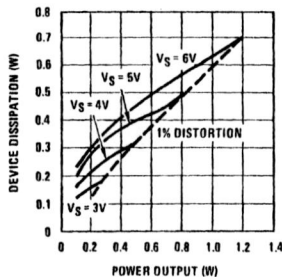

FIGURE 4.13.36 Device Dissipation
– 2Ω Load
LM2001 Only

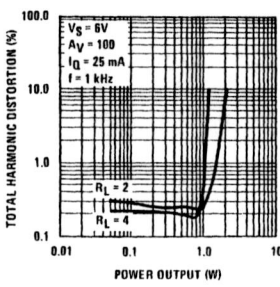

FIGURE 4.13.37 Distortion vs.
Output Power

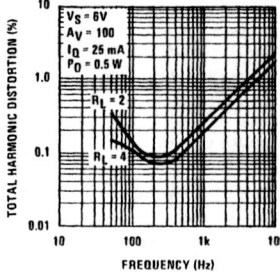

FIGURE 4.13.38 Distortion vs. Frequency

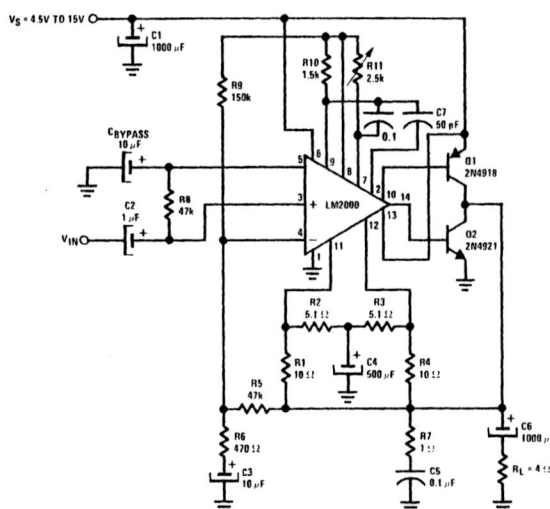

FIGURE 4.13.39 Complementary Output Amplifier

4.14 HEATSINKING

Insufficient heatsinking accounts for many phone calls made to complain about power ICs not meeting published specs. This problem may be avoided by proper application of the material presented in this section. Heatsinking is not difficult, although the first time through it may seem confusing.

If testing a breadboarded power IC results in premature waveform clipping, or a "truncated shape," or a "melting down" of the positive peaks, the IC is probably in thermal shutdown and requires more heatsinking. The following information is provided to make proper heat sink selection easier and help take the "black magic" out of package power dissipation.

4.14.1 Heat Flow

Heat can be transferred from the IC package by three methods, as described and characterized in Table 4.14.1.

TABLE 4.14.1 Methods of Heat Flow

METHOD	DESCRIBING PARAMETERS
Conduction is the heat transfer method most effective in moving heat from junction to case and case to heat sink.	Thermal resistance θ_{JL} and θ_{LS}. Cross section, length and temperature difference across the conducting medium.
Convection is the effective method of heat transfer from case to ambient and heat sink to ambient.	Thermal resistance θ_{SA} and θ_{LA}. Surface condition, type of convecting fluid, velocity and character of the fluid flow (e.g., turbulent or laminar), and temperature difference between surface and fluid.
Radiation is important in transferring heat from cooling fins.	Surface emissivity and area. Temperature difference between radiating and adjacent objects or space. See Table 4.14.2 for values of emissivity.

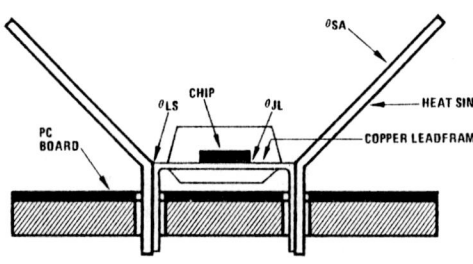

(a) Mechanical Diagram

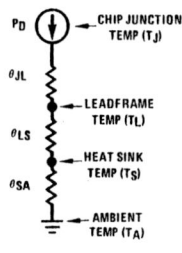

(b) Electrical Equivalent

Symbols and Definitions

θ = Thermal Resistance (°C/W)
θ_{JL} = Junction to Leadframe
θ_{LS} = Leadframe to Heat Sink
θ_{SA} = Heat Sink to Ambient
θ_{JS} = Junction to Heat Sink = $\theta_{JL} + \theta_{LS}$
θ_{JA} = Junction to Ambient = $\theta_{JL} + \theta_{LS} + \theta_{SA}$
T_J = Junction Temperature (maximum) (°C)
T_A = Ambient Temperature
P_D = Power Dissipated (W)

(c) Symbols and Definitions

FIGURE 4.14.1 Heat Flow Model

4.14.2 Thermal Resistance

Thermal resistance is nothing more than a useful figure-of-merit for heat transfer. It is simply temperature drop divided by power dissipated, under steady state conditions. The units are usually °C/W and the symbol most used is θ_{AB}. (Subscripts denote heat flowing from A to B.)

The thermal resistance between two points of a conductive system is expressed as:

$$\theta_{12} = \frac{T_1 - T_2}{P_D} \text{ °C/W} \qquad (4.14.1)$$

4.14.3 Modeling Heat Flow

An analogy may be made between thermal characteristics and electrical characteristics which makes modeling straightforward:

T — temperature differential is analogous to V (voltage)
θ — thermal resistance is analogous to R (resistance)
P — power dissipated is analogous to I (current)

Observe that just as R = V/I, so is its analog θ = T/P. The model follows from this analog.

A simplified heat transfer circuit for a power IC and heat sink system is shown in Figure 4.14.1. The circuit is valid only if the system is in thermal equilibrium (constant heat flow) and there are, indeed, single specific temperatures T_J, T_L, and T_S (no temperature distribution in junction, case, or heat sink). Nevertheless, this is a reasonable approximation of actual performance.

4.14.4 Where to Find Parameters

P_D

Package dissipation is read directly from the "Power Dissipation vs. Power Output" curves that are found on all of the audio amp data sheets. Most data sheets provide separate curves for either 4, 8 or 16Ω loads. Figure 4.14.2 shows the 8Ω characteristics of the LM378.

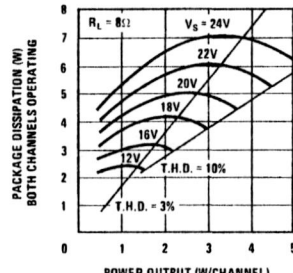

FIGURE 4.14.2 Power Dissipation vs. Power Output

Note: For P_o = 2W and V_S = 18V, $P_{D(max)}$ = 4.1W, while the same P_o with V_S = 24V gives $P_{D(max)}$ = 6.5W — 50% greater! This point cannot be stressed too strongly: *For minimum P_D, V_S must be selected for the minimum value necessary to give the required power out.*

For loads other than those covered by the data sheet curves, max power dissipation may be calculated from Equation (4.14.2). (See Section 4.12.)

$$P_{D(max)} = \frac{V_S^2}{20\,R_L} \qquad (4.14.2)$$

Equation (4.14.2) is for each channel when applied to duals.

When used for bridge configurations, package dissipation will be twice that found from Figure 4.14.2

θ_{LS}

The thermal resistance between lead frame and heat sink is a function of how close the bond can be made. For the D.I.P., soldering to the ground pins with 60/40 solder is recommended. When soldered, θ_{LS} may be neglected or a value of θ_{LS} = 0.25°C/W may be used. Where the package style permits bolting to the heat sink, θ_{LS} will depend on whether a heat sink compound and/or an insulating washer is used. For a TO-3 case style 0.1°C/W is obtained with compound, increasing to 0.4°C/W with a 3 mil mica washer. The TO-220 case style used by the LM383 has corresponding values for θ_{LS} between 1.6°C/W and 2.6°C/W.

$T_{J(max)}$

Maximum junction temperature for each device is 150°C.

θ_{JL}

Thermal resistance between junction to lead frame (or junction to heat sink if θ_{LS} is ignored) is read, directly from the "Maximum Dissipation vs. Ambient Temperature" curve found on the data sheet. Figure 4.14.3 shows a typical curve for the LM378.

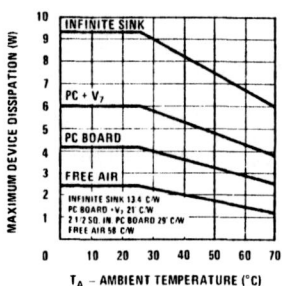

FIGURE 4.14.3 Maximum Dissipation vs. Ambient Temperature

Note: θ_{JL} is the slope of the curve labeled "Infinite Sink." It is also $\theta_{JA(best)}$, while $\theta_{JA(worst)}$ is the slope of the "Free Air" curve, i.e., infinite heat sink and no heat sink respectively.

So, what does it mean? Simply that with no heat sink you can only dissipate

$$\frac{150°C - 25°C}{58°C/W} = 2.16W.$$

And with the best heat sink possible, the maximum dissipation is

$$\frac{150°C - 25°C}{13.4°C/W} = 9.33W$$

Or, for you formula lovers:

$$\text{Max Allowable } P_D = \frac{T_{J(max)} - T_A}{\theta_{JA}} \qquad (4.14.3)$$

4.14.5 Procedure for Selecting Heat Sink

1. Determine $P_{D(max)}$ from curve or Equation (4.14.2).
2. Neglect θ_{LS} if soldering; if not, θ_{LS} must be considered.
3. Determine θ_{JL} from curve.
4. Calculate θ_{JA} from Equation (4.14.3).
5. Calculate θ_{SA} for necessary heat sink by subtracting (2) and (3) from (4) above, i.e., $\theta_{SA} = \theta_{JA} - \theta_{JL} - \theta_{LS}$ (4.14.4)

For example, calculate heat sink required for an LM378 used with V_S = 24V, R_L = 8Ω, P_o = 4W/channel and T_A = 25°C:

1. From Figure 4.14.2, P_D = 7W.
2. Heat sink will be soldered, so θ_{LS} is neglected.
3. From Figure 4.14.3, θ_{JL} = 13.4°C/W.
4. From Equation (4.14.3),

$$\theta_{JA} = \frac{150°C - 25°C}{7W} = 17.9°C/W.$$

5. From Equation (4.14.4),

$$\theta_{SA} = 17.9°C/W - 13.4°C/W = 4.5°C/W.$$

Therefore, a heat sink with a thermal resistance of 4.5°C/W is required. Examination of Figure 4.14.3 shows this to be substantial heatsinking, requiring forethought as to board space, sink cost, etc.

Results modeled:

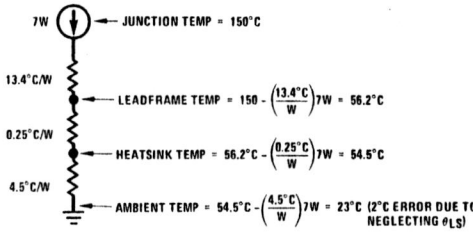

FIGURE 4.14.4 Heat Flow Model for LM378 Example

4.14.6 Custom Heat Sink Design

The required θ_{SA} was determined in Section 4.14.5. Even though many heat sinks are commercially available, it is sometimes more practical, more convenient, or more economical to mount the device to chassis, to an aluminum extrusion, or to a custom heat sink. In such cases, design a simple heat sink.

Simple Rules

1. Mount cooling fin vertically where practical for best conductive heat flow.
2. Anodize, oxidize, or paint the fin surface for better radiation heat flow; see Table 4.14.2 for emissivity data. However, note that although paint increases the emissivity of a surface, the paint itself has a high thermal resistance and should be removed where the semiconductor device is attached to the heat sink. (This will also apply to anodized and oxidized surfaces.)
3. Use 1/16" or thicker fins to provide low thermal resistance at the IC mounting where total fin cross-section is least.

Fin Thermal Resistance

The heat sink-to-ambient thermal resistance of a vertically mounted symmetrical square or round fin (see Figure 4.4.5) in still air is:

$$\theta_{SA} = \frac{1}{2H^2 \eta (h_c + h_r)} °C/W \qquad (4.14.5)$$

where: H = height of vertical plate in inches
η = fin effectiveness factor
h_c = convection heat transfer coefficient
h_r = radiation heat transfer coefficient

$$h_c = 2.21 \times 10^{-3} \left(\frac{T_S - T_A}{H}\right)^{1/4} W/in^2°C \qquad (4.14.6)$$

$$h_r = 1.47 \times 10^{-10} E \left(\frac{T_S + T_A}{2} + 273\right)^3 W/in^2°C \qquad (4.14.7)$$

where: T_S = temperature of heat sink at IC mounting, in °C
T_A = ambient temperature in °C
E = surface emissivity (see Table 4.14.2)

Fin effectiveness factor η includes the effects of fin thickness, shape, thermal conduction, etc. It may be determined from the nomogram of Figure 4.14.6.

TABLE 4.14.2 Emissivity Values for Various Surface Treatments

SURFACE	EMISSIVITY, E
Polished Aluminum	0.05
Polished Copper	0.07
Rolled Sheet Steel	0.66
Oxidized Copper	0.70
Black Anodized Aluminum	0.7 - 0.9
Black Air Drying Enamel	0.85 - 0.91
Dark Varnish	0.89 - 0.93
Black Oil Paint	0.92 - 0.96

For untreated copper and aluminum surfaces, E can be approximated to about 0.2.

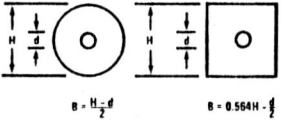

Note: For H >> d, using B = H/2 is a satisfactory approximation for either square or round fins.

FIGURE 4.14.5 Symmetrical Fin Shapes

The procedure for use of the nomogram of Figure 4.14.6 is as follows:

1. Specify fin height H as first approximation.
2. Calculate $h = h_r + h_c$ from Equations (4.14.6) and (4.14.7).
3. Determine α from values of h and fin thickness x (line a).
4. Determine η from values of B (from Figure 4.14.5) and α (line b).

The value of η thus determined is valid for vertically mounted symmetrical square or round fins (with H ≫ d) in still air. For other conditions, η must be modified as follows:

Horizontal mounting — multiply h_c by 0.7.

Horizontal mounting where only one side is effective — multiply η by 0.5 and h_c by 0.94.

For 2:1 rectangular fins — multiply h by 0.8.

For non-symmetrical fins where the IC is mounted at the bottom of a vertical fin — multiply η by 0.7.

Fin Design

1. Establish initial conditions, T_A and desired θ_{SA} as determined in Section 4.14.5.
2. Determine T_S at contact point with the IC by rewriting Equation (4.14.1):

$$\theta_{JL} + \theta_{LS} = \frac{T_J - T_S}{P_D} \qquad (4.14.8)$$

$$T_S = T_J - (\theta_{JL} + \theta_{LS})(P_D) \qquad (4.14.9)$$

$$\approx T_J - \theta_{JL} P_D$$

3. Select fin thickness, $x > 0.0625''$ and fin height, H.
4. Determine h_c and h_r from Equations (4.14.6) and (4.14.7).
5. Find fin effectiveness factor η from Figure 4.14.6.
6. Calculate θ_{SA} from Equation (4.14.5).
7. If θ_{SA} is too large or unnecessarily small, choose a different height and repeat steps (3) through (6).

Design Example

Design a symmetrical square vertical fin of 1/16" thick black anodized aluminum to be bolted onto an LM379 delivering a maximum power of 4W/Ch into 8Ω from a 28V supply.

1. LM379 operating conditions are:

   $T_J = 150°C_{(MAX)}$, $T_A = 55°C_{(MAX)}$

   From Figure 4.4.9, $\theta_{JL} = 6°C/W$

   From Figure 4.4.8, $P_{D(MAX)} = 9.5W$

2. From Equation 4.4.3

   $$\theta_{JA} = \frac{150°C - 55°C}{9.5W} = 10°C/W$$

   From Equation 4.14.4 (neglect θ_{LS})

   $\theta_{SA} = 10°C - 6°C/W = 4°C/W$

3. $T_S = 150°C - 6°C/W(9.5W) = 93°C$

4. $X = 0.0625''$ from initial conditions

   $E = 0.9$ from Table 4.14.2.

   Select $H = 3.5''$ for first trial (experience will simplify this step).

5. From Equation 4.14.6

   $$h_c = 2.21 \times 10^{-3} \frac{93 - 60}{3.5}^{1/4}$$

   $= 3.87 \times 10^{-3}$ W/°C in^2

   From Equation 4.14.7

   $$h_r = 1.47 \times 10^{-10} \times 0.9 \frac{93 + 60 + 273}{2}^3$$

   $= 5.6 \times 10^{-3}$ W/°C in^2

   $h = h_r + h_c = 9.47 \times 10^{-3}$ W/°C in^2

6. From h and f_{in} thickness use Figure 4.14.6 to find α (line a)

   $\alpha = 0.24$

7. From Figure 4.14.5

   $B = 1.91$ inches

8. From Figure 4.14.6 (line b)

   $\eta = 0.85$

9. From Equation 4.14.5

   $$\theta_{SA} = \frac{10^3}{2 \times 12.25 \times 0.85 \times 9.46} = 5.1°C/W$$

   Since the required heatsink thermal resistance is 4°C/W a larger f_{in} will be needed. A 4.25" square will increase the area by about 40% and a new calculation is made.

5'. $h_c = 3.7 \times 10^{-3}$ W/°C in^2

$h_r = 5.6 \times 10^{-3}$ W/°C in^2

$h = 9.3 \times 10^{-3}$ W/°C in^2

6'. $\alpha = 0.24$

7'. $B = 2.4$

8'. $\eta = 0.73$

9'. $\theta_{SA} = 4.08°C/W$ which is satisfactory.

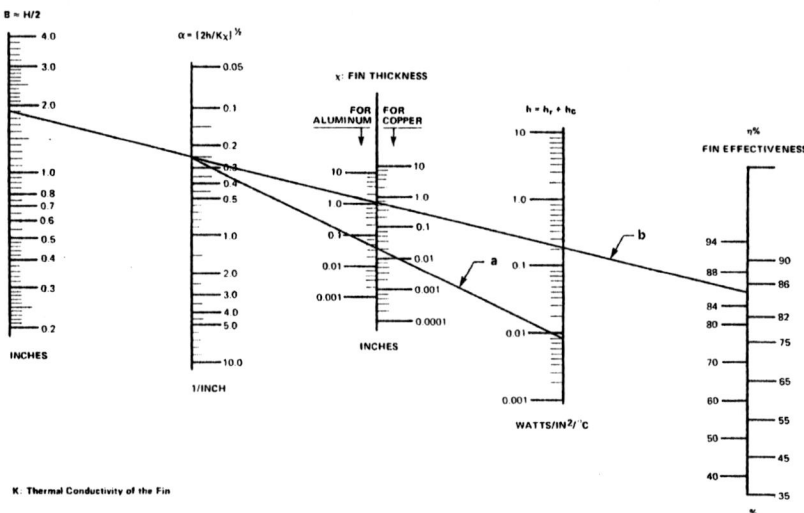

FIGURE 4.14.6 Fin Effectiveness Nomogram for Symmetrical, Flat, Uniformly-Thick, Vertically Mounted Fins

Although the above design procedure will specify the dimensions of the required heatsink, any design should be thoroughly tested under actual operating conditions to ensure that the maximum device case temperature does not exceed the rating for worst case thermal and load conditions.

4.14.7 Heatsinking with PC Board Foil

National Semiconductor's use of copper leadframes in packaging power ICs, where the center three pins on either side of the device are used for heatsinking, allows for economical heat sinks via the copper foil that exists on the printed circuit board. Adequate heatsinking may be obtained for many designs from single-sided boards constructed with 2 oz. copper. Other, more stringent, designs may require two-sided boards, where the top side is used entirely for heatsinking. Figure 4.14.7 allows easy design of PC board heat sinks once the desired thermal resistance has been calculated from Section 4.14.5.

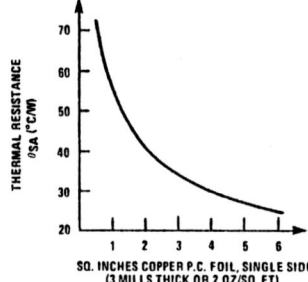

FIGURE 4.14.7 Thermal Resistance vs. Square Inches of Copper Foil

5.0 Floobydust

5.1 BIAMPLIFICATION

The most common method of amplifying the output of a preamplifier into the large signal required to drive a speaker system is with one large wideband amplifier having a flat frequency response over the entire audio band. An alternate method is to employ two amplifiers, or biamplification, where each amplifier is committed to amplifying only one part of the frequency spectrum. Biamping requires splitting up the audio band into two sections and routing these signals to each amplifier. This process is accomplished by using an active crossover network as discussed in the next section.

The most common application of biamping is found in conjunction with speaker systems. Due to the difficulty of manufacturing a single speaker capable of reproducing the entire audio band, multiple speakers are used, where each speaker is designed only to reproduce one section of frequencies. In conventional systems using one power amplifier the separation of the audio signal is done by passive high and low pass filters located within the speaker enclosure as diagrammed in Figure 5.1.1. These filters must be capable of processing high power signals and are therefore troublesome to design, requiring large inductors and capacitors.

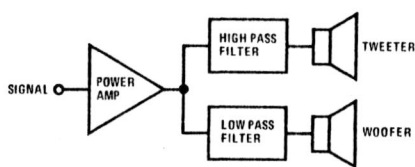

FIGURE 5.1.1 Passive Crossover, Single Amp System

Biamping with active crossover networks (Figure 5.1.2) allows a more flexible and easier design. It also sounds better. Listening tests demonstrate that biamped systems have audibly lower distortion.[4] This is due chiefly to two effects. The first results from the consequence of bass transient clipping. Low frequency signals tend to have much higher transient amplitudes than do high frequencies, so amplifier overloading normally occurs for bass signals. By separating the spectrum one immediately cleans up half of it and greatly improves the other half, in that the low frequency speaker will not allow high frequency components generated by transient clipping of the bass amplifier to pass, resulting in cleaner sound. Second is a high frequency masking effect, where the low level high frequency distortion components of a clipped low frequency signal are "covered up" (i.e., masked) by high level undistorted high frequencies. The final advantage of biamping is allowing the use of smaller power amplifiers to achieve the same sound pressure levels.

5.2 ACTIVE CROSSOVER NETWORKS

An active crossover network is a system of active filters (usually two) used to divide the audio frequency band into separate sections for individual signal processing by biamped systems. Active crossovers are audibly desirable because they give better speaker damping and improved transient response, and minimize midrange modulation distortion.

5.2.1 Filter Choice

The choice of filter type is based upon the need for good transient and frequency response. Bessel filters offer excellent phase and transient response but suffer from frequency response change in the crossover region, being too slow for easy speaker reproduction. Chebyshev filters have excellent frequency division but possess unacceptable instabilities in their transient response. Butterworth characteristics fall between Bessel and Chebyshev and offer the best compromise for active crossover design.

5.2.2 Number of Poles (Filter Order)

Intuitively it is reasonable that if the audio spectrum is split into two sections, their sum should exactly equal the original signal, i.e., without change in phase or magnitude (vector sum must equal unity). This is known as a constant voltage design. Also it is reasonable to want the same power delivered to each of the drivers (speakers). This is known as constant power design. What is required, therefore, is a filter that exhibits constant voltage and constant power. Having decided upon a Butterworth filter, it remains to

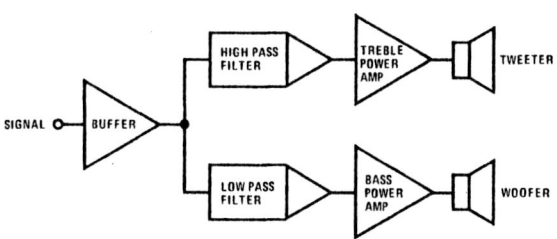

FIGURE 5.1.2 Active Crossover, Biamp System

determine an optimum order of the filter (the number of poles found in its transfer function) satisfying constant voltage and constant power.

Both active and passive realizations of a Butterworth filter have identical transfer functions, so a good place to start is with conventional passive crossover networks. Passive crossovers exhibit a single pole (1st order) response and have a transfer function given by Equations (5.2.1) and (5.2.2) (normalized to $\omega_0 = 1$).

$$T_L(S) = \frac{1}{S+1} \quad (5.2.1)$$

$$T_H(S) = \frac{S}{S+1} \quad (5.2.2)$$

where $T_L(S)$ equals low pass transfer function and $T_H(S)$ equals high pass transfer function. This filter exhibits constant voltage (hence, constant power) as follows:

require $T_L(S) + T_H(S) = 1 \quad (5.2.3)$

Inspection of Equations (5.2.1) and (5.2.2) shows this to be true.

The problem with a single pole system is that the rolloff beyond the crossover point is only −6dB/octave and requires the speakers to operate linearly for two additional octaves if distortion is to be avoided.[6]

The 2nd order system exhibits transfer functions:

$$T_L(S) = \frac{1}{S^2 + \sqrt{2}S + 1} \quad (5.2.4)$$

$$T_H(S) = \frac{S^2}{S^2 + \sqrt{2}S + 1} \quad (5.2.5)$$

These transfer functions exhibit constant power but not constant voltage. This is demonstrated by applying Equation (5.2.3), yielding:

$$T_L(S) + T_H(S) = \frac{S^2 + 1}{S^2 + \sqrt{2}S + 1} \quad (5.2.6)$$

At crossover, $S = -j\omega_0 = -j$ (since $\omega_0 = 1$); substitution into Equation (5.2.6) equals zero. This means that at the crossover frequency there exists a "hole," or a frequency that is not reproduced by either speaker. Ashley[1] demonstrated that this hole is audible. A commonly seen solution to this problem is to invert the polarity of one speaker in the system. Mathematically this changes the sign of the transfer function and effectively subtracts the two terms rather than adds them. This does eliminate the hole, but it creates a new problem of severe phase shifting at the crossover point which Ashley also demonstrated to be audible, making consideration of 3rd order Butterworth filters necessary.

The transfer functions for 3 pole Butterworth filters are given as Equations (5.2.7) and (5.2.8).

$$T_L(S) = \frac{1}{S^3 + 2S^2 + 2S + 1} \quad (5.2.7)$$

$$T_H(S) = \frac{S^3}{S^3 + 2S^2 + 2S + 1} \quad (5.2.8)$$

Applying Equation (5.2.3) yields:

$$T_L(S) + T_H(S) = \frac{S^3 + 1}{S^3 + 2S^2 + 2S + 1} \quad (5.2.9)$$

which at $S = -j\omega_0$ gives

$$T_L(-j\omega_0) + T_H(-j\omega_0) = -1 \quad (5.2.10)$$

Equation (5.2.10) satisfies constant voltage and constant power with one nagging annoyance — the phase has been inverted. Examination of the phase characteristics of Equation (5.2.9) shows that there is a gradual phase shift from $0°$ to $-360°$ as the frequency is swept through the filter sections, being $-180°$ at ω_0. Is it audible? Ashley[2] demonstrated that the ear cannot detect this gradual phase shift when it is not accompanied by ripple in the magnitude characteristic. (It turns out that all odd ordered Butterworth filters exhibit this effect with increasing amounts of phase shift, e.g., 5th order gives 0 to $-720°$, etc.)

The conclusion is that the best compromise is to use a 3rd order Butterworth filter. It will exhibit maximally flat magnitude response, i.e., no peaking (which minimizes the work required by the speakers); it has sharp cutoff characteristics of −18dB/octave (which minimizes speakers being required to reproduce beyond the crossover point); and it has flat voltage and power frequency response with a gradual change in phase across the band.

5.2.3 Design Procedure for 3rd Order Butterworth Active Crossovers

Many circuit topologies are possible to yield a 3rd order Butterworth response. Out of these the infinite-gain, multiple-feedback approach offers the best tradeoffs in circuit complexity, component spread and sensitivities. Figure 5.2.1 shows the general admittance form for any 3rd order active filter. The general transfer function is given by Equation (5.2.11).

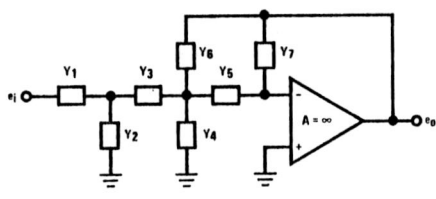

FIGURE 5.2.1 General Admittance Form for 3rd Order Filter

By substituting resistors and capacitors for the admittances per Figures 5.2.2 and 5.2.3, low and high pass active filters are created.

$$\frac{e_o}{e_i} = -\frac{Y_1 Y_3 Y_5}{(Y_5 Y_6 + Y_3 Y_7 + Y_4 Y_7 + Y_5 Y_7 + Y_6 Y_7)(Y_1 + Y_2 + Y_3) - Y_7 Y_3^2} \quad (5.2.11)$$

Low Pass:

$$\frac{e_{oL}}{e_i} = -\frac{\dfrac{1}{R_1 R_3 R_5 C_2 C_4 C_7}}{S^3 + \left(\dfrac{R_5 R_6 + R_3 R_6 + R_3 R_5}{R_3 R_5 R_6 C_4} + \dfrac{R_1 + R_3}{R_1 R_3 C_2}\right) S^2 + \left(\dfrac{1}{R_5 R_6 C_4 C_7} + \dfrac{R_5 R_6 + R_3 R_6 + R_3 R_5 + R_1 R_5 + R_1 R_6}{R_1 R_3 R_5 R_6 C_2 C_4}\right) S + \dfrac{R_1 + R_3}{R_1 R_3 R_5 R_6 C_2 C_4 C_7}}$$

(5.2.12)

High Pass:

$$\frac{e_{oH}}{e_i} = -\frac{\dfrac{C_1 C_3}{C_6 (C_1 + C_3)} S^3}{S^3 + \left(\dfrac{C_1(C_3 + C_5 + C_6) + C_3(C_5 + C_6)}{R_7 C_5 C_6 (C_1 + C_3)} + \dfrac{1}{(C_1 + C_3) R_2}\right) S^2 + \left(\dfrac{1}{C_5 C_6 R_4 R_7} + \dfrac{C_3 + C_5 + C_6}{C_5 C_6 (C_1 + C_3) R_2 R_7}\right) S + \dfrac{1}{C_5 C_6 (C_1 + C_3) R_2 R_4 R_7}}$$

(5.2.13)

Substitution of the appropriate admittances shown in Figures 5.2.2 and 5.2.3 into Equation 5.2.11 gives the general equation for a 3rd order low pass (Equation (5.2.12)) and for a 3rd order high pass (Equation (5.2.13)):

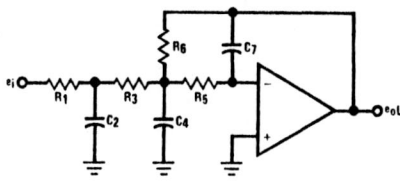

FIGURE 5.2.2 General 3rd Order Low Pass Active Filter

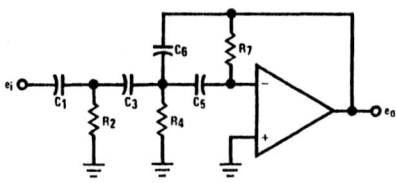

FIGURE 5.2.3 General 3rd Order High Pass Active Filter

Equation (5.2.12) is of form

$$\frac{K\omega_o^3}{S^3 + aS^2 + bS + \omega_o^3}$$

where: K = passband gain = 1

Letting $a = b = 2$ and normalizing $\omega_o^3 = 1$ gives the 3rd order Butterworth response of Equation (5.2.7).

Similarly, Equation (5.2.13) is of form

$$\frac{KS^3}{S^3 + aS^2 + bS + \omega_o^3}$$

and corresponds to Equation (5.2.8).

By letting $R_1 = R_3 = R_5 = R$ and $R_6 = 2R$ and equating coefficients between Equations (5.2.12) and (5.2.7), it is possible to solve for the capacitor values in terms of R. Doing so yields the relationships shown in Figure 5.2.4. For the high pass section, let $C_1 = C_3 = C_5 = C$ and $C_6 = C/2$ and equate coefficients to get the resistor values in terms of C. The high pass results also appear in Figure 5.2.4, which shows the complete 3rd order Butterworth crossover network.

Example 5.2.1

Design an active crossover network with −18dB/octave rolloff (3rd order), maximally flat (Butterworth) characteristics having a crossover frequency equal to 500Hz.

1. Let: R = 10k (1%)

2. Calculate C_2, C_4 and C_7 from Figure 5.2.4:

$$C_2 = \frac{2.4553}{(2\pi)(500)(10K)} = 7.82 \times 10^{-8}$$

Use $C_2 = 0.082\mu F$, 2%.

$$C_4 = \frac{2.1089}{(2\pi)(500)(10K)} = 6.71 \times 10^{-8}$$

Use $C_4 = 0.068\mu F$, 2%.

$$C_7 = \frac{0.1931}{(2\pi)(500)(10K)} = 6.51 \times 10^{-9}$$

Use $C_7 = 0.0056\mu F$, 2%.

3. Select C for high pass section to have same impedance level as R_{IN} for low pass, i.e., 20kΩ:

Let C = 0.015μF, 2%

C/2 = 0.0082μF, 2%.

4. Calculate R_2, R_4 and R_7 from Figure 5.2.4:

$$R_2 = \frac{0.4074}{(2\pi)(500)(1.592 \times 10^{-8})} = 8148$$

Use $R_2 = 8.06K$, 1%.

$$R_4 = \frac{0.4742}{(2\pi)(500)(1.592 \times 10^{-8})} = 9484$$

Use $R_4 = 9.53K$, 1%.

$$R_7 = \frac{5.1766}{(2\pi)(500)(1.592 \times 10^{-8})} = 103532$$

Use $R_7 = 102K$, 1%.

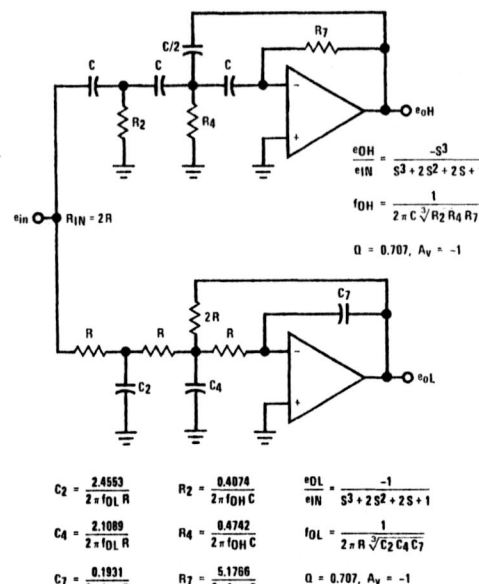

$$\frac{e_{OH}}{e_{IN}} = \frac{-s^3}{s^3 + 2s^2 + 2s + 1}$$

$$f_{OH} = \frac{1}{2\pi C \sqrt[3]{R_2 R_4 R_7}}$$

$$Q = 0.707, \quad A_v = -1$$

$$\frac{e_{OL}}{e_{IN}} = \frac{-1}{s^3 + 2s^2 + 2s + 1}$$

$$f_{OL} = \frac{1}{2\pi R \sqrt[3]{C_2 C_4 C_7}}$$

$$Q = 0.707, \quad A_v = -1$$

$$C_2 = \frac{2.4553}{2\pi f_{OL} R} \quad R_2 = \frac{0.4074}{2\pi f_{OH} C}$$

$$C_4 = \frac{2.1089}{2\pi f_{OL} R} \quad R_4 = \frac{0.4742}{2\pi f_{OH} C}$$

$$C_7 = \frac{0.1931}{2\pi f_{OL} R} \quad R_7 = \frac{5.1766}{2\pi f_{OH} C}$$

FIGURE 5.2.4 Complete 3rd Order Butterworth Crossover Network

The completed design is shown in Figure 5.2.5 using LF356 op amps for the active devices. LF356 devices were chosen for their very high input impedances, fast slew and extremely stable operation into capacitive loads. A buffer is used to drive the crossover network for two reasons: it guarantees low driving impedance which active filters require, and it gives another phase inversion so that the outputs are in phase with the inputs. Power supplies are ±15V, decoupled with 0.1 ceramic capacitors located close to the integrated circuits (not shown). Figure 5.2.6 gives the frequency response of Figure 5.2.5.

Figure 5.2.7 can be used to "look up" values for standard crossover frequencies of 100Hz to 5kHz.

5.2.4 Alternate Design for Active Crossovers

The example of Figure 5.2.5 is known as a symmetrical filter since both high and low pass sections are symmetrical about the crossover point (see Figure 5.2.6). An interesting alternate design is known as the asymmetrical filter (since the high and low pass sections are asymmetrical about the crossover point). This design is based upon the simple realization that if the output of a high pass filter is subtracted from the original signal then the result is a low pass.[3] Constant voltage is assured since the sum of low and high pass are always equal to unity (with no phase funnies). But, as always, there are tradeoffs and this time they are not obvious.

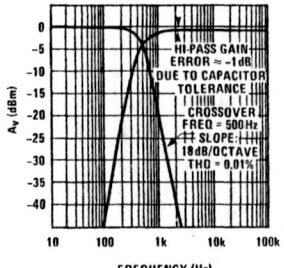

FIGURE 5.2.6 Active Crossover Frequency Response for Typical Example of Figure 5.2.5

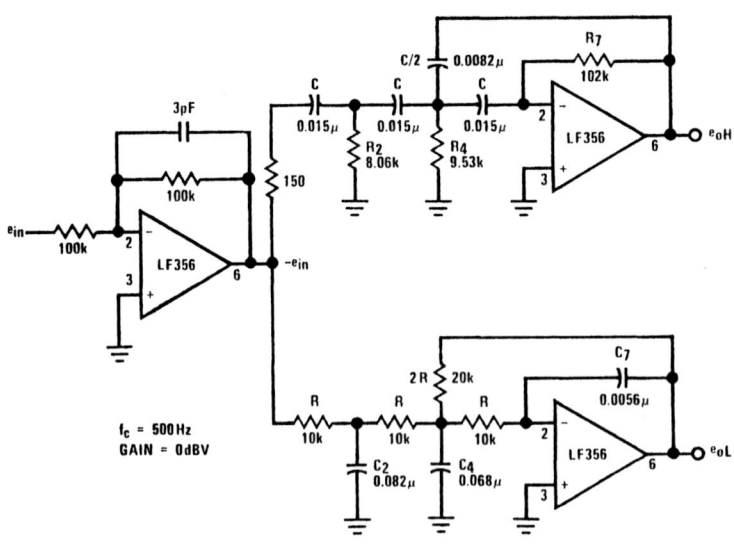

FIGURE 5.2.5 Typical Active Crossover Network Example

f_c Hz	C μF	R_2 Ω	R_4 Ω	R_7 Ω	C_2 μF	C_4 μF	C_7 μF
100	0.080	8148	9484	103532	0.391	0.336	0.0307
200	0.040				0.195	0.168	0.0154
300	0.027				0.130	0.112	0.0102
400	0.020				0.0977	0.0839	0.00768
500	0.016				0.0782	0.0671	0.00615
600	0.013				0.0651	0.0559	0.00512
700	0.011				0.0558	0.0479	0.00439
800	0.010				0.0488	0.0420	0.00384
900	0.0088				0.0434	0.0373	0.00341
1k	0.008				0.0391	0.0336	0.00307
2k	0.004				0.0195	0.0168	0.00154
3k	0.0027				0.0130	0.0112	0.00102
4k	0.002	↓	↓	↓	0.00977	0.00839	768 pF
5k	0.0016				0.00782	0.00671	615 pF

* Assumes R = 10k

FIGURE 5.2.7 Precomputed Values for Active Crossover Circuit Shown in Figure 5.2.4 (Use nearest available value.)

Referring back to Equation (5.2.8) for the transfer function of a 3rd order high pass and subtracting it from the original signal yields the following:

$$T_L(S) = 1 - T_H(S) \quad (5.2.14)$$

$$T_L(S) = 1 - \frac{S^3}{S^3 + 2S^2 + 2S + 1}$$

$$T_L(S) = \frac{2S^2 + 2S + 1}{S^3 + 2S^2 + 2S + 1} \quad (5.2.15)$$

Analysis of Equation (5.2.15) shows it has two zeros and three poles. The two zeros are in close proximity to two of the poles and near cancellation occurs. The net result is a low pass filter that exhibits only −6dB rolloff and rather severe peaking (~ +4dB) at the crossover point. For low frequency drivers with extended frequency response, this is an attractive design offering lower parts count, easy adjustment, no crossover hole and without gradual phase shift.

Figure 5.2.8 shows the circuit design for an asymmetrical filter, and Figure 5.2.9 gives its frequency response.

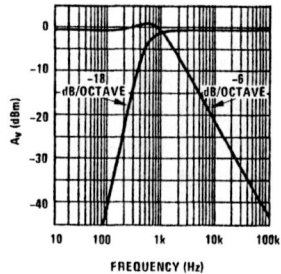

FIGURE 5.2.9 Frequency Response of Asymmetrical Filter Shown in Figure 5.2.8

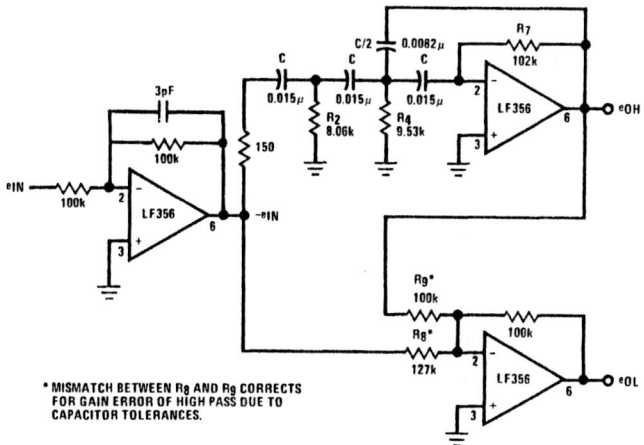

* MISMATCH BETWEEN R₈ AND R₉ CORRECTS FOR GAIN ERROR OF HIGH PASS DUE TO CAPACITOR TOLERANCES.

FIGURE 5.2.8 Asymmetrical 3rd Order Butterworth Active Crossover Network

5.2.5 Use of Crossover Networks and Biamping

Symbolically, Figure 5.2.5 can be represented as shown in Figure 5.2.10:

Figures 5.2.11-5.2.14 use Figure 5.2.10 to show several speaker systems employing active crossover networks and biamping.

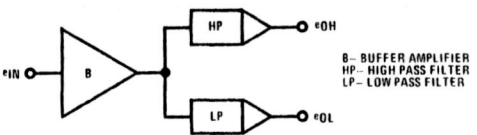

FIGURE 5.2.10 Symbolic Representation of Figure 5.2.5

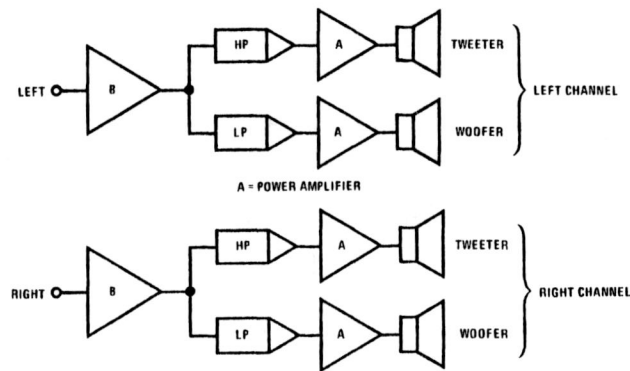

FIGURE 5.2.11 Stereo 2-Way System (Typical crossover points from 800 to 1600 Hz)

Cascading low pass (LP) and high pass (HP) active filters creates a bandpass and allows system triamping as follows:

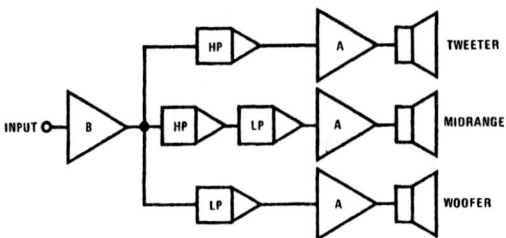

FIGURE 5.2.12 Single Channel 3-Way System
(Duplicate for Stereo)
(Typical crossover points: LP = 200 Hz, HP = 1200 Hz)

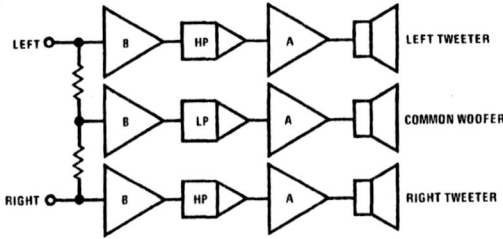

FIGURE 5.2.13 Common Woofer 2-Way Stereo System[5]
(Stereo-to-mono crossover point typically 150 Hz)

5-6

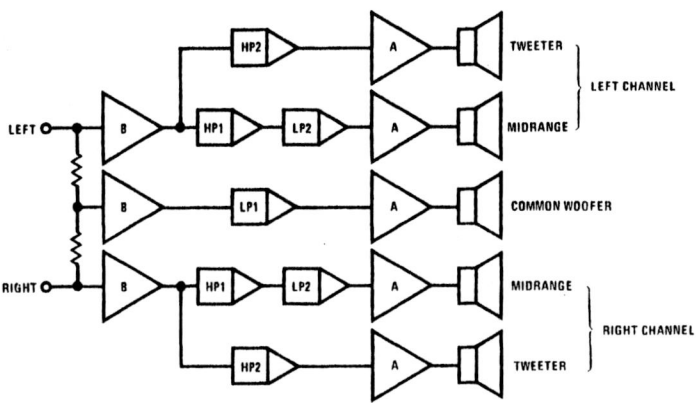

FIGURE 5.2.14 Common Woofer 3-Way Stereo System (Typically LP1 = HP1 = 150 Hz, LP2 = HP2 = 2500 Hz)

REFERENCES

1. Ashley, J. R., "On the Transient Response of Ideal Crossover Networks," *Jour. Aud. Eng. Soc.*, vol. 10, no. 3, July 1962, pp. 241-244.
2. Ashley, J. R. and Henne, L. M., "Operational Amplifier Implementation of Ideal Electronic Crossover Networks," *Jour. Aud. Eng. Soc.*, vol. 19, no. 1, January 1971, pp. 7-11.
3. Ashley, J. R. and Kaminsky, A. L., "Active and Passive Filters as Loudspeaker Crossover Networks," *Jour. Aud. Eng. Soc.*, vol. 19, no. 6, June 1971, pp. 494-501.
4. Lovda, J. M. and Muchow, S., "Bi-Amplification – Power vs. Program Material vs. Crossover Frequency," *AUDIO*, vol. 59, no. 9, September 1975, pp. 20-28.
5. Read, D. C., "Active Crossover Networks," *Wireless World*, vol. 80, no. 1467, November 1974, pp. 443-448.
6. Small, R. H., "Constant-Voltage Crossover Network Design," *Jour. Aud. Eng. Soc.*, vol. 19, no. 1, January 1971, pp. 12-19.

5.3 REVERBERATION

Reverberation is the name applied to the echo effect associated with a sound after it has stopped being generated. It is due to the reflection and re-reflection of the sound off the walls, floor and ceiling of a listening environment and under certain conditions will act enhance the sound. It is the main ingredient of concert hall ambient sound and accounts for the richness of "live" versus "canned" music. By using electro-mechanical devices, it is possible to add artificial reverberation to existing music systems and enhance their performance. The most common reverberation units use two precise springs that act as mechanical delay lines, each delaying the audio signal at slightly different rates. (Typical delay times are $\sim$ 30 milliseconds for one spring and $\sim$ 40 milliseconds for the other, with total decay times being around 2 seconds.) The electrical signal is applied to the input transducer where it is translated into a torsional force via two small cylindrical magnets attached to the springs. This "twisting" of one end of each spring slowly propogates along the length of the unit until it arrives at the other end, where similar magnets convert it back into an electrical signal. (Reflection also occurs, which creates the long decay time, relative to the delay time.)

5.3.1 Design Considerations for Driver and Recovery Amplifiers

Since the reverb driver is applying an electrical signal to a coil, its load is essentially inductive and as such has a rising impedance vs. frequency characteristic of +6dB/octave. Further, since the spring assembly operates best at a fixed value of ampere/turns (independent of frequency), it becomes desirable to drive the transducer with constant current. Constant current can be achieved in two ways: (1) by incorporating the input transducer as part of the negative feedback network, or (2) by creating a rising output voltage response as a function of frequency to follow the corresponding rise in output impedance. Method (1) precludes the use of grounded input transducers, which tend to be quieter and less susceptible to noise transients. (While grounded load, constant current sources exist, they require more parts to implement.) For this reason method (2) is preferred and will be used as a typical design example.

A high slew rate ($\sim$ 2 V/μs) amplifier should be used since the rising amplitude characteristic necessitates full output swing at the maximum frequency of interest (typical spring assemblies have a frequency response of 100 Hz-5 kHz), thereby allowing enough headroom to prevent transient clipping. It is also advisable to roll the amplifier off at high frequencies as a further aid in headroom. "Booming" at low frequencies is controlled by rolling off low frequencies below 100 Hz.

The requirements of the recovery amplifier are determined by the recovered signal. Typical voltage levels at the transducer output are in the range of 1-5 mV, therefore requiring a low noise, high gain preamp. Hum and noise need to be minimized by using shielding cable, mounting the reverb assembly and preamp away from the power supply transformer, and using good single point ground techniques to avoid ground loops. Equalization is not necessary if a constant current drive amplifier is used since the output voltage is constant with frequency.

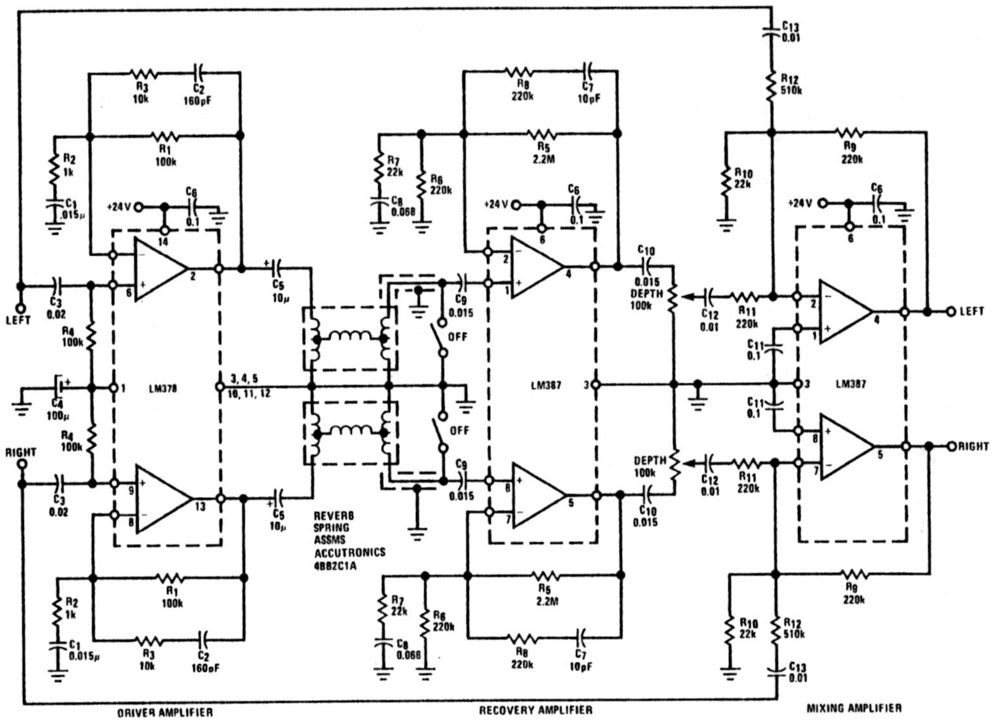

FIGURE 5.3.1 Stereo Reverb System

5.3.2 Stereo Reverb System

A complete stereo reverb system is shown in Figure 5.3.1, with its idealized "straightline" frequency response appearing as Figure 5.3.2.

The LM378 dual power amplifier is used as the spring driver because of its ability to deliver large currents into inductive loads. Some reverb assemblies have input transducer impedance as low as 8Ω and require drive currents of ~ 30mA. (There is a preference among certain users of reverbs to drive the inputs with as much as several hundred milliamps.) The recovery amplifier is easily done by using the LM387 low noise dual preamplifier which gives better than 75dB signal-to-noise performance at 1kHz (10mV recovered signal). Mixing of the delayed signal with the original is done with another LM387 used in an inverting summing configuration.

Figure 5.3.2 shows the desired frequency shaping for the driver and recovery amplifiers. The overall low frequency response is set by f_0 and occurs when the reactance of the coupling capacitors equals the input impedance of the next stage. For example, the driver stage low frequency corner f_0 is found from Equation (5.3.1).

$$f_0 = \frac{1}{2\pi R_4 C_3} \approx 80\,\text{Hz (as shown)} \quad (5.3.1)$$

The +6dB/octave response is achieved by proper selection of R_1, R_2 and C_1 as follows:

$$f_1 = \frac{1}{2\pi (R_1 + R_2) C_1} \approx 100\,\text{Hz (as shown)} \quad (5.3.2)$$

$$f_2 = \frac{1}{2\pi R_2 C_1} \approx 10\,\text{kHz (as shown)} \quad (5.3.3)$$

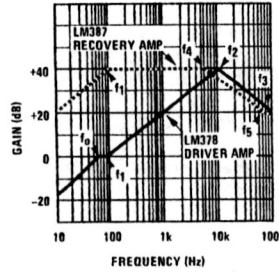

FIGURE 5.3.2 Straightline Frequency Response of Reverb Driver and Recovery Amplifiers

Ultimate gain is given by the ratio of R_2 and R_1:

$$A_o = 1 + \frac{R_2}{R_1} \text{ (gain beyond } f_2 \text{ corner)} \quad (5.3.4)$$

High frequency rolloff is accomplished with R_3 and C_2, beginning at f_2 and stopping at f_3.

$$f_2 = \frac{1}{2\pi R_1 C_2} \approx 10\,\text{kHz (as shown)} \quad (5.3.5)$$

$$f_3 = \frac{1}{2\pi R_3 C_2} \approx 100\,\text{kHz (as shown)} \quad (5.3.6)$$

Stopping high frequency rolloff at f_3 is necessary so the gain of the amplifier does not drop lower than 20dB, thereby preserving stability. (LM378 is not unity gain stable.) Resistors R_5 and R_6 are selected to bias the output of the LM387 at half-supply. (See Section 2.8.) Low frequency corner f_1 is fixed by R_7 and C_8:

$$f_1 = \frac{1}{2\pi R_7 C_8} \approx 100\,\text{Hz (as shown)} \quad (5.3.7)$$

High frequency rolloff is done similar to the LM377 by R_8 and C_7:

$$f_4 = \frac{1}{2\pi R_5 C_7} \approx 7\,\text{kHz (as shown)} \quad (5.3.8)$$

$$f_5 = \frac{1}{2\pi R_8 C_7} \approx 70\,\text{kHz (as shown)} \quad (5.3.9)$$

The same stability requirements hold for the LM387 as for the LM378.

Resistors R_9 and R_{10} are used to bias the LM387 summing amplifier. The output of the summer will be the scaled sum of the original signal and the delayed signal. Scaling factors are adjusted per Equation (5.3.10).

$$-V_{OUT} = \frac{R_9}{R_{12}} V_s + \frac{R_9}{R_{11}} V_D \quad (5.3.10)$$

where: V_s = original signal
V_D = delayed signal

As shown, the output is the sum of approximately one half of the original signal and all of the delayed signal.

5.3.3 Stereo Reverb Enhancement System

The system shown in Figure 5.3.3 can be used to synthesize a stereo effect from a monaural source such as AM radio or FM-mono broadcast, or it can be added to an existing stereo (or quad) system where it produces an exciting "opening up" spacial effect that is truly impressive.

The driver and recovery sections are as in Figure 5.3.1 with the exception that only one spring assembly is required. The second half of the LM387 recovery amplifier is used as an inverter and a new LM387 is added to mix both channels together. The outputs are inverted, scaled sums of the original and delayed signals such that the left output is composed of LEFT minus DELAY and the right output is composed of RIGHT plus DELAY.

When applied to mono source material, both inputs are tied together and the two outputs become INPUT minus DELAY and INPUT plus DELAY, respectively. If the outputs are to be used to drive speakers directly (as in an automotive application, or small home systems), then the LM387 may be replaced by one of the LM1877/378/379 dual 2W/4W/6W amplifier family wired as an inverting power summer per Figure 5.3.4.

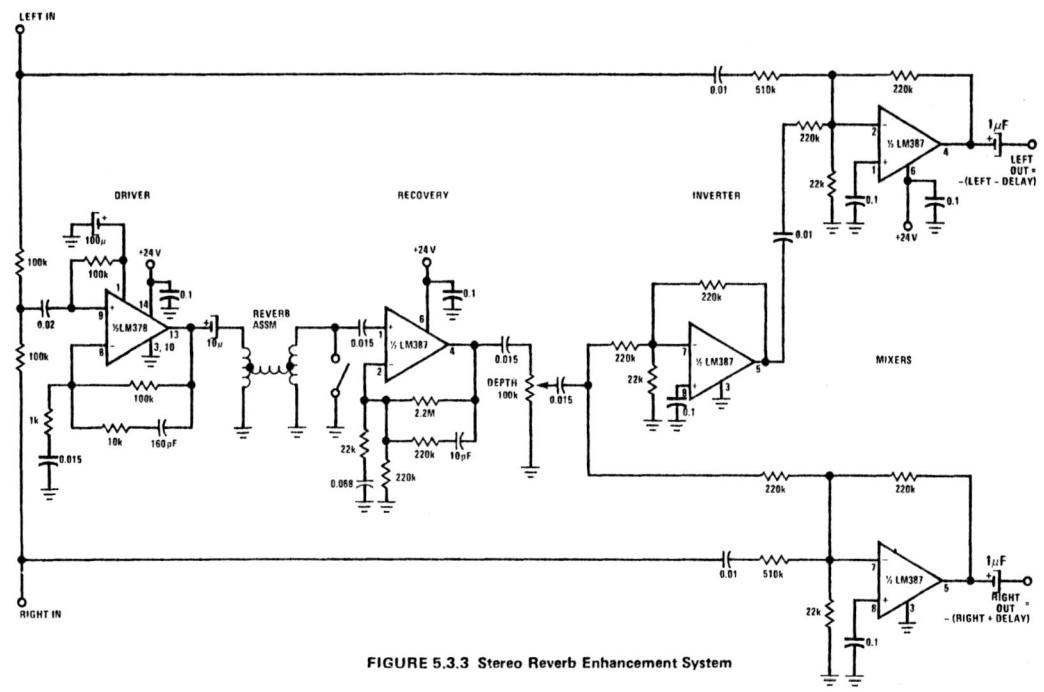

FIGURE 5.3.3 Stereo Reverb Enhancement System

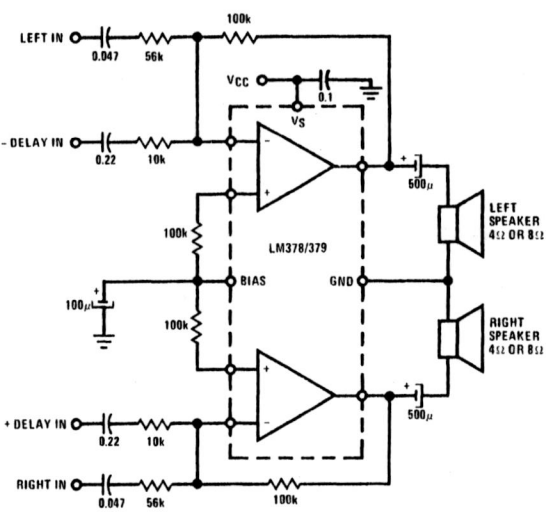

FIGURE 5.3.4 Alternate Output Stage for Driving Speakers Directly Using LM378/379 Family of Power Amplifiers

REFERENCES

1. "Application of Accutronic's Reverberation Devices," Technical paper available from Accutronics, Geneva, Ill.

2. "What Is Reverberation?," Technical paper available from Accutronics, Geneva, Ill.

5.4 PHASE SHIFTER

A popular musical instrument special effect circuit called a "phase shifter" can be designed with minimum parts by using two quad op amps, two quad JFET devices and one LM741 op amp (Figure 5.4.1). The sound effect produced is similar to a rotating speaker, or Doppler phase shift characteristic, giving a whirling, ethereal, "inside out" type of sound. The method used by recording studios is called "flanging," where two tape recorders playing the same material are summed together while varying the speed of one by pressing on the tape reel "flange." The time delay introduced will cause some signals to be summed out of phase and cancellation will occur. This phase cancellation produces the special effect and when viewed in the frequency domain is akin to a comb filter with variable rejection frequencies.[1] The phase shift stage used (Figure 5.4.1) is a standard configuration[2] displaying constant magnitude and a varying phase shift of 0-180° as a function of the resistance between the positive input and ground. Each stage shifts 90° at the frequency given by $1/(2\pi RC)$, where C is the positive input capacitor and R is the resistance to ground. Six phase shift stages are used, each spaced one octave apart, distributed about the center of the audio spectrum (160Hz-3.2kHz). JFETs are used to shift the frequency at which there is 90° delay by using them as voltage adjustable resistors. As shown, the resistance varies from 100Ω (FET full ON) to 10kΩ (FET full OFF), allowing a wide variation of frequency shift (relative to the 90° phase shift point). The gate voltage is adjusted from 5V to 8V (optimum for the AM9709CN), either manually (via foot operated rheostat) or automatically by the LM741 triangle wave generator. Rate is adjustable from as slow as 0.05Hz to a maximum of 5Hz. The output of the phase shift stages is proportionally summed back with the input in the output summing stage.

REFERENCES

1. Bartlett, B., "A Scientific Explanation of Phasing (Flanging)," *Jour. Aud. Eng. Soc.*, vol. 18, no. 6, December 1970, pp. 674-675.

2. Graeme, J. G., *Applications of Operational Amplifiers*, McGraw-Hill, New York, 1973, pp. 102-104.

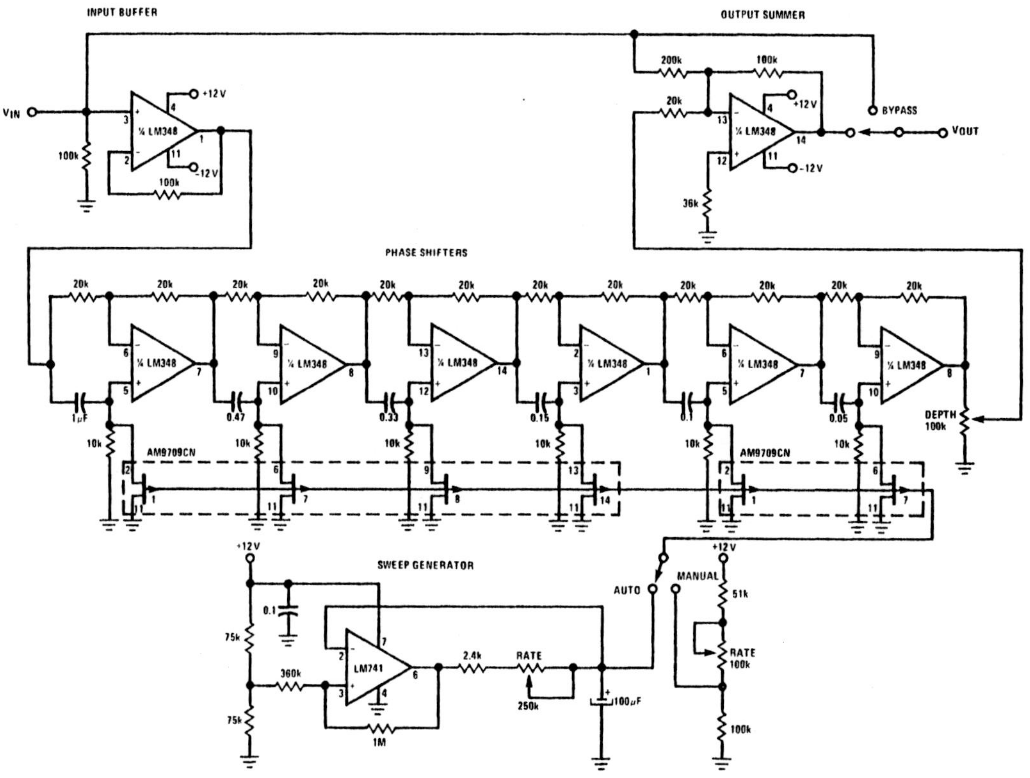

FIGURE 5.4.1 Phase Shifter

5.5 FUZZ

Two diodes in the feedback of a LM324 create the musical instrument effect known as "fuzz" (Figure 5.5.1). The diodes limit the output swing to ±0.7V by clipping the output waveform. The resultant square wave contains predominantly odd-ordered harmonics and sounds similar to a clarinet. The level at which clipping begins is controlled by the Fuzz Depth pot while the output level is determined by Fuzz Intensity.

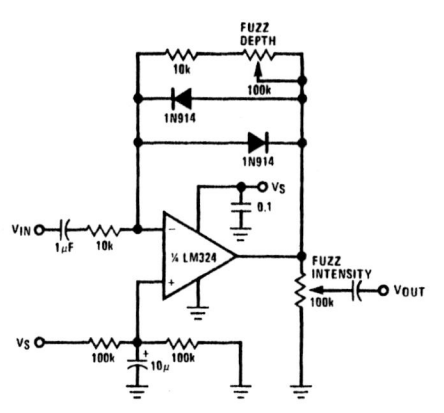

FIGURE 5.5.1 Fuzz Circuit

5.6 TREMOLO

Tremolo is amplitude modulation of the incoming signal by a low frequency oscillator. A phase shift oscillator (Figure 5.6.1) using the LM324 operates at an adjustable rate (5-10 Hz) set by the SPEED pot. A portion of the oscillator output is taken from the DEPTH pot and used to modulate the "ON" resistance of two 1N914 diodes operating as voltage controlled attenuators. Care must be taken to restrict the incoming signal level to less than $0.6 V_{p-p}$ or undesirable clipping will occur. (For signals greater than 25mV, THD will be high but is usually acceptable. Applications requiring low THD require the use of a light detecting resistor (LDR) or a voltage-controlled gain block. See Figure 4.8.9.)

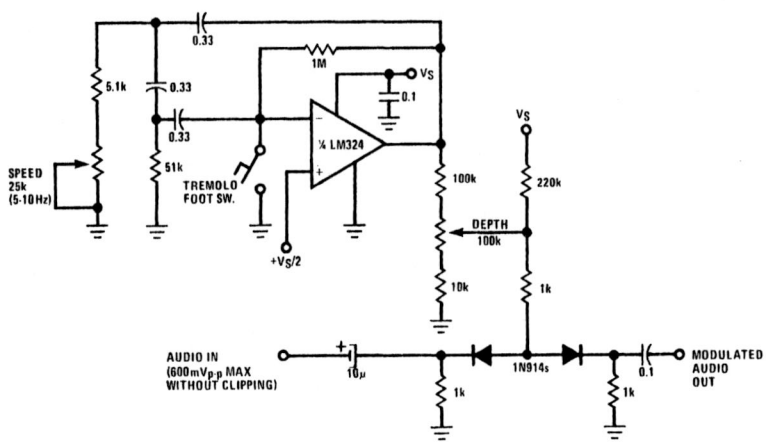

FIGURE 5.6.1 Tremolo Circuit

5.7 ACOUSTIC PICKUP PREAMP

Contact pickups designed for detection of vibrations produced by acoustic stringed musical instruments (e.g., guitar, violin, dulcimer, etc.) require preamplification for optimum performance. Figure 5.7.1 shows the LM387 configured as an acoustic pickup preamp, with Bass/Treble tone control, volume control, and switchable ±10dB gain select. The pickup used is the Ibanez "Bug," which is a flat response piezo-ceramic contact unit that is easy to use, inexpensive, and has excellent tone response. By using one half of the LM387 as the controllable gain stage and the other half as an active two-band tone control block, the complete circuit is done with only one 8-pin IC and requires very little space, allowing custom built-in designs where desired.

The tone control circuit is as described in Section 2.14.8. Addition of the midrange tone control (Section 2.14.9) is possible, making tone modification even more flexible.

Switchable gain control of ±10dB is achieved using a DPDT, center off, switch to add appropriate paralleling resistors around the main gain setting resistors R_8 and R_6. Resistor R_9 is capacitively coupled (C_{14}) so as not to disturb dc conditions set up by R_8 and R_{10}.

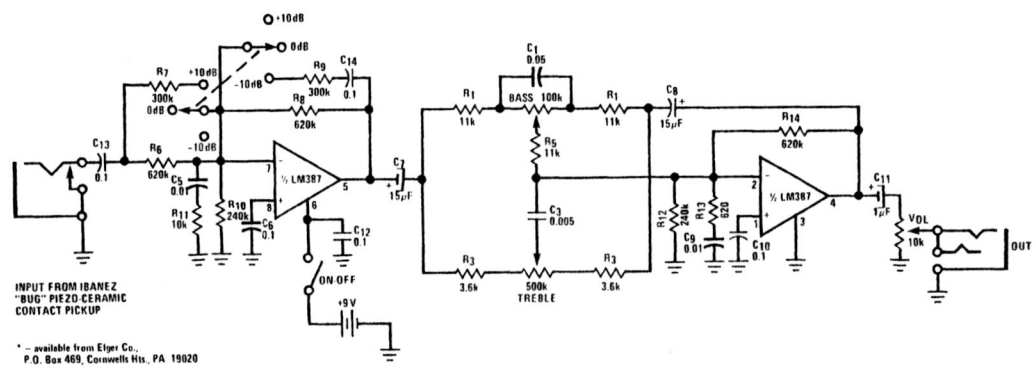

FIGURE 5.7.1 Acoustic Pickup Preamp

LM13600 – DUAL TRANSCONDUCTANCE AMPLIFIER

The LM13600 is similar to the more familiar op-amp with the major exception that the output is a current, the magnitude and polarity of which is defined by the product of the amplifier transconductance and the input voltage (i.e. $I_{OUT} = gm V_{IN}$). This output circuit is characterized as an infinite impedance current generator rather than the zero impedance voltage generator that represents the output of the conventional op-amp. The schematic for one half of the LM13600 is shown in Figure 5.8.1 and the circuit has a differential input stage with a tail current defined by the current injected into pin 1 (16). This current I_{ABC} controls the input stage transconductance, and the differential components of this current in Q_4 and Q_5 are mirrored into the output stage such that

$$I_{OUT} = \frac{V_{IN} I_{ABCq}}{2KT} = gm V_{IN} \qquad (5.8.1)$$

where $gm = 19.2 \, I_{ABC}$ at room temperature.

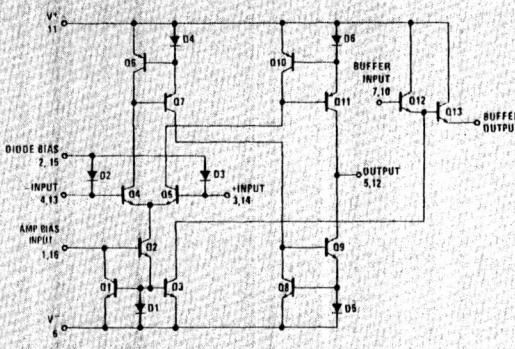

FIGURE 5.8.1 LM13600 Schematic

To use either section of the LM13600 as a low pass filter, we can configure it as shown in Figure 5.8.2

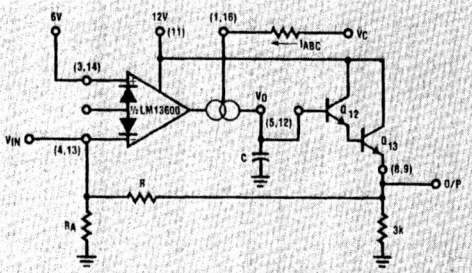

FIGURE 5.8.2 Voltage Controlled Low Pass Filter

By using the output Darlington buffer transistors Q_{12}, Q_{13}, the signal voltage V_{OUT} that appears at the capacitor C, is fed back to the amplifier inverting input, attenuated by the feedback resistors R_A and R, such that

$$V_{IN} = \frac{V_{OUT} R_A}{(R_A + R)} \qquad (5.8.2)$$

This input voltage will produce an output current I_{OUT} dependent on the control current magnitude I_{ABC}. From Equation (5.8.1)

$$I_{OUT} = gm V_{IN}$$
$$= gm V_{OUT} \frac{R_A}{(R_A + R)}$$

Therefore the amplifier output resistance R_O is given by

$$\frac{V_{OUT}}{I_{OUT}} \text{(Pin 5, 12)} = \frac{V_{OUT}}{gm V_{OUT}} \frac{(R_A + R)}{R_A}$$

i.e. $R_{OUT} = \frac{(R_A + R)}{R_A gm} \qquad (5.8.3)$

Since gm is controlled by I_{ABC}, the amplifier appears as a variable resistance R_{OUT} driving a capacitance C, which is a low pass filter configuration with a –3dB corner frequency given by

$$f_c = \frac{1}{2\pi R_{OUT} C} \qquad (5.8.4)$$

As R_{OUT} is changed by I_{ABC}, the corner frequency is changed by the same amount.

5.8 NON-COMPLEMENTARY NOISE REDUCTION

One of the many contributors to the success of cassette recorders in becoming part of component hi-fi systems has been the Dolby B-type noise reduction scheme. This is a complementary system — i.e. the original material is encoded in such a way before recording that the complementary decoding process reduces the noise that can be added by the tape recorder. A weighted 9dB S/N ratio improvement is obtained without affecting the fidelity of the source. Unfortunately the Dolby B system cannot improve the S/N ratio of the original material — it simply prevents further degradation by the recorder. So what can we do about old and favorite recordings made before Dolby circuits were widely available and whose value is marred by the ever present tape hiss? Also, for many of us, FM broadcasts still leave something to be desired in the attainment of low background noise levels. In either case the alternative is a non-complementary noise reduction system that operates to remove the noise *already present* in the source.

This can be done by restricting the system bandwidth (down to about 800Hz) in the absence of programme material. For a typical cassette source this will improve the S/N ratio by about 14dB (CCIR/ARM weighted). When programme material is present, with sufficient amplitude in the appropriate frequency range to mask the noise, the system bandwidth is automatically opened up. The degree to which the bandwidth can be opened depends largely on the masking effect of the programme material which, in turn, depends on the pitch and loudness of the noise. For this reason, the detector circuit used to determine the signal amplitude for which the audio bandwidth can be opened should include frequency response shaping networks. While several such audio processing systems have been built, Ref. 1, 2, 3, with varying degrees of complexity, the introduction of a dual transconductance amplifier, the LM13600 (see box), has made the implementation of automatically variable filters both simple and economical.

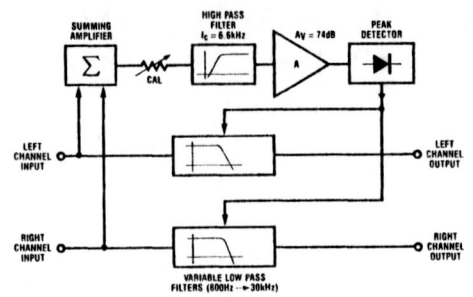

FIGURE 5.8.3 Noise Processing System Block Diagram

A block diagram containing the necessary functions is shown in Figure 5.8.3. This format is suitable for a stereo system with two unity gain current controlled filters operating with a common signal. A common control signal path, with frequency shaping, is used to prevent possible loss of stereo image which could occur if the bandwidth of the L and R channels were different.

A single LM13600 is used for both variable filters with the advantage that both channels will be inherently well matched and no set-up adjustments are required. When the control current to each filter from the peak detector is 4µA, the audio bandwidth is 800Hz, increasing to 20kHz, when the control current is 100µA. Since both filters are equivalent to single section RC low pass filters, they have a 6dB/octave roll-off slope above the cut-off frequency.

The response times of the filters for a bandwidth change are determined by the detector circuit time constants — in this case a 1msec attack time is used to obtain rapid opening of the bandwidth with programme transients, and a 50msec decay time to prevent the filters cutting off the natural reverberation following a music transient.

The control path sums the L & R inputs into a high pass filter. this filter has a corner frequency of 6.6kHz and a 12dB/octave roll off slope to ensure that proper weighting is given to the programme material in terms of its noise masking ability. A single adjustment for the system, a sensitivity control, also precedes the high pass filter and sets the summed input level such that the noise in the source (during a blank period in the programme) is just beginning to open the audio channel filters.

A practical circuit for the noise processor is shown in Figure 5.8.4 and is designed to be included in the tape monitor loop of a hi-fi system. The gain in the audio channel is unity, with an 88dB S/N ratio for a 775mV$_{rms}$ input level and a 30kHz system bandwidth.

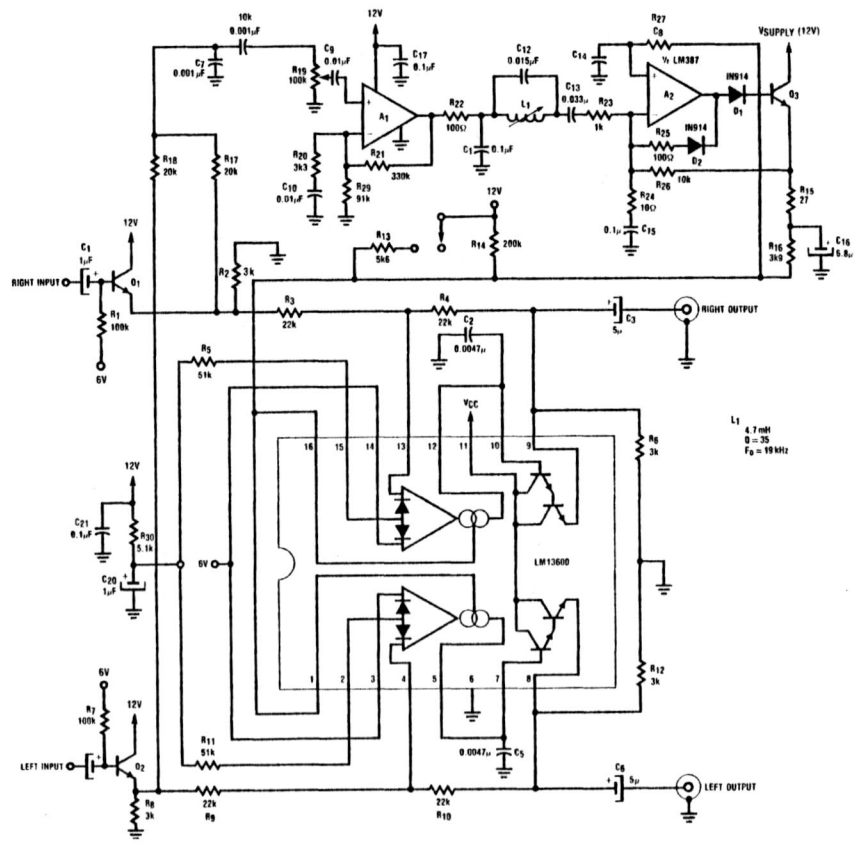

FIGURE 5.8.4 Stereo Noise Reduction Circuit

Two input buffer transistors Q_1 and Q_2 are used to provide a high input impedance and to allow the audio signals to be summed into the sensitivity potentiomenter R_{19} with minimum crosstalk. To ensure that the programme signals above the noise floor are capable of opening the audio filters, an LM387 is used to give two stages of gain before the detector capacitor C_{16}. These amplifier stages also perform necessary detector signal filtering so that masking is obtained. To obtain a -12dB/octave slope, two RC sections are used, $R_{23}C_{13}$ and $R_{20}C_{10}$. For a half power frequency f_c of 6.6kHz with two stages having identical corner frequencies f_o, we use Equation (5.8.5)

$$f_c = f_o \sqrt{10^{0.3/n} - 1} \quad (5.8.5)$$

For n = 2

$f_o = 0.643 f_c = 4.24\text{kHz}$

Therefore $R_{23}C_{13} = R_{20}C_{10} = 37.5\mu\text{S}$.
Above the filter corner frequency the midband gain is obtained with stage gains of 40dB for a total of 80dB.

$$|A_{V1}| = 1 + \frac{R_{21}}{R_{20}} \quad (5.8.6)$$

$$|A_{V2}| = \frac{R_{26}}{R_{24}} \quad (5.8.7)$$

The detector time constants are set by charging C_{16} through the resistor R_{15} and discharging C_{16} through the resistor R_{16} connected to the filter control pins (pins 1 & 16) of the LM13600. To bypass the noise reduction effect, a 5.6kΩ resistor R_{13} is switched into the control path forcing the filter to a fixed B-W in excess of 200kHz.

As shown, the circuit gives a 14dB S/N ratio improvement (weighted) with a distortion level of 0.13% with the rated input level (0.775mV$_{\text{rms}}$).

To display the action of the noise processor, the circuit in Figure 5.8.5(a) can be used. When connected to the detector capacitor C_{16}, the LM3915 will illuminate successive L.E.D.s for each 3dB increase in detected signal level (floating dot mode), providing a dynamic display of the instantaneous audio bandwidth. A suitable power supply, utilizing a 200mA filament transformer, is shown in Figure 5.8.5(b).

REFERENCES:

1. Burwen, Richard S., "A Dynamic Noise Filter For Mastering," *Audio*, June 1972, page 29.
2. Hellyer, H. W., "Noise Reduction Techniques," *Audio*, October 1972, page 18.
3. Scott, H. H., "Dynamic Noise Processor," *Electronics*, December 1947, page 96.

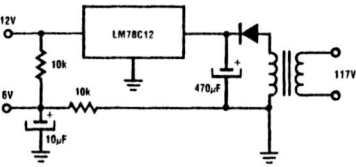

FIGURE 5.8.5(b) Power Supply

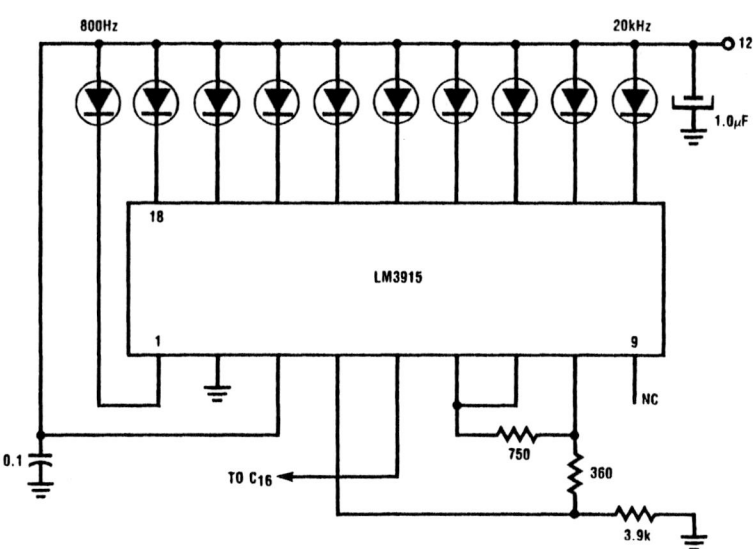

FIGURE 5.8.5(a) Audio Bandwidth Display

Section 6.0
Appendices

6.0 Appendices

A1.0 POWER SUPPLY DESIGN

A1.1 Introduction

One of the nebulous areas of power IC data sheets involves the interpretation of "absolute maximum ratings" as opposed to "operating conditions." The fact that parameters are specified at an operating voltage quite a few volts below the absolute maximum is not nearly so important in "garden variety" op amps as in power amps — because a key spec of any power amplifier is how much power it can deliver, a spec that is a strong function of the supply. Indeed P_O is approximately proportional to the square of supply voltage. Since many audio ICs are powered from a step down transformer off the 120 V_{AC} line, the "absolute maximum voltage" is an attempt to spec the highest value the supply might ever reach under power company overvoltages, transformer tolerances, etc. This spec says the IC will not die if taken to its "absolute maximum rating." Operating voltage, on the other hand, should be approximately what a nominal supply will sag under load at normal power company voltages. Some audio amplifiers are improperly specified at their "absolute maximum voltages" in order to give the illusion of large output power capability. However, since few customers regulate the supply voltage in their applications of audio ICs, this sort of "specsmanship" can only be termed deceptive.

A1.2 General

This section presents supply and filter design methods and aids for half-wave, full-wave center tap, and bridge rectifier power supplies. The treatment is sufficiently detailed to allow even those unfamiliar with power supply design to specify filters, rectifier diodes and transformers for single-phase supplies. A general treatment referring to Figure A1.1 is given, followed by a design example. No attempt is made to cover multiphase circuits or voltage multipliers. For maximum applicability a regulator is included, but may be omitted where required.

A1.3 Load Requirements

The voltage, current, and ripple requirements of the load must be fully described prior to filter and supply design. Actually, so far as the filter and supply are concerned, the load requirements are those at the regulator input. (See Figure A1.1.) Therefore, V_{IN} and I_{IN} become the governing conditions, where:

$I_{IN} = I_O + I_Q$, output current plus regulator quiescent current

$I_{IN(MAX)} \approx I_{O(MAX)}$, full-load operating current

$I_{IN(MIN)} \leqslant I_Q$, no-load or minimum operating current; could be near zero

$V_{IN(PK)} = V_M$, maximum permissible instantaneous no-load filter output voltage equal to peak value of transformer secondary voltage at highest design line voltage V_{PRI}; limited by absolute maximum regulator input voltage

$V_{IN} > V_O$, nominal DC voltage input to the regulator, usually 2 to 15V higher than V_O

$V_{IN(MIN)} \approx V_O + 2V$, minimum instantaneous full-load filter output voltage including ripple voltage; limited by minimum regulator input voltage to insure satisfactory regulation ($V_O + V_{dropout}$) or minimum regulator input voltage to allow regulator start-up under full load or upon removal of a load short circuit

r_f RMS ripple factor at filter output expressed as a percentage of V_{IN}; limited by maximum permissible ripple at load as modified by the ripple rejection characteristics of the regulator

A1.4 Filter Selection, Capacitor or Inductor-Input

For power supplies using voltage regulators, the filter will most often use capacitor input; therefore, emphasis will be placed upon that type of filter in following discussions. Notable differences between the two types of filters are that the capacitor input filter exhibits:

1. Higher DC output voltage
2. Poorer output voltage regulation with load variation
3. Higher peak to average diode forward currents

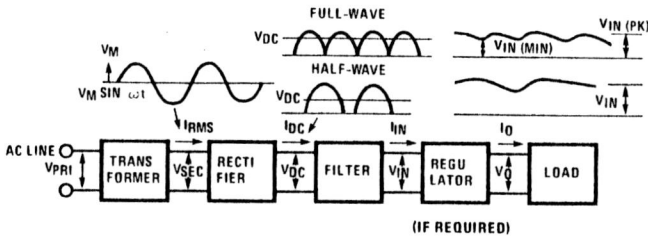

FIGURE A1.1 Power Supply Block Diagram, General Case

TABLE A1.1 Summary of Significant Rectifier Circuit Characteristics, Single Phase Circuits
Capacitive Data is for $\omega C R_L = 100$ & $R_S/R_L = 2\%$ (higher values)
and for $\omega C R_L = 10$ & $R_S/R_L = 10\%$ (lower values)

Rectifier Circuit Connection	Single Phase Half Wave			Single Phase Full Wave Center Tap			Single Phase Full Wave Bridge		
Voltage Waveshape to Load of Filter									
CHARACTERISTIC LOAD	R	L	C	R	L	C	R	L	C
Average Diode Current $I_{F(AVG)}/I_{O(DC)}$	1	1	1	0.5	0.5	0.5	0.5	0.5	0.5
Peak Diode Current $I_{FM}/I_{F(AVG)}$	3.14	-	8 / 5.2	3.14	2	10 / 6.2	3.14	2	10 / 6.2
Diode Current Form Factor, $F = I_{F(RMS)}/I_{F(AVG)}$	1.57	-	2.7 / 2	1.57	1.41	3 / 2.2	1.57	1.41	3 / 2.2
RMS Diode Current $I_{F(RMS)}/I_{O(DC)}$	1.57	-	2.7 / 2	0.785	0.707	1.35 / 1.1	0.785	0.707	1.35 / 1.1
RMS Input Voltage per Transformer Leg $V_{SEC}/V_{IN(DC)}$	2.22	2.22	0.707	1.11	1.11	0.707	1.11	1.11	0.707
Transformer Primary VA Rating VA/P_{DC}	3.49	-	-	1.23	1.11	-	1.23	1.11	-
Transformer Secondary VA Rating VA/P_{DC}	3.49	-	-	1.75	1.57	-	1.23	1.11	-
Total RMS Ripple %	121	-	-	48.2	-	-	48.2	-	-
Rectification Ratio (Conversion Efficiency) %	40.6	-	-	81.2	100	-	81.2	100	-

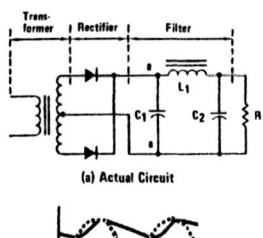

(a) Actual Circuit

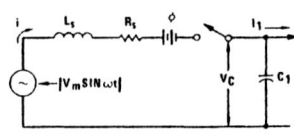

(b) Equivalent Circuit

(c) Voltage Across Input Capacitor C_1

(d) Current Through Diodes

FIGURE A1.2 Actual and Equivalent Circuits of Capacitor-Input Rectifier System, Together with Oscillograms of Voltage and Current for a Typical Operating Condition

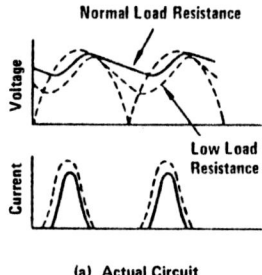
(a) Actual Circuit

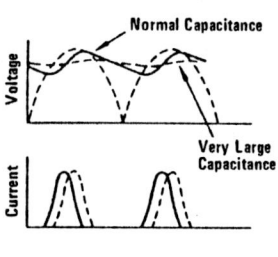
(b) Equivalent Circuit

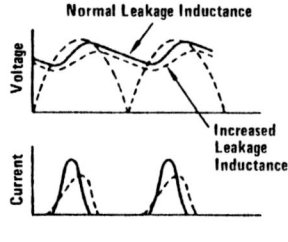
(c) Voltage Across Input Capacitor C_1

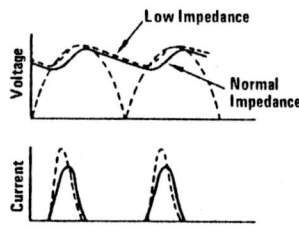
(d) Current Through Diodes

FIGURE A1.3 Effects of Circuit Constants and Operating Conditions on Behavior of Rectifier Operated with Capacitor-Input Filter

4. Lower diode PIV rating requirements
5. Very high diode surge current at turn-on
6. Higher peak to average transformer currents

The voltage regulator overcomes disadvantage (2) while semiconductor diodes of moderate price meet most of the peak and surge requirements except in supplies handling many amperes. Still, it may be necessary to balance increased diode and transformer cost against the alternative of a choke-input filter. In power supply designs employing voltage regulators, it is assumed that only moderate filter output regulation and ripple are required. Therefore, a capacitor input filter would exhibit peak currents considerably lower than indicated in the comparison of Table A1.1.

A1.5 Filter Design, Capacitor-Input

Figure A1.2 shows a full-wave, capacitor-input (filter) rectifier system with typical voltage and current waveforms. Note that ripple is inevitable as the capacitor discharges approximately linearly between voltage peaks. Figure A1.3 shows the effects on DC voltage, ripple, and peak diode current under varying conditions of load resistance, input capacitance, series diode and transformer resistance R_S, and transformer leakage inductance. The most practical design procedure for capacitor-input filters is to use the graphs of Figures A1.4-A1.7. Note, however, that these include the effects of diode dynamic resistance within R_S. Diode forward drop is not included, and must be subtracted from the transformer secondary voltage. A good rule of thumb is to subtract 0.7V from the transformer voltage and assume diode dynamic resistance is insignificant (0.02Ω at I_F = 1A, 0.26Ω at I_F = 100mA); ordinarily the transformer resistance will overshadow diode dynamic resistance.

Figures A1.4 and A1.5 show the relationship between peak AC input voltage and DC output voltage as a relation to load resistance R_L, series circuit resistance R_S, and filter input capacitance C. Figure A1.4 is for half-wave rectifiers and Figure A1.5 is for full-wave rectifiers. Note that the horizontal axis is labeled in units of $\omega C R_L$ where:

ω = AC line frequency in Hertz x 2π

C = value of input capacitor in Farads

$R_L = V_{IN}/I_{IN} \approx V_O/I_O$, equivalent load resistance in Ohms

R_S = total of diode dynamic resistance, transformer secondary resistance, reflected transformer primary resistance, and any added series surge limiting resistance

The major design trade-off encountered in designing capacitor-input filters is that between achieving good voltage regulation with low ripple and achieving low cost.

Referring to Figures A1.4 and A1.5:

1. Good regulation means $\omega C R_L \approx 10$.
2. Low ripple may mean $\omega C R_L > 40$.
3. High efficiency means $R_S/R_L < 0.02$.
4. Low cost usually means low surge currents and small C.
5. Good transformer utilization means low VA ratings, best with full-wave bridge FWB circuit, followed by full-wave center tap FWCT circuit.

In most cases, a minimum capacitance accomplishing a reasonable full-load to no-load regulation is preferable for low cost. To achieve this, use an intercept with the upper

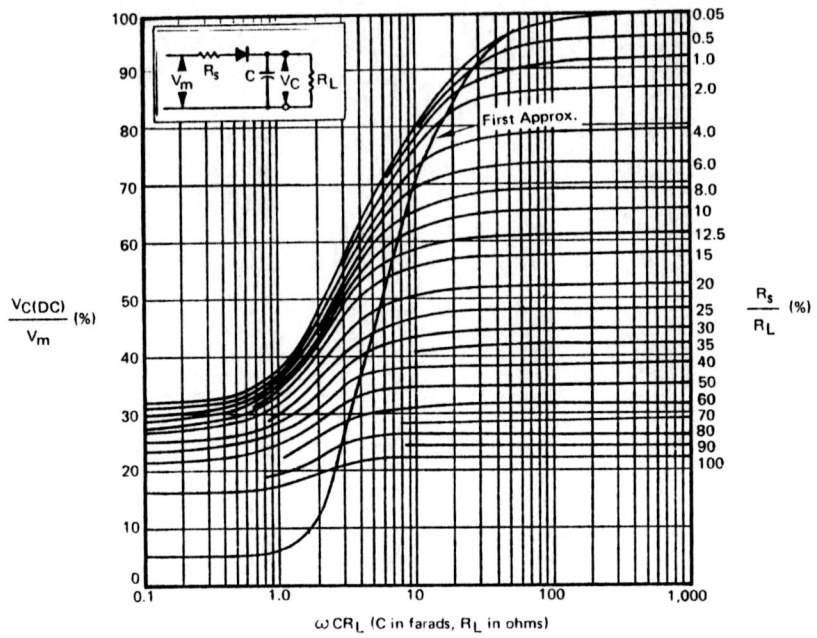

FIGURE A1.4 Relation of Applied Alternating Peak Voltage to Direct Output Voltage in Half-Wave Capacitor-Input Circuits (From O. H. Schade, Proc. IRE, vol. 31, p. 356, 1943.)

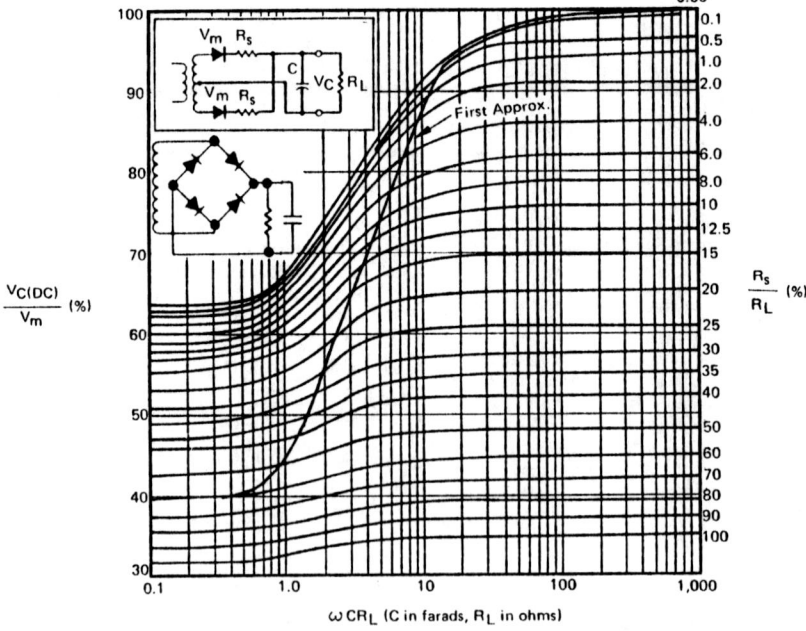

FIGURE A1.5 Relation of Applied Alternating Peak Voltage to Direct Output Voltage in Full-Wave Capacitor-Input Circuits (From O. H. Schade, Proc. IRE, vol. 31, p. 356, 1943.)

knee of the curves in Figures A1.4 and A1.5. Occasionally, a minimum value filter capacitor will not result in a lower cost system. For example, increasing the value of C may allow higher R_S/R_L to result in lower surge and RMS currents, thus allowing lower cost transformers and diodes. Be sure that capacitors used have adequate ripple current ratings.

Design procedure is as follows:

1. Assuming that V_O, I_O, ω, and load ripple factor r_f have been established and an appropriate voltage regulator has been selected, we know or can determine:

   $\omega = 2\pi f = 377$ rad/sec for 60 Hz line

   $r_{f(in)} = r_{f(out)} \times$ ripple reduction factor of selected regulator

   $V_{IN(PK)} \leq$ Max V_{IN} for the selected regulator; allow for highest line voltage likely to be encountered

   $V_{IN(MIN)} \approx V_O + 2V$; allow for lowest line voltage

   $V_{IN(DC)}^+ = V_{IN}$, usually 2-15 V above V_O; if chosen midway between $V_{IN(PK)}$ and $V_{IN(MIN)}$ or slightly below that point, will allow for greatest ripple voltage

   $I_{IN} \approx I_O$ for full load

   $I_{IN(MIN)} = I_Q$ for open load

   $R_L = V_{IN(DC)}/I_{IN}$

   $R_{L(MIN)} = V_{IN(MIN)}/I_{IN}$

2. Set $V_M \leq V_{IN(PK)}$ and calculate $V_{IN(DC)}/V_{IN}$. Enter the graph of Figure A1.4 or A1.5 at the calculated $V_{IN(DC)}/V_M$ to intercept one of the R_S/R_L = constant lines. Either estimate R_S at this time or intercept the curve marked "First Approximation."

3. Drop vertically from the intercept of Step (2) to the horizontal axis and read $\omega C R_L$. Calculate C, allowing for usual commercial tolerance on capacitors of +100, −50%.
   If $V_{IN(DC)}$ is midway between $V_{IN(PK)}$ and $V_{IN(MIN)}$, the supply can present maximum ripple to the regulator. A low value of C is then practical. If $V_{IN(DC)}$ is near $V_{IN(MIN)}$, regulator power dissipation is low and supply efficiency is high; however, ripple must be low, requiring large C.

4. Determine ripple factor r_f from Figure A1.6. Make certain that the ripple voltage does not drop instantaneous V_{IN} below $V_{IN(MIN)}$.

   The ripple factor could determine minimum required C if ripple is the limiting factor instead of voltage regulation. Again, allow for −50% tolerance on the capacitor.

   $V_{ripple(pk)} = \sqrt{2} \dfrac{r_f}{100} V_{IN(DC)}$

A1.6 Diode Specification

Find diode requirements as follows:

1. $I_{F(AVG)} = I_{IN(DC)}$ for half-wave rectification

   $= I_{IN(DC)}/2$ for full-wave rectification

2. Determine peak diode current ratio from Figure A1.7; remember to allow for highest operating line voltage and +100% capacitor tolerance.

   $I_{FM} = I_{FM}/I_{F(AVG)} \times I_{IN(DC)}$ for half-wave

   $= I_{FM}/I_{F(AVG)} \times I_{IN(DC)}/2$ for full-wave

3. Determine diode surge current requirement at turn-on of a fully discharged supply when connected at the peak of the highest expected AC line waveform. Surge current is:

   $$I_{SURGE} = \dfrac{V_M}{R_S + ESR}$$

   where ESR = effective series resistance of capacitor.

4. Find required diode PIV rating from Figure A1.8. Actually, required PIV may be considerably more than the value thus obtained due to noise spikes on the line. See Section A1.9 for details on transient protection. Remember that the PIV for the diodes in the FWB configuration are one half that of diodes as found in FWCT or HW rectifier circuits.

The diodes may now be selected from diode manufacturers' data sheets. If calculated surge current rating or peak current ratings are impractically high, return to Step A1.5(2) and choose a higher R_S/R_L or lower C. Conversely, it may be practical to choose lower R_S/R_L or higher C if diode current ratings can be practically increased without adverse effect on transformer cost; the result will be higher supply efficiency.

A1.7 Transformer Specification

A decision may have been made at Step A1.5(2) as to using half-wave or full-wave rectification. The half-wave circuit is often all that is required for low current regulated supplies; it is rarely used at currents over 1 A, as large capacitors and/or high surge currents are dictated. Transformer utilization is also quite low, meaning that higher VA rating is required of the transformer in HW circuits than in FW circuits. (See VA ratings of Table A1.1.)

Half-wave circuits are characterized by low $V_{IN(DC)}/V_M$ ratio, or very large C required (about 4 times that required for FW circuits, high ripple, high peak to average diode and transformer current ratios, and poor transformer utilization). They do, however, require only one diode.

Full-wave circuits are characterized by high $V_{IN(DC)}/V_M$ ratio, low C value required, low ripple, low peak to average diode and transformer current ratios, and good transformer utilization. They do require two diodes in the center-tap version, while the bridge configuration with its very high transformer utilization requires four diodes.

The information necessary to specify the transformer is:

1. Half-wave, full-wave CT or full-wave bridge circuit

2. Secondary V_{RMS} per transformer leg, $(V_M + 0.7^*)/\sqrt{2}$, from Section A1.5

3. Total equivalent secondary resistance including reflected primary resistance from Section A1.5

4. Peak, average, and RMS diode or winding currents from Sections A1.6(1) and -(2), and VA ratings.

*1.4 for full-wave bridge circuit.

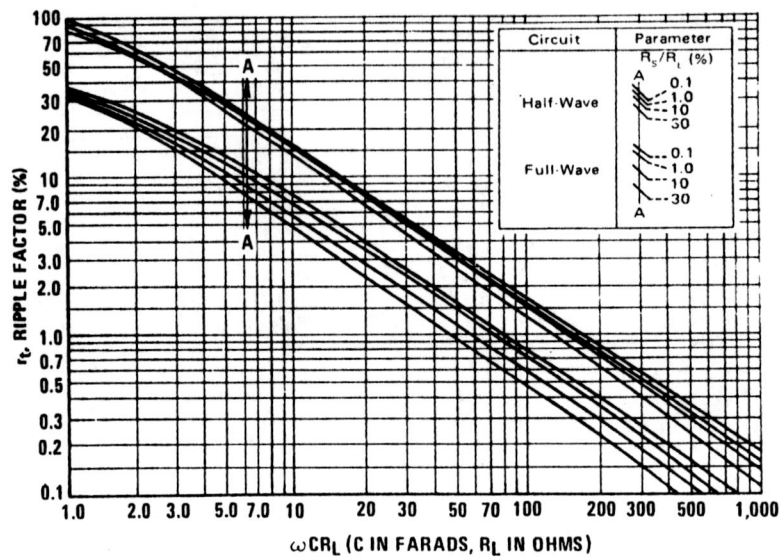

FIGURE A1.6 Root-Mean-Square Ripple Voltage for Capacitor-Input Circuits
(From O. H. Schade, Proc. IRE, vol. 31, p. 356, 1943.)

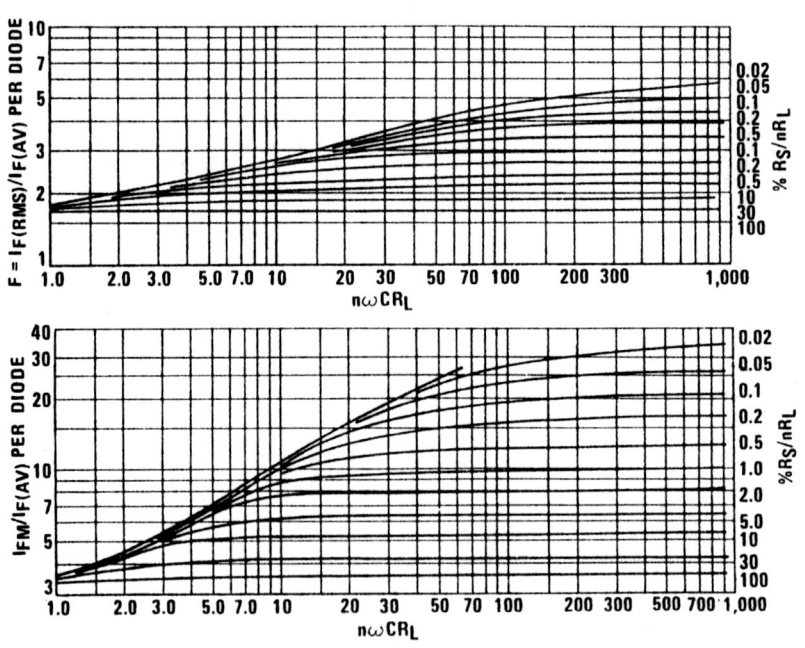

FIGURE A1.7 Relation of RMS and Peak-to-Average Diode Current in Capacitor-Input Circuits
(From O. H. Schade, Proc. IRE, vol. 31, p. 356, 1943.)

Transformer VA rating and secondary current ratings are determined as follows:

	FWB	FWCT	HW
$I_{RMS(SEC)}$ =	$I_{IN(DC)} F/\sqrt{2}$	$I_{IN(DC)} F/2$	$I_{IN(DC)} F$
VA_{SEC} =	$V_{RMS} I_{RMS}$	$2 V_{RMS} I_{RMS}$	$V_{RMS} I_{RMS}$
VA_{PRI} =	VA_{SEC}	$VA_{SEC}/\sqrt{2}$	VA_{SEC}

where: $F = I_{F(RMS)}/I_{IN(DC)}$
 = form factor from Figure A1.7
V_{RMS} = secondary RMS voltage per leg

A1.8 Additional Filter Sections

Occasionally, it is desirable to add an additional filter to reduce ripple. When this is done, an LC filter section is cascaded with the single C section filter already designed. If the inductor is of low resistance, the effect on output voltage is small. The additional ripple reduction may be determined from Figure A1.9.

A1.9 Transient Protection

Often the PIV rating of the rectifier diodes must be considerably greater than the minimum value determined from Figure A1.8. This is due to the likely presence of high-voltage transients on the line. These transients may be as high as 400 V on a 115 V line. The transients are a result of switching inductive loads on the power line. Such loads could be motors, transformers, or could even be caused by SCR lamp dimmers or switching-type voltage regulators, or the reverse recovery transients in rectifying diodes. As the transients appearing on the transformer primary are coupled to the secondary, the rectifier diodes may see rather high peak voltages. A simple method of protecting against these transients is to use diodes with very high PIV. However, high-current diodes with very high PIV ratings can be expensive.

There are several alternate methods of protecting the rectifier diodes. All rely on the existence of some line impedance, primary transformer resistance or secondary circuit resistance. See Figure A1.10 for the system circuit.

The several methods of transient protection rely on shunting the transient around the rectifier diodes to dissipate the transient energy in the series circuit resistance and the protective device. The usual protection methods are:

1. Series resistor at the primary with shunt capacitor across the primary winding — see Figure A1.10

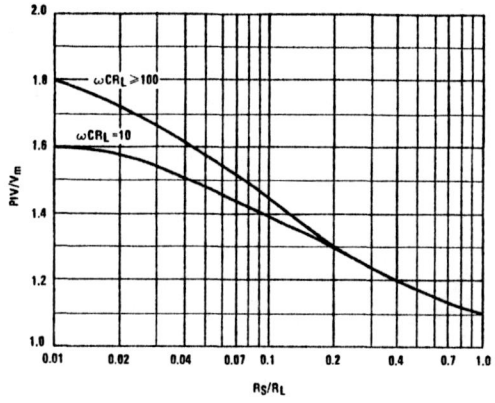

FIGURE A1.8 Ratio of Operating Peak Inverse Voltage to Peak Applied AC for Rectifiers Used in Capacitor-Input, Single-Phase, Filter Circuits

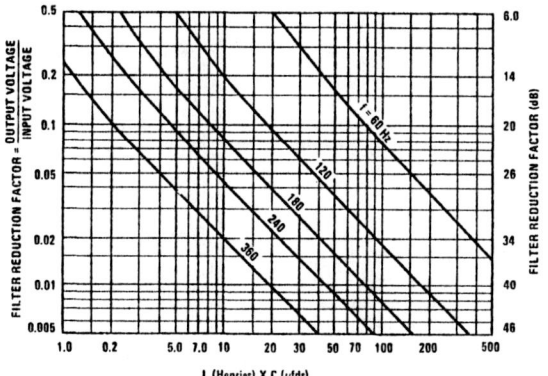

FIGURE A1.9 Reduction in Ripple Voltage Produced by a Single Section Inductance-Capacitance Filter at Various Ripple Frequencies

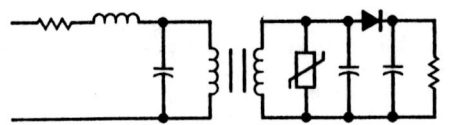

FIGURE A1.10 Transformer/Filter Circuit Showing Placement of Transient Protection Components

2. Series inductance at the primary, possibly with a shunt capacitor across the primary — see Figure A1.10
3. Shunt capacitor on the secondary — see Figure A1.10
4. Capacitor shunt on the rectifier diode — transient power is thus dissipated in circuit series resistance.

5. Surge suppression varactor shunt on the rectifier diode — this scheme is quite effective, but costly.

6. Dynamic clipper shunt on the rectifier diode — the clipper consists of an R, a C and a diode.

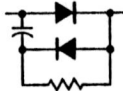

7. Zener shunt on the rectifier diode — may also include a series resistance.

8. Shunt varistor (e.g., GE MOVs) on the secondary — see Figure A1.10.

Of the several protective circuits:

- (1), (2), (3) and (4) are least costly, but are limited in their utility to incomplete protection.
- (4) is probably the circuit providing the most protection for the money and is all that may be required in low-current regulated supplies.
- (5), (6), (7) and (8) are most costly, but provide greatest protection. Their use is most worthwhile on high current supplies where high PIV ratings on high-current diodes is costly, or where very high transient voltages are encountered.

A1.10 Voltage Doublers

Occasionally, a voltage doubler is required to increase the voltage output from an existing transformer. Although the doubler circuits will provide increased output voltage, this is accomplished at the expense of an increased component count. Specifically, two filter capacitors are required. There are two basic types of doubler circuits as indicated in Figure A1.11. Figure A1.11a is the conventional full-wave doubler circuit wherein two capacitors connected in series are charged on alternate half cycles of the line waveform.

Figure A1.11b is a half-wave doubler circuit wherein C_2 is partially charged on one half cycle and then on the second half cycle the input voltage is added to provide a doubling effect. C_1 is normally considerably larger than C_2. The advantage of the half-wave circuit is that there is a common input and output terminal; disadvantages are high ripple, low I_O capability, and low V_{OUT}.

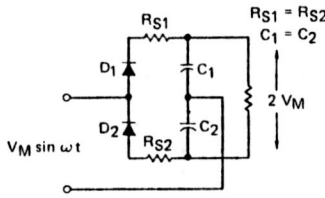

(a) Conventional Full-Wave Voltage Doubling Circuit

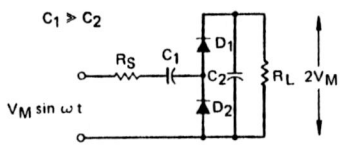

(b) Cascade (Half-Wave) Voltage Doubling Circuit

FIGURE A1.11 Voltage Doubler Circuits

These rectifying circuits, being capacitively loaded, exhibit high peak currents when energy is transferred to the capacitors. Filter design for the doubler circuits is similar to that of the conventional capacitor filter circuits. Figures A1.12, A1.13 and A1.14 provide the necessary design aids for full-wave voltage doubler circuits. They are used in the same way as Figures A1.5, A1.6 and A1.7.

A1.11 Design Example

Design a 5V, 3A regulated supply using an LM123K. Determine the filter values and transformer and diode specifications. Ripple should be less than $7\,mV_{RMS}$. Assume 60 dB ripple reduction from typical curves.

1. Establish operating conditions:

$\omega = 377$ rad/sec

$V_{IN(PK)} = 18V$ and 10% high line voltage; this allows some 2V headroom before reaching the 20V absolute maximum V_{IN} rating of the LM123K

$V_{IN(MIN)} = 7.5V$ at 10% low line voltage including effects of ripple voltage

$V_{IN(DC)} = 11V$ at nominal line voltage; chosen to exceed $V_{IN(MIN)}$ + peak ripple voltage

$V_{ripple(out)} \leq 7\,mV_{RMS}$

$V_{ripple(in)} \leq 7\,V_{RMS}$

$r_{f(in)} \leq 7V/11V = 63.5\%$

$I_{IN} = 3A$

$I_{IN(MIN)} = I_Q = 20\,mA$

$R_L = 11V/3A = 3.67\,\Omega$

$R_{L(MIN)} = 7.5V/3A = 2.5\,\Omega$

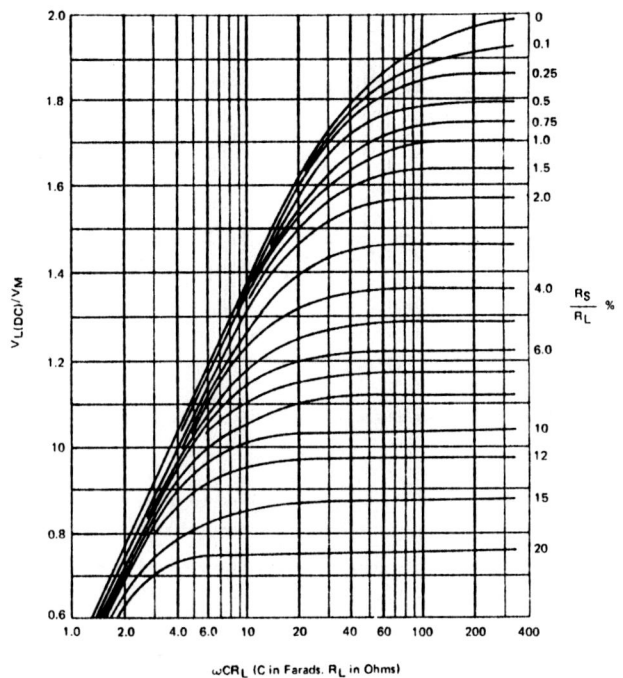

FIGURE A1.12 Output Voltage as a Function of Filter Constants for Full-Wave Voltage Doubler for Full-Wave Voltage Doubler

2. Set:

   V_M = 16.3 V nominal, which is 18 V − 10% line variation

   $V_{IN(DC)}/V_M$ = 11/16.3 = 0.67

   Assume full-wave bridge rectification because of the high current load. Enter the graph of Figure A1.5 at $V_{IN(DC)}/V_M$ = 0.67 to intercept the "First Approximation" curve.

3. Drop down to the horizontal axis to find $\omega C R_L$ = 3.33. Thus, $R_S/R_L \approx$ 13%, or R_S = 0.4 Ω is allowable.

   $$C = \frac{3.33}{3.67 \times 377} = 2400 \mu F$$

   (4800 μF allowing for −50% capacitor tolerance)

4. Ripple factor is 15% from Figure A1.6. Ripple is then

   $V_{ripple(pk)} = \sqrt{2} \times 0.15 \times 11$ = 2.33 V pk.

5. Checking for $V_{IN(MIN)}$:

   V_M = 16.3 V or, allowing for 10% low line voltage, 14.8 V

   $V_{IN(DC)}$ = 14.8 × 0.67 = 9.91 V

   Subtracting peak ripple, $V_{IN(MIN)}$ = 9.91 − 2.33 = 7.6 V which is within specifications

   In fact, all requirements have been met.

6. Diode specifications are:

   $$I_{F(AVG)} = \frac{I_{IN(DC)}}{2} = 1.5 A \text{ for FW rectifiers}$$

   I_{FM} = 8 × 1.5 A = 12 A, from figure A1.7, allowing C = 100% high, for commercial tolerances

   I_{SURGE} = 18 V/0.48 Ω = 37.5 A, worst case with 10% high line, neglecting capacitor ESR

   $I_{F(RMS)}$ = 2.1 × 1.5 A = 3.15 A, from Figure A1.7, allowing for 100% high tolerance on C

7. Transformer specifications are:

   $$V_{SEC(RMS)} = \frac{16.3 + 1.4}{\sqrt{2}} = 12.6 \text{ for FWB}$$

   (24 VCT for FWCT)

   R_S = 0.48 Ω including reflected primary resistance, but subtract 2 × diode resistance

   $I_{AVG} = I_{IN(DC)}$ = 3 A

   $$I_{SEC(RMS)} = \frac{I_{IN(DC)} \times F}{\sqrt{2}} = \frac{3 A \times 2.1}{1.414} = 4.45 A$$

   VA rating = 4.45 A × 12.6 = 56 VA, or 62 VA, allowing for 10% high line.

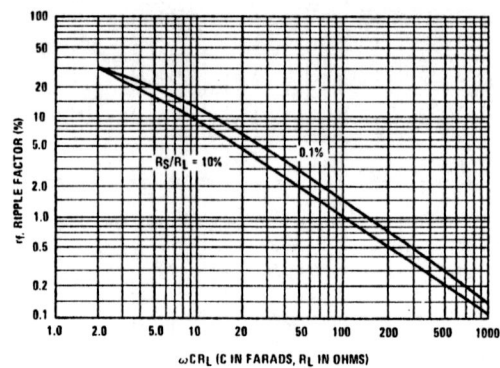

FIGURE A1.13 Ripple as a Function of Filter Constants for Full-Wave Voltage Doubler

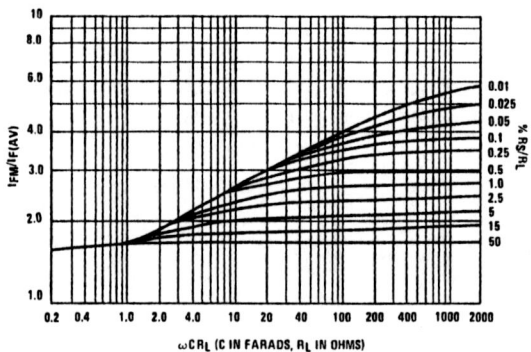

RMS Rectifier Current as a Function of Filter Constants for Full-Wave Voltage Doubler

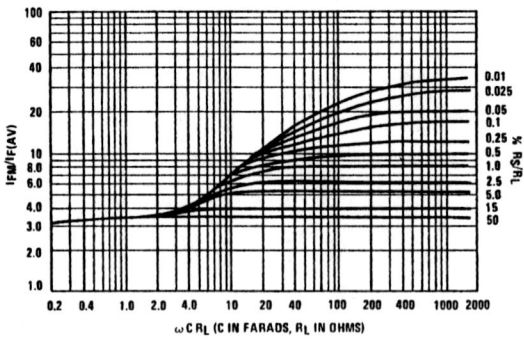

FIGURE A1.14 Relation of RMS to Peak and Average Diode Currents

A2.0 DECIBEL CONVERSION

A2.1 Definitions

The decibel (dB) is the unit for comparing relative levels of sound waves or of voltage or power signals in amplifiers.

The number of dB by which two power outputs P_1 and P_2 (in Watts) may differ is expressed by:

$$10 \log \frac{P_1}{P_2}$$

or, in terms of volts:

$$20 \log \frac{E_1}{E_2} \quad \text{(Figure A2.1)}$$

or, in current:

$$20 \log \frac{I_1}{I_2}$$

While power ratios are independent of source and load impedance values, voltage and current ratios in these formulas hold true only when the source and load impedances Z_1 and Z_2 are equal. In circuits where these impedances differ, voltage and current ratios are expressed by:

$$dB = 20 \log \frac{E_1 \sqrt{Z_2}}{E_2 \sqrt{Z_1}} \quad \text{or} \quad 20 \log \frac{I_1 \sqrt{Z_1}}{I_2 \sqrt{Z_2}}$$

Specific reference levels, i.e., the 0dB point, are denoted by a suffix letter following the abbreviation dB. Common suffixes and their definitions follow:

dBm — referenced to 1mW of power
dBV — referenced to 1V
dBW — referenced to 1W

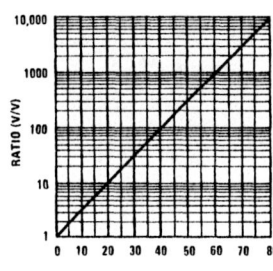

FIGURE A2.1 Gain Ratio to Decibel Conversion Graph
(Note: For negative values of decibels, i.e., gain attenuation, simply invert the ratio number. For example, −20dB = 1/10 V/V.)

A2.2 Relationship Between dB/Octave and dB/Decade

dB/Octave	dB/Decade
3	10
6	20
9	30
10	33.3
12	40
15	50
18	60

A3.0 WYE-DELTA TRANSFORMATION

Wye-delta transformation techniques (and the converse, delta-wye) are very powerful analytical tools for use in understanding feedback networks. Known also as tee-pi and pi-tee transformations, their equivalencies are given below.

A3.1 Wye-Delta (Tee-Pi)

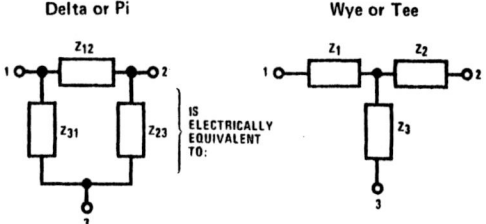

where:

$$Z_{12} = Z_1 + Z_2 + \frac{Z_1 Z_2}{Z_3} \quad (A3.1.1)$$

$$Z_{23} = Z_2 + Z_3 + \frac{Z_2 Z_3}{Z_1} \quad (A3.1.2)$$

$$Z_{31} = Z_3 + Z_1 + \frac{Z_3 Z_1}{Z_2} \quad (A3.1.3)$$

A3.2 Delta-Wye (Pi-Tee)

where:

$$Z_1 = \frac{Z_{12} Z_{31}}{Z_{12} + Z_{23} + Z_{31}} \quad (A3.2.1)$$

$$Z_2 = \frac{Z_{12} Z_{23}}{Z_{12} + Z_{23} + Z_{31}} \quad (A3.2.2)$$

$$Z_3 = \frac{Z_{31} Z_{23}}{Z_{12} + Z_{23} + Z_{31}} \quad (A3.2.3)$$

A4.0 STANDARD BUILDING BLOCK CIRCUITS

Definitions:

A_V = Closed Loop AC Gain
f_o = Low Frequency −3dB Corner
R_{in} = Input Impedance

General Comments:

Power supply connections omitted for clarity.
Split supplies assumed.
Single supply biasing per A4.9 or A4.10.

A4.1 Non-Inverting AC Amplifier

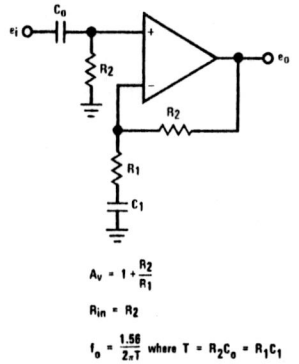

$A_V = 1 + \dfrac{R_2}{R_1}$

$R_{in} = R_2$

$f_o = \dfrac{1.56}{2\pi T}$ where $T = R_2 C_o = R_1 C_1$

A4.2 Inverting AC Amplifier

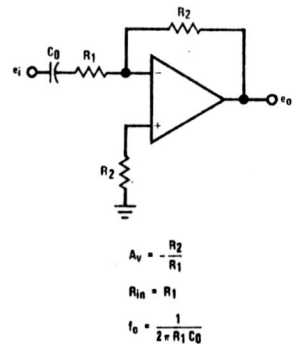

$A_V = -\dfrac{R_2}{R_1}$

$R_{in} = R_1$

$f_o = \dfrac{1}{2\pi R_1 C_0}$

A4.3 Inverting Summing Amplifier

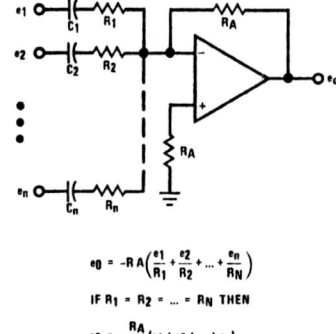

$e_0 = -R_A \left(\dfrac{e_1}{R_1} + \dfrac{e_2}{R_2} + ... + \dfrac{e_n}{R_N} \right)$

IF $R_1 = R_2 = ... = R_N$ THEN

$e_0 = -\dfrac{R_A}{R_1}(e_1 + e_2 + ... + e_n)$

A4.4 Non-Inverting Buffer

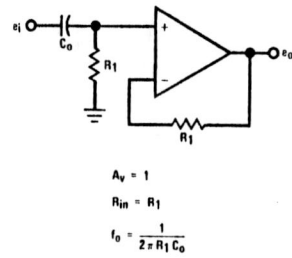

$A_V = 1$

$R_{in} = R_1$

$f_o = \dfrac{1}{2\pi R_1 C_0}$

A4.5 Inverting Buffer

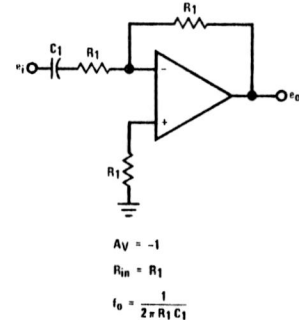

$A_V = -1$

$R_{in} = R_1$

$f_o = \dfrac{1}{2\pi R_1 C_1}$

A4.6 Difference Amplifier

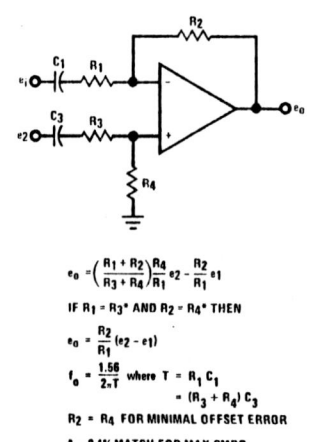

$e_0 = \left(\dfrac{R_1 + R_2}{R_3 + R_4} \right) \dfrac{R_4}{R_1} e_2 - \dfrac{R_2}{R_1} e_1$

IF $R_1 = R_3$* AND $R_2 = R_4$* THEN

$e_0 = \dfrac{R_2}{R_1}(e_2 - e_1)$

$f_o = \dfrac{1.56}{2\pi T}$ where $T = R_1 C_1$
$= (R_3 + R_4) C_3$

$R_2 = R_4$ FOR MINIMAL OFFSET ERROR

* — 0.1% MATCH FOR MAX CMRR

A4.7 Variable Gain AC Amplifier

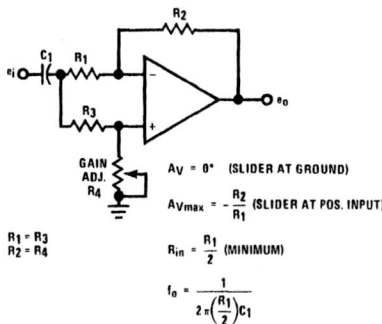

$A_V = 0$ (SLIDER AT GROUND)
$A_{Vmax} = -\frac{R_2}{R_1}$ (SLIDER AT POS. INPUT)

$R_1 = R_3$
$R_2 = R_4$

$R_{in} = \frac{R_1}{2}$ (MINIMUM)

$f_o = \frac{1}{2\pi \left(\frac{R_1}{2}\right) C_1}$

*LIMITED BY CMRR OF AMPLIFIER AND MATCH OF $R_1 = R_3$, $R_2 = R_4$, e.g., LF356 AND 0.1% MATCH EQUALS > 80 dB FOR A_{Vmax} = 20 dB.

A4.8 Switch Hitter (Polarity Switcher, or 4-Quadrant Gain Control)

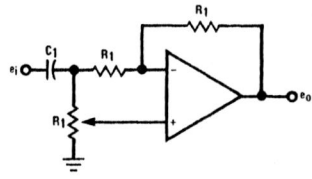

$A_v = +1$ (SLIDER AT C_1)
$A_v = 0^*$ (SLIDER MIDPOSITION)
$A_v = -1$ (SLIDER AT GROUND)

$R_{in} = \frac{R_1}{2}$ (MINIMUM)

$f_o = \frac{1}{2\pi \left(\frac{R_1}{2}\right) C_1}$

*WITHIN CMRR OF AMPLIFIER

A4.9 Single Supply Biasing of Non-Inverting AC Amplifier

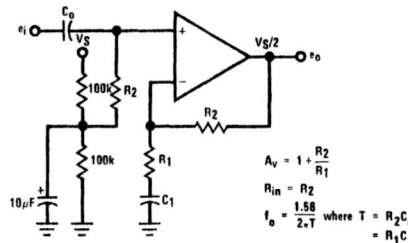

$A_v = 1 + \frac{R_2}{R_1}$
$R_{in} = R_2$
$f_o = \frac{1.58}{2\pi T}$ where $T = R_2 C_o$
$= R_1 C_1$

A4.10 Single Supply Biasing of Inverting AC Amplifier

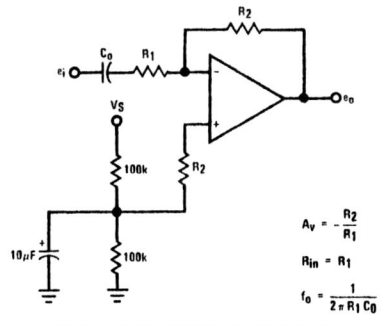

$A_v = -\frac{R_2}{R_1}$
$R_{in} = R_1$
$f_o = \frac{1}{2\pi R_1 C_o}$

A5.0 MAGNETIC PHONO CARTRIDGE NOISE ANALYSIS

A5.1 Introduction

Present methods of measuring signal-to-noise (S/N) ratios do not represent the true noise performance of phono preamps under real operating conditions. Noise measurements with the input shorted are only a measure of the preamp noise voltage, ignoring the two other noise sources: the preamp current noise and the noise of the phono cartridge.

Modern phono preamps have typical S/N ratios in the 70 dB range (below 2 mV @ 1 kHz), which corresponds to an input noise voltage of $0.64 \mu V$, which looks impressive but is quite meaningless. The noise of the cartridge[1] and input network is typically *greater* than the preamp noise voltage, ultimately limiting S/N ratios. This must be considered when specifying preamplifier noise performance. A method of analyzing the noise of complex networks will be presented and then used in an example problem.

A5.2 Review of Noise Basics

The noise of a passive network is thermal, generated by the real part of the complex impedance, as given by Nyquist's Relation:

$$\overline{V_n^2} = 4kT\,Re(Z)\,\Delta f \quad (A5.2.1)$$

where: $\overline{V_n^2}$ = mean square noise voltage
k = Boltzmann's constant (1.38×10^{-23} W-sec/°K)
T = absolute temperature (°K)
$Re(Z)$ = real part of complex impedance (Ω)
Δf = noise bandwidth (Hz)

The total noise voltage over a frequency band can be readily calculated if it is white noise (i.e., $Re(Z)$ is frequency independent). This is not the case with phono cartridges or most real world noise problems. Rapidly changing cartridge network impedance and the RIAA equalization of the preamplifier combine to complicate the issue. The total input noise in a non-ideal case can be calculated by breaking the noise spectrum into several small bands where the noise is nearly white and calculating the noise of each band. The total input noise is the RMS sum of the noise in each of the bands $N_1, N_2, \ldots, N_n$.

$$V_{noise} = (V_{N1}^2 + V_{N2}^2 + \ldots + V_{Nn}^2)^{\frac{1}{2}} \quad (A5.2.2)$$

This expression does not take into account gain variations of the preamp, which will also change the character of the noise at the preamp output. By reflecting the RIAA equalization to the preamp input and normalizing the gain to 0 dB at 1 kHz, the equalized cartridge noise may then be calculated.

$$V_{EQ} = (|A_1|^2 V_{N1}^2 + |A_2|^2 V_{N2}^2 + \ldots + |A_n|^2 V_{Nn}^2)^{\frac{1}{2}}$$

$$(A5.2.3)$$

where: V_{EQ} = equalized preamp input noise
$|A_n|$ = magnitude of the equalized gain at the center of each noise band (V/V)

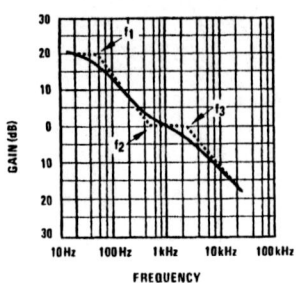

FIGURE A5.1 Normalized RIAA Gain

$R = R_A \| R_p$
$L = L_p$
$C = C_s + C_c$

A5.3 Cartridge Impedance

The simplified lumped model of a phono cartridge consists of a series inductance and resistance shunted by a small capacitor. Each cartridge has a recommended load consisting of a specified shunt resistance and capacitor. A model for the cartridge and preamp input network is shown in Figure A5.2.

The impedance relations for this network are:

$$Re(Z) = \frac{R X_L^2 X_C^2}{(R X_L - R X_C)^2 + X_L^2 X_C^2}$$ (A5.3.2)

$$|Z| = \frac{R X_L X_C}{[(R X_L - R X_C)^2 + X_L^2 X_C^2]^{1/2}}$$

A5.4 Example

Calculations of the RIAA equalized phono input noise are done using Equations (A5.2.1)-(A5.3.2). Center frequencies and frequency bands must be chosen: values of R_p, L_p, $Re(Z)$, $|Z|$ and noise calculated for each band, then summed for the total noise. Octave bandwidths starting at 25 Hz will be adequate for approximating the noise.

An ADC27 phono cartridge is used in this example, loaded with $C = 250$ pF and $R_A = 47$ kΩ, as specified by the manufacturer, with cartridge constants of $R_s = 1.13$ kΩ and $L_s = 0.75$ H. (C_c may be neglected.) Table A5.1 shows a summary of the calculations required for this example.

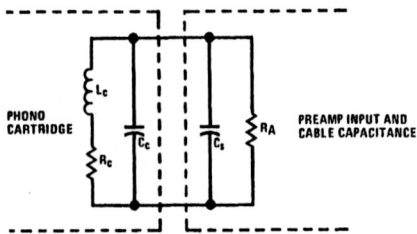

FIGURE A5.2 Phono Cartridge and Preamp Input Network

This seemingly simple circuit is quite formidable to analyze and needs further simplification. Through the use of Q equations,[2] a series L-R is transformed to a parallel L-R.

A5.5 Conclusions

The RIAA equalized noise of the ADC27 phono cartridge and preamp input network was $0.75 \mu V$ for the audio band. This is the limit for S/N ratios if the preamp was noiseless, but zero noise amplifiers do not exist. If the preamp noise voltage was $0.64 \mu V$ then the actual noise of the system is $0.99 \mu V$ ($[0.64^2 + 0.75^2]^{1/2} \mu V$) or 66 dB S/N ratio (re 2 mV @ 1 kHz input). This is a 4 dB loss and the preamp current noise will degrade this even more.

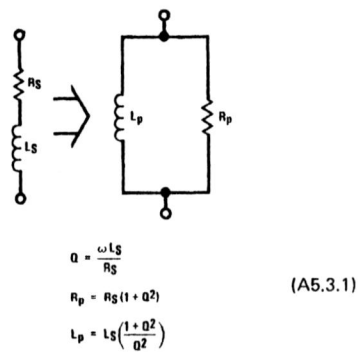

$$Q = \frac{\omega L_s}{R_s}$$

$$R_p = R_s (1 + Q^2)$$

$$L_p = L_s \left(\frac{1 + Q^2}{Q^2}\right)$$

(A5.3.1)

Simplifying the input network,

TABLE A5.1 Summary of Calculations

f Range (Hz)	25 - 50	50 - 100	100 - 200	200 - 400	400 - 800	800 - 1.6k	1.6k - 3.2k	3.2k - 6.4k	6.4k - 12.8k	12.8k - 20k
f Center (Hz)	37.5	75	150	300	600	1200	2400	4800	9600	16.4k
f_{BW} (Hz)	25	50	100	200	400	800	1600	3200	6400	7.2k
$Q = \frac{\omega L_s}{R_s}$	0.156	0.313	0.625	1.25	2.5	5	10	20	40	68.4
Q^2	0.0244	0.098	0.391	1.56	6.25	25	100	400	1600	4678.6
$1 + Q^2$	1.0244	1.098	1.391	2.56	7.25	26	101	401	1601	4679.6
$\frac{1+Q^2}{Q^2}$	42	11.24	3.56	1.64	1.16	1.04	1.01	1.0	1.0	1.0
R_p (Ω)	1.16k	1.24k	1.57k	2.9k	8.2k	29.4k	114k	454k	1.8M	5.29M
L_p (H)	31.5	8.43	2.67	1.23	0.87	0.78	0.76	0.75	0.75	0.75
$R_p \| R_A$ (Ω)	1.13k	1.21k	1.52k	2.74k	7k	18.1k	32.9k	42.6k	45.8k	46.6k
X_L (Ω)	7.42k	3.97k	2.52k	2.32k	3.28k	5.88k	11.45k	22.6k	45.2k	77.2k
X_C (Ω)	17M	8.48M	4.24M	2.12M	1.06M	0.53M	0.265M	0.133M	66.3k	38.8k
$R_e(Z)$ (Ω)	1.11k	1.11k	1.11k	1.15k	1.26k	1.73k	3.86k	12.4k	41.5k	34k
$|Z|$ (Ω)	1.12k	1.15k	1.3k	1.77k	2.97k	5.59k	11.7k	24.4k	43.6k	40.1k
e_{nz} (nV/√Hz)	4.1	4.1	4.1	4.1	4.3	5.1	7.3	14	26	23
V_N (nV)	20.5	29	41	58	86	144.2	292	792	2080	1952
V_n^2 (nV²)	420.3	840.5	1681	3362	7396	20.8k	85.3k	627.7k	4.33M	3.81M
A^2	63.04	31.6	10	3.17	1.59	0.89	0.45	0.159	0.05	0.025
$A^2 V_n^2$ (nV²)	26.5k	26.6k	16.8k	10.7k	11.8k	18.5k	38.1k	99.7k	216.3k	95.2k

$(\Sigma V_n^2)^{1/2} = 2.98 \mu V$ unequalized noise.
$(\Sigma |A_n|^2 V_n^2)^{1/2} = 0.75 \mu V$ RIAA equalized noise.

Thus it is apparent that present phono preamp S/N ratio measurement methods are inadequate for defining actual system performance, and that a new method should be used — one that more accurately reflects true performance.

REFERENCES

1. Hallgren, B. I., "On the Noise Performance of a Magnetic Phonograph Pickup," *Jour. Aud. Eng. Soc.*, vol. 23, September 1975, pp. 546-552.
2. Fristoe, H. T., "The Use of Q Equations to Solve Complex Electrical Networks," *Engineering Research Bulletin*, Oklahoma State University, 1964.
3. Korn, G. A. and Korn, T. M., *Basic Tables in Electrical Engineering*, McGraw-Hill, New York, 1965.
4. Maxwell, J., *The Low Noise JFET — The Noise Problem Solver*, Application Note AN-151, National Semiconductor, 1975.

A6.0 GENERAL PURPOSE OP AMPS USEFUL FOR AUDIO

National Semiconductor's line of integrated circuits designed specifically for audio applications consists of 4 dual preamplifiers, 3 dual power amplifiers, and 6 mono power amplifiers. All devices are discussed in detail through most of this handbook; there are, however, other devices also useful for general purpose audio design, a few of which appear in Table A6.1. Functionally, most of these parts find their usefulness between the preamplifier and power amplifier, where line level signal processing may be required. The actual selection of any one part will be dictated by its actual function.

TABLE A6.1 General Purpose Op Amps Useful for Audio

Device[1]	Single	Dual	Quad	Compensated	Decompensated[2]	Uncompensated	Slew Rate[3] (V/μs)	Supply Voltage Typical (Min. → Max.)	Supply Current Maximum (mA)	General Features of Audio Application Interest
LM301A	X					X	5[4]	±3 → ±18	3	Low THD.
LM310	X			X			30	±5 → ±18	5.5	Fast unity-gain buffer.
LM318	X			X			50	±5 → ±18	10	High slew rate.
LM324			X	X			0.3	3 → 30 (±1.5 → ±15)	2	Low supply current quad.
LM343	X			X			2.5	±4 → ±34	5	High supply voltage.
LM344	X					X	30	±4 → ±34	5	Fast LM343.
LM348			X	X			0.5	±5 → ±18	4.5	Quad LM741.
LM349			X		X		2	±5 → ±18	4.5	Fast LM348.
LF355	X			X			5	±5 → ±18	4	Low supply current LF356.
LF356[5]	X			X			12	±5 → ±18	10	Fast, JFET input, low noise.
LF357	X				X		50	±5 → ±18	10	Higher slew rate LF356.
LM358		X		X			0.3	3 → 30 (±1.5 → ±15)	1.2	Dual LM324.
LM394	—	—	—	—	—	—	—	—	—	Supermatch low noise transistor pair.
LM741	X			X			0.5	±3 → ±18	2.8	Workhorse of the industry.
LM747		X		X			0.5	±3 → ±18	5.6	Dual LM741 (14 pin).
LM1458		X		X			0.2	±3 → ±18	5.6	Dual LM741 (8 pin).
LM3900			X	X			0.5	4 → 30 (±2 → ±15)	10	Quad current differencing amp.
LM4250	X			X			0.03	±1 → ±18	0.1	Micropower.

1. Commercial devices shown (0°C-70°C); extended temperature ranges available.
2. Decompensated devices stable above a minimum gain of 5 V/V.
3. A_V = 1 V/V unless otherwise specified.
4. Compensation capacitor = 3 pF; A_V = 10 V/V minimum.
5. Highly recommended as general purpose audio building block.

A7.0 FEEDBACK RESISTORS AND AMPLIFIER NOISE

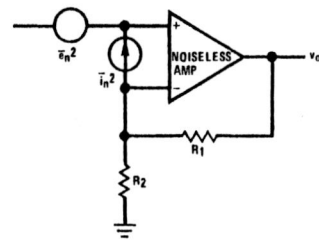

FIGURE A7.1 Practical Feedback Amplifier

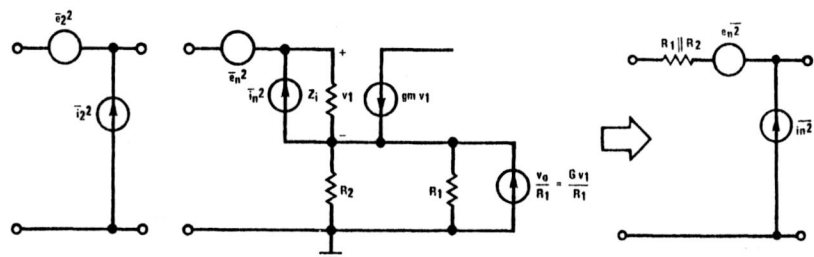

FIGURE A7.2 Model of First Stage of Amplifier

To see the effect of the feedback resistors on amplifier noise, model the amplifier of Figure A7.1 as shown in Figure A7.2, and neglect thermal noise.

We must now show that the intrinsic noise generators $\overline{e_n}^2$ and $\overline{i_n}^2$ are related to the noise generators outside the feedback loop, $\overline{e_2}^2$ and $\overline{i_2}^2$. In addition, the output noise at v_0 can be related to v_1 by the open loop gain of the amplifier G, i.e.,

$$v_0 = v_1 G$$

Thus v_1 is a direct measure of the noise behavior of the amplifier. Open circuit the amplifier and equate the effects of the two noise current generators. By superposition:

$$v_1 = i_2 Z_i$$

also $\quad v_1 = i_n Z_i$

$$\therefore \overline{i_n}^2 = \overline{i_2}^2$$

Short circuit the input of the amplifier to determine the effect of the noise voltage generators. To do this, short the amplifier at $\overline{e_2}^2$ and determine the value of v_1, then short circuit the input at $\overline{e_n}^2$ and find the value of v_1.

$$e_2 = v_1 + R_1 \| R_2 \left(gm\, v_1 + \frac{G v_1}{R_1} \right)$$

$$v_1 = e_2 \frac{1}{1 + gm\, R_1 \| R_2 + G \frac{R_1 \| R_2}{R_1}} \quad , \quad \text{(A7.1)}$$

Now short the input at $\overline{e_n}^2$; $\overline{e_n}^2$ and $\overline{i_n}^2$ both affect v_1.

$\overline{e_n}^2$ gives:

$$v_1 = e_n \frac{1}{1 + gm\, R_1 \| R_2 + G \frac{R_1 \| R_2}{R_1}} \quad \text{(A7.2)}$$

$\overline{i_n}^2$ gives:

$$-v_1 = Z_i \| R_1 \| R_2 \left(gm\, v_1 + G \frac{v_1}{R_1} - i_n \right)$$

Assume $Z_i \gg R_1 \| R_2$

$$v_1 = \frac{i_n R_1 \| R_2}{1 + gm\, R_1 \| R_2 + G \frac{R_1 \| R_2}{R_1}} \quad \text{(A7.3)}$$

Add Equations (A7.2) and (A7.3) and equate to Equation (A7.1):

$$\frac{\overline{e_n}^2 + \overline{i_n}^2 (R_1 \| R_2)^2}{\left(1 + gm\, R_1 \| R_2 + G \frac{R_1 \| R_2}{R_1}\right)^2} =$$

$$\frac{\overline{e_2}^2}{\left(1 + gm\, R_1 \| R_2 + G \frac{R_1 \| R_2}{R_1}\right)^2}$$

$$\therefore \overline{e_2}^2 = \overline{e_n}^2 + \overline{i_n}^2 (R_1 \| R_2)^2.$$

A8.0 RELIABILITY

A8.1 Consumer plus program

National's Consumer Plus Program is a comprehensive program that assures high quality and reliability of molded integrated circuits. The C+ Program improves both the quality and reliability of National's consumer products. It is intended for the manufacturing user who cannot perform 100% inspection of his ICs, or does not wish to do so, yet needs significantly-better-than-usual incoming quality and reliability levels for his ICs.

Integrated circuit users who specify Consumer Plus processed parts will find that the program:

- eliminates 100% the need for incoming electrical inspection
- eliminates the need for, and thus the costs of, independent testing laboratories
- reduces the cost of reworking assembled boards
- reduces field failures
- reduces equipment downtime

Reliability Saves You Money

With the increased population of integrated circuits in modern consumer products has come an increased concern with IC failures, and rightly so, for at least two major reasons. First of all, the effect of component reliability on system reliability can be quite dramatic. For example, suppose that you, as a color TV manufacturer, were to choose ICs that are 99% reliable. You would find that if your TV system used only seven such ICs, the overall reliability of IC portion would be only 50% for one out of each ten sets produced. In other words, only nine out of your ten systems would operate. The result? Very costly to produce and probably very difficult to sell. Secondly, whether the system is large or small, you cannot afford to be hounded by the spectre of unnecessary maintenance costs, not only because labor, repair or rework costs have risen — and promise to continue to rise — but also because field replacement may be prohibitively expensive.

Reliability vis-a-vis Quality

The words "reliability" and "quality" are often used interchangeably as though they connoted identical facets of a product's merit. But reliability and quality are different and IC users must understand the difference to evaluate various vendors' programs for product improvement that are generally available, and National's Consumer Plus Program in particular.

The concept of *quality* gives us information about the population of faulty IC devices among good devices, and generally relates to the number of faulty devices that arrive at a user's plant. But looked at in another way, quality can instead relate to the number of faulty ICs that escape detection at the IC vendor's plant.

It is the function of a vendor's Quality Control arm to monitor the degree of success of that vendor in reducing the number of faulty ICs that escape detection. QC does this by testing the outgoing parts on a sampled basis. The Acceptable Quality Level (AQL) in turn determines the stringency of the sampling. As the AQL decreases it becomes more difficult for bad parts to escape detection; thus the quality of the shipped parts increases.

The concept of *reliability*, on the other hand, refers to how well a part that is initially good will withstand its environment. Reliability is measured by the percentage of parts that fail in a given period of time.

Thus ICs of high quality may, in fact, be of low reliability, while those of low quality may be of high reliability.

Improving the Reliability of Shipped Parts

The most important factor that affects a part's reliability is its construction: the materials used and the method by which they are assembled.

It's true that reliability cannot be tested into a part, but there are tests and procedures that can be implemented which subject the IC to stresses in excess of those that it will endure in actual use. These will eliminate most marginal parts.

In any test for reliability the weaker parts will normally fail first. Stress tests will accelerate the failure of the weak parts. Because the stress tests cause weak parts to fail prior to shipment to the user, the population of shipped parts will in fact demonstrate a higher reliability in use.

Quality Improvement

When an IC vendor specifies 100% final testing of his parts, every shipped part should be a good part. However, in any population of mass-produced items there does exist some small percentage of defective parts.

One of the best ways to reduce the number of such faulty parts is, simply, to retest the parts prior to shipment. Thus, if there is a 1% chance that a bad part will escape detection initially, retesting the parts reduces that probability to only 0.01%. (A comparable tightening of the QC group's sampled-test plan ensures this.)

National's Consumer Plus Program Gets It All Together

We've stated that the C+ Program improves both the quality and reliability of National's molded integrated circuits, and pointed out the difference between these two concepts. Now, how do we bring them together? The answer is in the C+ Program processing, which is a continuum of stress and double testing. With the exception of the final QC inspection, which is sampled, all steps of the C+ processes are performed on 100% of the program parts. The following flow chart shows how we do it.

Epoxy B Processing for All Molded Parts

At National, all molded semiconductors, including ICs, have been built by this process for some time now. All processing steps, inspections and QC monitoring are designed to provide highly reliable products. (A reliability report is available that gives, in detail, the background of Epoxy B, the reason for its selection at National and reliability data that proves its success.)

Six Hour, 150°C Bake

This stress places the die bond and all wire bonds into a tensile and shear stress mode, and helps eliminate marginal bonds and connections.

Five Temperature Cycles (0°C to 100°C)

Exercising the circuits over a 100°C temperature range generally eliminates any marginal bonds missed during the bake.

High Temperature (100°C) Functional Electrical Test

A high-temperature test such as this with voltages applied places the die under the most severe stress possible. The test is actually performed at 100°C — 30°C higher than the commercial ambient limit. All devices are thoroughly exercised at the 100°C ambient. (Even though Epoxy B has virtually eliminated thermal intermittents, we perform this test to insure against even the remote possibility of such a problem. Remember, the emphasis in the C+ Program is on the elimination of those marginally performing devices that would otherwise lower field reliability of the parts.)

DC Functional and Parametric Tests

These room-temperature functional and parametric tests are the normal, final tests through which all National products pass.

Tighter-Than-Normal QC Inspection Plans

Most vendors sample inspect outgoing parts to a 0.65% AQL. Some use even a looser 1% AQL. However, not only do we sample your parts to a 0.28% AQL for all data-sheet DC parameters, but they receive a 0.14% AQL for functionality as well. Functional failures — not parameter shifts beyond spec — cause most system failures. Thus, the five-times to seven-times tightening of the sampling procedure (from 0.65%-1% to 0.14% AQL) gives a substantially higher quality to your C+ parts. And you can rely on the integrity of your received ICs without incoming tests.

Ship Parts

Here are the QC sampling plans used in our Consumer Plus test program:

Test	Temperature	AQL
Electrical Functionality	25°C	0.14%
Parametric, DC	25°C	0.28%
Parametric, DC	100°C	1%
Parametric, AC	25°C	1%
Major Mechanical	—	0.25%
Minor Mechanical	—	1%

A8.2 Operating Junction Temperatures

For steady state operation within the operating junction temperature range of the part, most failure modes are due to die surface related effects such as zener voltage drift due to field effect changes caused by movement of ions in the oxide. After extensive life testing, National Semiconductor has developed some average "acceleration factors" relating increased surface related failure rates to increased junction temperature. For example: an IC device operating steady state at $T_J = 125°C$ for 500 hours will experience approximately the same failure rate as if operated at $T_J = 70°C$ for 72,500 hours. The acceleration factor from 70°C to 125°C (T_J) would be 145. From 125°C to 150°C (T_J) the acceleration factor is 6.3. This indicates the greatly increased part lifetime the user can realize by maintaining the part at a low operating junction temperature.

A9.0 AUDIO RADIO GLOSSARY

"A" Line Filter

The L-C filter used in the power supply lead of an automotive radio to suppress transient voltage spikes. A series inductance from 0.2mHz to 2mHz with a $1000\mu F$ to $2000\mu F$ capacitor to ground is typical.

AB-Bias

A technique used in class B audio amplifiers to prevent crossover distortion. The complementary output devices of the amplifier are biased "on" so that a small amount of current runs in them in the absence of a signal allowing a smooth transition from a positive signal swing to a negative signal swing and vice-versa. This AB-Bias current is typically from 1 — 30mA and is the major component of the amplifier quiescent current and stand-by power dissipation.

Adjacent Channel Rejection

A measure of AM receiver selectivity — the ratio of the detected signal level of a desired r.f. carrier to the detected signal level of an undesired carrier of similar strength located ±10kHz from the desired carrier. Usually > 20dB. (see also selectivity)

Ambience

The indirect sounds heard in a concert hall or other large listening area that contribute to the overall auditory effect obtained when listening to live performances.

Amplifier

Class A

A class A transistor audio amplifier refers to an amplifier with a single output device that has a collector flowing for the full 360° of the input cycle.

Class B

The most common type of audio amplifier that basically consists of two output devices each of which conducts for 180° of the input cycle (see AB-Bias however).

Class C

In a class C amplifier the collector current flows for less than 180°. Although highly efficient, high distortion results and the load is frequently tuned to minimize this distortion (primarily used in R.F. power amplifiers).

Class D

A switching or sampling amplifier with extremely high efficiency (approaching 100%). The output devices are used as switches, voltage appearing across them only while they are off, and current flowing only when they are saturated.

AM Rejection
(AM Suppression)

The ratio of the recovered audio output produced by a desired FM signal of specified level and deviation to the recovered audio output produced by an unwanted AM signal of specified amplitude and modulation index. Usually the AMR of a system is measured over a range of input signal levels with 100% FM and 30% AM, 1kHz modulating frequency.

High quality tuner receiver: AMR > 50dB
Mid quality/multi-band/TV sound: AMR > 40dB

Anechoic Chamber

A derived term for a room or enclosure that is designed to be echo-free over a specified frequency range. Any sound reflections within this frequency range must be less than 10% of the source sound pressure.

A.F.C.

Automatic Frequency Control — controlling the frequency of the local oscillator of a superheterodyne receiver at the value required to produce the desired intermediate frequency. An AFC system will correct for mistuning and oscillator frequency drifts caused by temperature/supply voltage variations and ageing. In higher quality receivers the AFC circuit output can be used to drive a display meter to facilitate tuning.

A.G.C.

Automatic Gain Control — an AGC system operates to maintain the output of an amplifier approximately constant despite input signal level variations, by changing the amplifier gain as the input signal changes. This allows tuning a radio from strong to weak signals without resetting the manual volume control.

AGC Figure of Merit — the widest possible range of input signal level required to make the output signal drop by a specified amount from the specified maximum output level. Typical F.O.M. numbers are from 40dB to 50dB, for domestic radios and about 60dB for automotive radios (for −10dB output level change).

A.L.C.

Automatic Level Control — a compressor circuit usually located at the microphone input of a tape recorder that operates to keep the recorded sound level within predetermined limits regardless of input sound level changes. A figure of merit can be measured similar to that for an A.G.C. system.

Average Power

The signal produced by amplifier into a given load with a given input signal. For sine-wave inputs the average power (also termed continuous or RMS power) is a measure of the amplifier capability to deliver peak outputs while delivering significant power at all levels below the peak. For a peak-to-peak sine-wave output signal Eo into a load R_L the power is given by

$$P_o = \frac{E_o^2}{8R_L}$$

Azimuth

The angle of a tape head's pole-piece slot relative to the direction of tape travel. Misalignment (Azimuth error) will cause a loss of high frequencies. For a track of width W and a recorded wavelength λ, and angle of misalignment θ (Azimuth error) will give a level loss of

$$20 \text{LOG}_{10} \left\{ \frac{\text{Sin} \frac{\pi W \theta}{\lambda}}{\frac{\pi W \theta}{\lambda}} \right\} \text{dB}$$

For example, at 5kHz a −3dB loss will be incurred at a tape speed of 1 7/8 I.P.S. by an error of 0.48 degrees.

Bandwidth

AM

The width of the band of frequencies over which the detector output amplitude does not drop to less than one half the center tuned response with a constant input signal strength. Because of the effect of the normal AM radio AGC circuit, the bandwidth is measured both before and after the onset of AGC action. Below the AGC threshold typical bandwidths are from 4kHz to 10kHz. Above the AGC threshold (signal input +40dB above level required for rated detector output) the I_F bandwidth is from 25kHz to 80kHz.

FM

The range of frequencies at the detector limited by the −3dB amplitude points. The measurement is made with an input signal that produces a detector output −3dB below the detector level obtained with a large r.f. input signal. For monophonic receivers the typical BW is 180kHz, and for stereo 225kHz.

Noise

Noise bandwidth is a term used in the design of phase locked loops (PLL) to describe the response of the loop to signals on either side of the desired locking frequency. It is not measured directly but is the equivalent bandwidth of the PLL derived by plotting the square of the loop amplitude/frequency response and deriving a rectangular pass band characteristic having the same peak value and enclosing the same area. For a loop filter with a single RC roll-off, the noise bandwidth of the filter is 1.57 f-3dB.

Power

The power bandwidth of an audio amplifier is the frequency range over which the amplifier voltage gain does not fall below 0.707 of the flat band voltage gain specified for a given load and output power.

Power bandwidth also can be measured by the frequencies at which a specified level of distortion is obtained while the amplifer delivers a power output 6dB below the rated output. For example, an amplifier rated at 60 watts with ≤ 0.25% THD, would make its power bandwidth measured as the difference between the upper and lower frequencies at which 0.25% distortion was obtained while the amplifier was delivering 30 watts.

Biamplification

The technique of splitting the audio frequency spectrum into two sections and using individual power amplifiers to drive a separate woofer and tweeter. Cross-over frequencies for the amplifiers usually vary between 500Hz and 1600Hz. "Biamping" has the advantages of allowing smaller power amps to produce a given sound pressure level and reducing distortion effects produced by overdrive in one part of the frequency spectrum affecting the other part.

Bias (Tape)

The magnetic coating on audio tapes exhibits non-linear regions in the magnetization characteristic at zero magnetization and at saturation levels. If a steady state magnetic field is applied to the tape during the recording process, the signal or audio information is restricted to the linear portion of the magnetization characteristic. This is called "Biasing" (analogous to the dc bias for solid state devices used to ensure operation in the linear region).

DC Bias

The simplest method of biasing a tape is to apply a steady state dc current to the recording head so that the tape is magnetized to a linear part of the characteristic.

AC Bias

An ultrasonic (50kHz – 110kHz) alternating current applied to the recording head so that ideal or "anhysteric" magnetization of the tape takes place. In the presence of the bias waveform, the signal (audio) magnetization characteristic is linearized enabling larger flux levels to be recorded, and improving the S/N ratio compared to dc bias.

Peak Bias

AC biasing also increases the tape sensitivity (larger recorded flux for given recording current). However, beyond a certain bias level, called the peak bias, the recorded flux level starts to decrease and distortion levels begin to increase. The required peak bias depends on the tape formulation used and is typically lower for ferric oxide tape than for chromium dioxide tapes.

Capstan

A motor driven spindle that feeds tape at a constant speed past the tape heads. The tape is held against the capstan by an idler or pinch wheel.

Capture Range

The capture range of a PLL is the frequency range, centered about the V.C.O. free running frequency, over which the loop can acquire lock with the input signal.

Capture Ratio

A measure of the ability of an FM tuner to select the stronger of two r.f. signals at or near the same frequency. It is the ratio of the signal strength of the carriers required for 30dB suppression of the audio from the weaker signal at the tuner output.

The rated capture ratio is measured by increasing the signal strength of an unmodulated carrier until there is a drop in the tuner audio output being obtained from a 100% modulated carrier at a 1mV signal level. The ratio in dB of the unmodulated carrier levels required to cause 1dB and 30dB drops in the audio output, divided by two, is the capture ratio.

Cartridge

A phonograph pick-up and stylus combination

Constant Amplitude Pick-Ups:

Known as ceramic or crystal cartridges, piezo electric pick-ups depend on the piezo-electric effect – i.e. when crystals (rochelle salt) or ceramics (barium titanate) are mechanically flexed, an EMF is developed directly proportional to the degree of flexure. Very popular in low fidelity sound systems, these cartridges have very high output levels from 100mV to 2V.

Constant Velocity Pick-ups:

Moving coil or moving magnet cartridges develop an output proportional to the velocity of the stylus motion and are used in high fidelity systems. Substantially lower output levels around 3mV to 5mV are obtained compared to crystal cartridges.

Channel Separation

The degree to which the signal in one amplifier is kept separate from an adjacent undriven amplifer. Channel separation for FM stereo decoders is typically >40dB whereas phono cartridge channel separation is typically between 20dB and 30dB.

C.C.I.R./A.R.M.

Literally: International Radio Consultative Committee/Average Responding Meter

This refers to a weighted noise measurement for a Dolby B type noise reduction system. A filter characteristic is used that gives a closer correlation of the measurement with the subjective annoyance of noise to the ear. Measurements made with this filter cannot necessarily be related to unweighted noise measurements by some fixed conversion factor since the answers obtained will depend on the spectrum of the noise source.

Coercivity

A measure of the magnetic field strength required to erase a tape to a state of zero magnetism. High coercivity tapes are harder to erase but suffer less from high frequency losses caused by self-erasure (see self-erasure).

Compandor

A complementary compression and expansion system used for audio noise reduction. Before recording or being transmitted the entire signal is compressed according to some fixed law and afterwards is expanded to its original dynamic range for replay.

Composite Signal

The stereo FM broadcast modulation signal consisting of a 19Hz pilot tone, (L+R) information and (L–R) information modulated on a suppressed 38kHz carrier and (if any) a 67kHz FM carrier with ±6kHz deviation S.C.A. channel.

Crest Factor

The ratio of the peak value of a waveform to its RMS value. For example, the crest factor for audio amplifier noise is from 3:1 to 5:1.

Crossmodulation

Crossmodulation is the name given to the phenomenon whereby information from an AM carrier is transferred to another carrier. For a reciever, non-linearities in the R.F. or mixer stages can cause the modulation from an adjacent undesired signal to modulate the desired carrier signal to which the receiver is tuned. For measurements typically 30% modulated carriers are used and the level of crossmodulation specified to be ≤1%.

Crossover Distortion

Distortion caused in the output stage of a class B amplifier. It can result from inadequate bias current (see AB Bias) allowing a dead zone where the output does not respond to the input as the input cycle goes through its zero crossing point. Also for I/Cs an inadequate frequency response of the output PNP device can cause a turn-on delay giving crossover distortion for negative going transition through zero at the higher audio frequencies.

Crossover Frequency

A frequency at which other frequencies above and below it are separated. A crossover network will separate the high and low frequencies in a tweeter/woofer speaker system for example, with a single crossover frequency between 1kHz and 3kHz.

dbx

An audio noise reduction system operating on the wide band companding principle. A true RMS detector controls the gain of an amplifier before recording with operator variable compression factor from 1.0 to 3.0. On playback the dynamic range is restored by a similar expansion process.

De-Emphasis

To reduce the effect of broadband noise in an FM broadcast the signal has pre-emphasis of the higher frequencies defined by a 75µS time constant (50µS in Europe). At the receiver, a 75µS de-emphasis restores the frequency/amplitude relationships while reducing the higher frequency noise added during the broadcast signal transmission.

Detector

The point in a receiver at which the modulating information is recovered from the carrier waveform.

Differential Peak Detector

An FM detector that operates by comparing the peak voltages detected on either side of a single-tuned circuit.

Quadrature Detector

Compares the relative phases of the I.F. signal on either side of a circuit tuned to give 90° phase shift at the intermediate frequency.

Synchronous Detector

A P.L.L. detector where the modulated signal is compared in a phase detector to a local oscillator signal.

Power Detector

An AM peak detector where the diode is the base-emitter junction of a transistor.

D.I.N.

Literally Deutsche Industrie Norm.

Designates a European performance standard or test procedure. It also describes a unitized audio connector plug and socket.

Deviation

The instantaneous frequency difference of an FM signal from the unmodulated carrier frequency.

Distortion

The effects produced by an electronic circuit when the signal output from the circuit does not exactly duplicate the input signal in all respects except magnitude (see intermodulation, THD, etc.).

Dolby B

Dolby B is a simplified version of the Dolby A professional quality noise reduction system. The amplitude of low level signals over a selected frequency range is increased prior to recording to enhance them above tape noise. On playback the original levels are restored causing a corresponding reduction in the audible tape noise. The major difference with Dolby A which used four frequency bands, is the use of a single variable frequency band with a cut-off frequency that increases in the presence of high level high frequency signals.

Dolby Level

Because of the complementary nature of the Dolby B noise reduction system, the audio channel between the encoder and the decoder must have a fixed gain such that the decoding signal level is within 2dB of the encoding signal level. Also if recordings are interchangeable the signals in the noise reduction system must be related to the levels in the audio channel. Dolby level provides this reference and corresponds to a specified tape flux density when recorded with a 400Hz tone. For reel to reel and eight track cartridge tapes this is 185nWb/m, and for cassettes Dolby level is 200nWb/m.

Equalization

The adjustment of the frequency or phase characteristics of a signal or audio device. Examples are the R.I.A.A. recording characteristic for phonograph discs and the N.A.B. recording playback characteristic for tape. Equalizers are audio equipment devices inserted into playback systems to compensate for signal variations, room acoustics and loudspeaker responses.

Erasure

The exposure of magnetic tape to a strong alternating magnetic field in order to leave the tape in a neutral state.

Self Erasure

The tendency for strongly magnetized sections of an audio tape to erase adjacent sections of opposite polarity magnetization. This is a significant cause of loss of high frequencies at lower tape speeds.

Excess Noise

A fudge factor to account for the extra noise components exhibited by passive electronic components that are not described by thermal noise effects.

Flutter

Rapidly repeating fluctuations in tape or turntable speed that give rise to warbling variations in the pitch of the reproduced sound. Flutter can be considered a higher frequency version of wow and typically measurements are made of wow and flutter combined.

Flux (Magnetic)

The magnetic force existing in the neighborhood of a magnetic pole (on an audio tape) can be represented by lines of force known as the magnetic flux. The recorded flux level of a tape is specified by the number of lines of force per unit track width of the tape and has units of nWb/m (nanowebers per meter). A typical reference flux level for tapes is 185nWb/m with less than 1% distortion being obtained at this level. In Europe the DIN reference level of 320nWb is used to calibrate equipment using peak reading program meters. (see also Dolby level).

Gap

The narrow slot between the pole pieces of the record or playback head of a tape machine.

Gap Length

The dimension of the gap in the direction of tape travel. When the gap length becomes comparable to the recorded signal wavelength there will be no output from the head. This is the most significant high frequency limitation of a tape recorder.

Gap Smear

Continual abrasion of the head by the tape causes head material to cold flow into the gap causing magnetic shorts.

Harmonic Distortion

A form of distortion characterized by the presence of spurious harmonics in the signal. It is usually measured by comparing the percentage amplitude of the spurious harmonics to the amplitude of the signal fundamental tone.

Head

A magnetic transducer used to record and/or playback tape signal.

Hyperbolic Head

A head, the pole pieces of which are shaped to follow a hyperbolic function. This helps to maintain good tape to gap contact — for separation d the loss is given by $55d/\lambda$

Note: The wavelength (λ) is the recorded wavelength. If d is measured in inches, the $\lambda = \dfrac{\text{tape speed (I.P.S.)}}{\text{Hz.}}$

Headroom

The margin between an actual signal operating level and the level that would cause substantial distortion. For a tape recorder this would be the level above zero VU that gives a (specified) distortion

Image Rejection

A superheterodyne receiver can usually respond to two frequencies whose difference from the local oscillator frequency is equal to the intermediate frequency. One is the desired frequency, the other is the image frequency. The receiver is tuned to the desired frequency and the RF level adjusted for a specified output (usually maximum sensitivity). The input signal is then switched to the image frequency and the RF level increased to obtain the same output. This change in levels is a measure (in dB) of the image rejection.

Input Sensitivity

A measure of a device's input signal requirement to produce a desired output. "High" sensitivity indicates a low input signal level whereas "Low" sensitivity implies a higher input signal requirement. Typical specifications on amplifier systems are: phono and mic, 2mV; auxiliary (radio) and tape, 200mV.

I.H.F.M.

Institute of High Fidelity Manufacturers

Intermodulation Distortion

Distortion characterized by the presence of the sum and the difference frequencies of the fundamentals and harmonics of two or more simultaneous tones being passed through a system.

The CCIR measurement procedure is to use a 1:1 ratio of test tones of nearly equal frequency with the distortion given by the amplitude of the beat note.

The SMPTE method used a 4:1 tone amplitude ratio.

Intermediate Frequency (I.F.)

In a superheterodyne receiver, the frequency to which the RF signal carrier frequency is converted by action of the local oscillator. Popular intermediate frequencies are: AM radio, 455kHz; AM automotive radio, 262.5kHz; FM radio 10.7mHz.

I.F. Rejection

A measure of the ability of a tuner to reject an RF signal at the intermediate frequency.

Limiter

An amplifier whose output signal has a constant amplitude when the input signal is above a certain specified level. Limiting amplifiers are used in FM IF strips to help eliminate spurious amplitude modulation of the signal. Limiters can also be found in audio equipment used to suppress short duration peak transients.

Limiting Threshold

The input signal level required for the amplifier output to be limited in amplitude. For an FM IF amplifier this is measured as the input signal level for which the output falls to 0.707 of the amplitude obtained with a strong signal input.

M.R.L.

Maximum recorded level. The signal level required at a given frequency to give a 3% distortion level on a tape (at mid-range frequencies — at high frequencies self-erasure determines the MRL). It is the performance ceiling for the particular tape with a given bias and equalization.

Medium Wave

A term applied by the CCIR to a frequency band between 300kHz and 3mHz.

In Europe this is popularly taken to mean the AM broadcast band encompassing carrier wavelengths between 580m and 190m. There is also a band designated long wave (LW) which includes wave lengths from 2000m to 1150m. (U.S. AM broadcast transmissions are in the MW band only).

Mixer

A circuit in which two separate frequencies, usually called the carrier and local oscillator signals, can be mixed or converted to the difference frequency between them (the intermediate frequency).

Microphone/Line Mixer

A device for adding two or more input signals in a linear fashion while exercising individual control over the amplitude of each.

Modulation Index (Modulation Percentage)

A measure of the degree of modulation of a carrier signal on a carrier waveform.

AM Modulation

For a sinusoidal modulation of a carrier waveform, if E_{MAX} is the peak waveform amplitude while the null amplitude is E_{MIN}, the modulation index m is given by the ratio

$$\frac{E_{MAX}/E_{MIN} - 1}{E_{MAX}/E_{MIN} + 1}$$

The modulation percentage is given by $m \times 100\%$.

FM Modulation

The modulation index for a frequency modulated wave (mf) is given by the ratio of the peak frequency deviation of the carrier to the modulating frequency, i.e.

$$mf = \frac{\Delta f}{fm}$$

For Broadcast FM signals, modulation percentage has quite a different meaning and 100% modulation refers to the peak deviation permitted by the particular broadcast standard. For example, 100% modulation can be:

Radio Broadcast (U.S.): ±75kHz
NTSC Television (U.S.): ±25kHz
PAL Television: ±50kHz

Motorboating

Audible spurious low frequency oscillations in a system usually caused by inadequate de-coupling of power supply leads. Signals in an output stage couple back through the common internal impedance of the power supply to the input stages.

Multi-Path

Multi-path describes a signal condition whereby the antenna of a receiving system receives not only the directly radiated signal (line of sight) but also delayed signals reflected from large buildings or hills. For an FM receiver in an automobile the delay time caused by the longer reflected signal path changes as the car moves causing a "fluttering" of the recovered audio at the rate of change of phase between the direct and reflected signals. This is sometimes called "Picket Fencing."

Music Power

A measurement of the peak output power capability of an amplifier with either a signal duration sufficiently short that the amplifier power supply does not sag during the measurement, or when high quality external power supplies are used. This measurement (an IHF standard) assumes that with normal music program material the amplifier power supplies will sag insignificantly.

Mute

To suppress the audio output of an amplifier in response to a command signal even though an input may be present. Used in stereo receivers to prevent the off-channel spurious response produced while tuning from reaching the speakers.

N.A.B.

National Association of Broadcasters — usually associated with various tape standards.

Noise

A term for unwanted electrical distrubances, other than crosstalk or distortion components, that occur at the output of a reproducing amplifier.

Noise Bandwidth (see Bandwidth)

Noise, Excess (see Excess Noise)

Noise Figure

The logarithmic ratio of the input signal to noise and the output signal to noise ratios.

Noise Current

The equivalent open circuit RMS noise current which occurs at the input of a noiseless amplifier due to current flowing at that input. It is measured by shunting an impedance across the input terminals and comparing this output noise obtained with the output noise obtained when the input is shorted (see below).

Noise Voltage

The equivalent short circuit input RMS noise voltage which occurs at the input of a noiseless amplifier if the input teminals are shorted. It is measured at the output, divided by the amplifier gain and the square root of the bandwidth over which the measurement is made to yield units of $nV/\sqrt{Hz}$.

Flicker Noise

1/f or flicker noise has a random amplitude similar to shot and thermal noise but with a 1/f spectral power sensity. This means that the noise increases at low frequencies and is associated with the level of direct current in the device.

Popcorn Noise

So named for the audible characteristic, popcorn noise is randomly occurring, random amplitude noise, lasting from a few microseconds to several seconds.

Shot Noise

The noise generated by a charge crossing a potential barrier. For medium and high frequencies it is the dominant noise mechanism in bipolar devices. Shot noise has a constant spectral density.

Thermal Noise

Also called Johnson Noise, this mechanism of noise voltage generation occurs spontaneously in all resistive devices and involves the random thermal agitation of electrons. It has a constant spectral density and the noise voltage is given by Nyquist's formula

$$V_N = \sqrt{4KTBR}$$

Modulation Noise

The noise produced on playback of a tape that is a function of the instantaneous amplitude of the signal. It is caused by poor particle dispersion and surface irregularities.

Pink Noise

Noise that has a constant mean squared voltage (or power) per octave, i.e. the mean squared noise voltage per unit bandwidth increases at 3dB per octave (10dB per decade) with falling frequency. Noise sources with this characteristic are popular in audio work since it allows correlation between successive octave equalizer stages by ensuring that the same voltage amplitude is available as a reference standard.

White Noise

Noise with a constant spectral density — the mean square noise voltage per unit bandwidth is independent of frequency. Resistor thermal noise has this characteristic.

Pan-Pot

A potentiometer used to adjust the stereo balance of a monophonic signal allowing it to be positioned anywhere across the stereo stage.

Phantom

A signal derived from two sources in such a way as to appear located from a third source. Stereo signals "appearing" between speakers are said to be "phantomed".

Print-Through

The transfer of signal through adjacent layers of tape on a reel. It causes faint echoes preceding or following loud passages.

Quad

Generally taken to mean quadrasonic or quadraphonic sound systems designed to give the impression of a field of sounds coming from an apparent 360° around the listener.

Quieting

A measure of the usable sensitivity of an FM tuner and is expressed as the least RF signal level (100% modulated with a 400 Hz tone) that reduces the receiver internal noise and distortion to 30 dB below the output level obtained with the modulated tone present (S + N/N = 30 dB, a null filter tuned to 400 Hz is used to remove the tone).

For an AM receiver the carrier is modulated by 30% and the field strength (μV/m) is measured that is necessary to provide a 20 dB S + N/N ratio.

R.I.A.A.

Record Industry Association of America

Usually referred to in connection with a phonograph disc recording equalization that helps limit the frequency and amplitude swings of a record cutting head over the audio frequency range. The reproducing amplifier has the inverse characteristic.

Retentivity

When a tape has a signal recorded on it the resulting magnetic field strength per unit coating cross section (width X thickness) is known as the retentivity of the tape.

Remanence

The magnetic field strength retained by a ¼" wide tape.

Reverberation

The persistence of sound in an enclosure after the original sound has ceased. Reverberation can be regarded as a series of multiple echoes closely spaced so as to appear continuous but gradually decaying in intensity. Electromechanical (or solid state) devices can be used to simulate reverberation with delay times from a few milliseconds and decay times up to 2 seconds using a frequency range from 100 Hz to 5 kHz.

Rumble

The name given to low frequency noise (below 100 Hz) caused by turntable and tape transport mechanisms.

S.O.A.

Safe operating area — of a solid state device. The curves displaying the collector current and collector voltage limits of the transistor that must be observed for reliable operation. Curves indicating instantaneous power dissipations are often shown as well as dc limits.

Saturation

The condition of a tape coating that has accepted its maximum degree of magnetization. It can also refer to an amplifier output that is at the point of clipping.

Selectivity

A tuner's ability to select a station in the presence of strong adjacent or alternate channel signals.

For FM, either 30% or 100% modulated signals can be used and involves the measurement of the ratio of the desired carrier at the usable sensitivity level (30 dB quieting) and the level of undesired carrier (0.2 mHz away for adjacent channel and 0.4 mHz away for alternate channel) needed to cause a 30 dB reduction in the tuner recovered audio level.

For AM receivers a 20 dB reduction is required with the undesired carrier located 10 kHz away (adjacent channel) or 20 kHz away (alternate channel).

Sensitivity

See Input Sensitivity

FM Sensitivity:

The radio frequency input signal (μV) required to produce 30 dB quieting of the recovered audio (also called usable sensitivity). The carrier is modulated to ±75 kHz deviation with a 400 Hz tone.

AM Sensitivity:

The radio frequency input field strength required to produce 20 dB quieting of the recovered audio. A 30% modulated carrier is used (IMF standard).

A popular technique with O.E.M.'s is to measure the field strength required to produce a given level at the speaker — for table radios this is 50 mW, for automotive radios it is 1 watt.

Microphone Sensitivity

The output voltage in dB referenced to 1 volt for an S.P.L. of 1 microbar (74 dB SPL).

A microphone with a sensitivity of −85 dBV will have an output of 5.6V for an S.P.L. of 174 dB (output = 174 dB − 74 dB − 85 dB = +15 dB above IV).

S/N

The ratio of a system's output signal level and the noise level obtained in the absence of signal. The reference signal level is either specified or measured as that which related to a specified distortion level.

Sinad

A measurement of the signal to noise ratio of a receiver system where the signal level measurement includes the system noise and distortion (S + N + D)/N.

Skating

The tendency of a pivoted tone arm to be pulled to the center spindle. It is caused by friction between the stylus and the record surface.

S.P.L.

Sound pressure level — usually measured with a microphone/meter combination calibrated to a pressure level of 0.0002μBars (approximately the threshold hearing level).

$$S.P.L. = 20 \log_{10} P/0.0002 dB$$

where P is the R.M.S. sound pressure in microbars. (1 Bar = 1 atmosphere = 14.5 lb/in^2 = 194 dB S.P.L.).

Squelch

An audio squelch is one that cuts off (or mutes) the output of the audio section of a receiver when there is no input signal. It is used to prevent listener fatigue on communications channels caused by noise in the absence of carrier signals.

S.C.A.

Literally subsidiary communications authorization. This applies to an additional modulation on the standard FM carrier (see composite signal) intended to provide commercial-free background music for stores, etc.

Thermal Resistance (R_{TH})

An analogy for heat transfer where the ability of a heat conductive system to transfer heat is described in similar terms to those used in an electrical system for power dissipated in a resistor with a given applied voltage. The thermal resistance is given by the temperature differential established when a given amount of power is being dissipated ($\theta = T1 - T2/P_D$) with units of °C/watt.

Tweeter

A loudspeaker designed for high frequencies (see cross-over frequency).

Ultrasonic Rejection

The level of rejection of the 19 kHz pilot tone and 38 kHz V.C.O. frequency in a stereo FM receiver. The intrinsic rejection of a stereo decoder is the logarithmic ratio of the level of 19 kHz and 38 kHz to a 1 kHz reference tone with only the standard de-emphasis filter at the decoder outputs.

VU

The abbreviation for Volume Unit, a form of decibel referenced to a standard value of 1 mW in a 600Ω load.

VU Meter

A recording level meter with a needle motion damped according to internationally recognized standards, which will respond to 99% of the input signal within 0.3 seconds and have less than 1.5% overshoot. The frequency response also has to be better than ±0.5 dB from 25 Hz to 16 kHz.

Woofer

Speaker designed to reproduce relatively low frequencies.

Wow

A slow variation in the pitch of a reproduced signal caused by tape or turntable speed variations (see flutter).

7.0 Index

AB Bias: 4-3, 6-19
A-Line Filter: 4-48
Absolute Maximum Ratings: 1-2, 6-1
Acoustic Pickup Preamp: 5-12
Active Crossover Networks
    Filter Choice: 5-1
    Filter Order: 5-1
    Table of Values: 5-5
    Third-Order Butterworth: 5-2
    Use of: 5-6
AGC: 3-5, 6-20
AM9709: 5-11
AM97C11: 2-68
Ambience, Rear Channel, Amplifier: 4-20
Amplifers
    AB Bias: 4-3, 6-19
    Bootstrapped: 4-4, 4-41, 4-45, 4-62
    Buffer: 6-12
    Class B: 4-2, 6-19
    Current Limit: 4-3
    Difference: 6-12
    Distortion: 4-1, 4-3, 6-21, 6-22, 6-23
    Frequency Response: 4-1
    gm: 4-1
    Inverting AC: 6-12
    Loop Gain: 2-1, 4-1
    Non-Inverting AC: 6-12
    Output Stages: 4-2
    Protection Circuits: 4-3
    RF Oscillation in: 4-2, 4-13, 4-25, 4-61
    Single Supply Biasing: 6-13
    Slew Rate: 1-1, 4-2
    Summing: 6-12
    Thermal Shutdown: 4-4, 4-53
    Transconductance: 4-1, 5-13
    Variable Gain: 6-13
Amplitude Modulation (see AM Radio): 6-23
AM Radio
    Field Strength Conversion: 3-1
    LM3820: 3-4
    Regenerative: 3-1
    Superheterodyne: 3-1
    Tuned RF: 3-1
    Typical Gain Stages: 3-4
AM Rejection Ratio: 3-18, 6-19
AM Suppression: 6-19
Analog Switching (see Switching, Noiseless)
Antenna Field Strength (see AM Radio)
Antennas
    Capacitive: 3-2
    Ferrite Rod: 3-1
AQL (Acceptable Quality Level): 6-18
Audio Rectification: 2-11
Audio Taper Potentiometer: 2-46
ALC Circuit (LM1818): 2-40, 4-38

Balance Control: 2-50, 4-19
Balanced Mic Preamp (see Mic Preamps)
Bandwidth: 1-2

Bass Control
    Active: 2-51, 2-53, 4-21, 4-33, 4-39, 5-12
    Passive: 2-46, 4-19
Baxandall Tone Control (see Tone Control, Active)
Biamplification: 5-1, 6-20
Bias Erasure: 2-30
Bias (Tape)
    AC: 2-38, 6-21
    DC: 2-38, 6-21
    Peak: 2-37, 6-21
Bias Trap: 2-32
Blend, Stereo/Monaural: 3-18, 3-20
Boosted Power Amplifiers
    Emitter Followers: 4-50
    LM391: 4-52
Bootstrapped Amplifiers (see Power Amplifiers, LM388, LM390)
Bootstrapping: 4-4, 4-41, 4-63
Bridge Amplifiers
    LM1877/LM378/LM379/LM1896: 4-15
    LM380: 4-26
    LM383: 4-48
    LM388: 4-39
    Power Dissipation of: 4-50
Buffer Amplifier: 6-12
Butterworth Filters: 2-56, 5-1

Capacitive Antenna (see Antennas, Capacitive)
Capture Ratio: 6-21
Cartridges (see Phono Cartridges)
Cassette Tape Preamplifier: 2-36
Ceramic Cartridge Compensation for R.I.A.A.: 4-38
Ceramic Cartridge Frequency Response: 4-35
Ceramic Phono Amplifier: 4-21, 4-25, 4-34, 4-39
Channel Separation: 6-21
Circuit Layout See Layout, Circuit)
Class B Output Stage: 4-2
Closed-Loop Gain: 2-1
C.C.I.R./ARM: 2-10, 6-21
CMRR in Mic Preamps: 2-45
Conduction: 4-65
Constant Amplitude Disc Recording: 2-24
Constant Current Tape Recording: 2-29
Constant Velocity Disc Recording: 2-24
Consumer Plus Program: 6-18
Contact Mic Preamp (see Acoustic Pickup Preamp)
Convection: 4-65
Crest Factor: 2-8, 6-21
Crossover Distortion (see Distortion)
Crossover Networks (see Active Crossover Networks)
Crystal Cartridge Frequency Response: 4-38
Current Amplifier: 2-67
Current Limit: 4-3
Cutover: 2-23

Decibel: 6-11
Decompensated Op Amp: 1-2
Delta-Wye Transformer: 6-11
Delta-V_{BE} Reference Voltage: 4-9

D.I.N. Cassette Tape Standard: 2-36, 6-22
Difference Amplifier: 2-44, 6-12
Disc (see Phono Disc)
Dissipation (see Power Dissipation)
Distortion
    Harmonic: 1-2, 4-1, 6-23
    Crossover: 4-3, 6-21
Dolby: 2-10, 2-42
Dynamic Range
    Phono Disc: 2-23
    Supply Voltage: 1-2

Emissivity: 4-67
Epoxy B: 6-18
Equalization (see RIAA or NAB Equalization)
Equalizer: 2-59
Equalizing Instrument: 2-62
Excess Noise: 2-3, 6-22

Feedback, Effects of
    Bandwidth: 2-1
    General: 2-1
    Harmonic Distortion: 2-1
    Input Impedance: 2-1
    Inverting Amplifier: 2-1
    Noise Gain: 2-1
    Non-Inverting Amplifier: 2-1
    Output Impedance: 2-1
    Series-Shunt: 2-1
    Shunt-Shunt: 2-1
Feedback Tone Control (see Tone Control, Active)
Ferrite Rod Antenna (see Antennas, Ferrite Rod)
Field Strength (see Antenna Field Strength)
Filters, Active
    Bandpass: 2-57, 2-58, 2-63
    High Pass: 2-55, 5-3, 5-14
    Low Pass: 2-55, 5-3, 5-13
    Parameter Definitions: 2-55
    Rumble: 2-55
    Scratch: 2-55
    Speech: 2-57
Flanging: 5-10
Flat Response: 2-46
Fletcher and Munson (see Loudness Control)
Flicker Noise: 2-4
FM Radio
    IF Amplifiers: 3-8
    LM1310: 3-14, 3-17
    LM1800: 3-14
    LM3089: 3-8
    LM3189: 3-13
    Stereo: 3-14
FM Scanner Power Amp: 4-44
FM Stereo Multiplex (see FM Radio, Stereo)
Form Factor: 6-7
Frequency Modulation (see FM Radio)
Full-Power Bandwidth: 1-1
Fuzz: 5-11

Gain-Bandwidth Product: 1-2
Gap Loss in Tape Heads: 2-29
General Purpose Op Amps: 6-16

Graphic Equalizer: 2-59
Groove Modulation: 2-23
Ground Loops: 2-1

Harmonic Distortion (see Distortion)
Head Gap (Width): 2-29, 6-22, 6-23
Headroom: 6-23
Heatsinking
    Custom Design: 4-67
    Heat Flow: 4-65
    LM1877/LM378/LM379: 4-11
    LM391: 4-55
    Modelling: 4-65
    PC Board Foil: 4-69
    Procedure: 4-66
    Staver V-7: 4-23
    Thermal Resistance: 4-65
    Where to Find Parameters: 4-66

Inductor Simulation: 2-60
IF Bandwidth: 3-14, 6-20
IF Selectivity: 6-25
Input Referred Ripple Rejection: 1-2
Input Sensitivity: 6-23
Instrumentation Amplifier: 2-45
Intercom: 4-27, 4-44
Inverse RIAA Response Generator: 2-38
Inverting AC Amplifiers: 6-12

JFET Switching: 2-68

Lag Compensation: 2-62
Large Signal Response: 1-1
Layout, Circuit: 2-1
LF356/LF357
    Active Crossover Network: 5-4, 5-5
    Mic Preamp: 2-44, 2-45
    Octave Equalizer: 2-61
LH0002: 2-67
Limiting Sensitivity: 6-25
Limiting Threshold: 6-23
Line Driver: 2-67
LM324: 5-11
LM348: 5-11, 2-61
LM349:
    Active Tone Control: 2-53, 2-55
    Equalizing Instrument: 2-64
LM378/LM379
    Boosted: 4-50
    Bridge Connection: 4-15
    Characteristics: 4-5
    Circuit Description: 4-7
    Comparison: 4-5
    Fast Turn-On Circuitry: 4-8
    Heatsinking: 4-11
    Inverting Amplifier: 4-9
    Layout: 4-13
    Non-Inverting Amplifier: 4-8, 4-10, 4-14
    Power Oscillator: 4-17
    Power Output: 4-11
    Power Summer: 5-10
    Proportional Speed Controller: 4-18

Rear Channel Ambience Amplifier: 4-20
Reverb Driver: 5-8, 5-9
Split Supply Operation: 4-13
Stabilization: 4-13
Stereo System: 4-19
Two-Phase Motor Drive: 4-18
Unity Gain Operation: 4-13

LM380
AC Equivalent Circuit: 4-23
Biasing: 4-24
Bridge: 4-26
Ceramic Phono: 4-25
Characteristics: 4-6
Circuit Description: 4-22
Common-Mode Tone Control: 4-25
Common-Mode Volume Control: 4-25
DC Equivalent Circuit: 4-22
Device Dissipation: 4-23
Dual Supply: 4-28
Heatsinking: 4-23
Intercom: 4-27
JFET Input: 4-28
Oscillation: 4-25
RF Precautions: 4-25
Siren: 4-29
Voltage-to-Current Converter: 4-28

LM381
Audio Rectification Correction: 2-11
Biasing: 2-13
Characteristics: 2-12
Circuit Description: 2-13
Equivalent Input Noise: 2-9
Inverting AC Amplifier: 2-16
Mic Preamp: 2-64
Non-Inverting AC Amplifier: 2-16
Split Supply Operation: 2-15
Tape Playback Preamp: 2-33
Tape Record Preamp: 2-32

LM381A
Characteristics: 2-12
Equivalent Input Noise: 2-9
General: 2-16
Mic Preamp: 2-43, 2-44
Phono Preamp: 2-50

LM382
Adjustable Gain for Non-Inverting Case: 2-20
Characteristics: 2-12
Equivalent Input Noise: 2-9
Internal Bias Override: 2-20
Inverting AC Amplifier: 2-21
Non-Inverting AC Amplifier: 2-19
Tape Preamp: 2-35, 4-20
Unity Gain Inverting Amplifer: 2-22

LM383
Bridge Amplifer: 4-48
Characteristics: 4-6
Circuit Description: 4-46
Heatsinking: 4-47
Layout: 4-47
Power Dissipation: 4-47

LM384
Characteristics: 4-6

Five Watt Amplifier: 4-30
General: 4-29

LM386
Bass Boost: 4-33
Biasing: 4-32
Characteristics: 4-6
Gain Control: 4-32
General: 4-31
Muting: 4-32
Non-Inverting Amplifier: 4-32, 4-33
Phono Amplifier (Minimum Parts): 4-35
Phono Power Supply Operation: 4-36
Sine Wave Oscillator: 4-34
Square Wave Oscillator: 4-39

LM387/LM387A
Acoustic Pickup Preamp: 5-12
Active Bandpass Filter: 2-59
Active Tone Control: 2-54
Adjustable Gain: 5-12
Characteristics: 2-12
Equivalent Input Noise: 2-9
Inverse RIAA Response Generator: 2-28
Inverter: 5-9
Inverting AC Amplifier: 2-17
Line Driver: 2-67
Mic Preamp: 2-44, 2-45
Mixer: 5-8, 5-9
Noise Reduction Circuit: 5-14
Noise, Measurement of: 2-8
Non-Inverting AC Amplifier: 2-17
Passive Tone Controls: 2-49
Reverb Recovery Amplifier: 5-8, 5-9
Rumble Filter: 2-56
Scratch Filter: 2-58
Speech Filter: 2-58
Summer: 5-8, 5-9
Tape Playback Preamp: 2-33
Tape Record Preamp: 2-32
Tone Control Amplifier: 2-18, 5-12
Two Channel Panning Circuit: 2-66
Unity Gain Inverting Amplifier: 2-17

LM388
Bootstrapping: 4-42
Bridge: 4-43
Characteristics: 4-6
FM Scanner Power Amp: 4-44
General: 4-41
Intercom: 4-44
Squelch: 4-45
Walkie Talkie Power Amp: 4-44

LM389
Ceramic Phono: 4-39
Characteristics: 4-6
General: 4-36
Logic Controlled Mute: 4-41
Muting: 4-37
Noise Generator: 4-40
Siren: 4-39
Tape Recorder: 4-38
Transistor Array: 4-37
Tremolo: 4-40
Voltage-Controlled Amplifier: 4-40

LM390
    Characteristics: 4-6
    General: 4-45
    One Watt, 6 Volt Amplifier: 4-45
LM391
    AB Bias: 4-52
    Characteristics: 4-36
    Circuit Description: 4-52
    Dual Slope Load Line: 4-57
    Non-Inverting Amplifier: 4-53, 4-57, 4-59
    Oscillations and Grounding: 4-61
    Output Device Heatsinks: 4-55
    Output Stage: 4-53
    Power Supply Requirements: 4-57
    Protection Circuits: 4-55
    Single Slope Load Line: 4-57
    Slew Rate: 4-52
    Thermal Shutdown: 4-53
    Transient Distortion: 4-61
    Turn-On Delay: 4-61
LM741: 5-11
LM1011: 2-42
LM1303
    Characteristics: 2-12
    Inverting AC Amplifier: 2-23
    Non-Inverting AC Amplifier: 2-23
    Tape Preamp: 2-36
LM1310: 3-23
LM1800: 3-14
LM1800A: 3-18
LM1818
    ALC Circuit: 2-40
    General Description: 2-37
    Meter Drive Circuit: 2-40
    Microphone Amplifier: 2-37
    Monitor Amplifier: 2-39
    Playback Amplifier: 2-37
LM1870 (see Blend)
    Application: 3-20
    Characteristics: 3-21
LM1877/2877
    Active Bass Tone Control Circuit: 4-21
    Characteristics: 4-5, 4-6
    Circuit Description: 4-9
    Comparator Operation: 4-9
    Inverting Amplifier: 4-10
    Non-Inverting Amplifier: 4-11
    Power Output: 4-11
    Reference Voltage: 4-9
    Single/Split Power Supply Operation: 4-14
    Stereo Phonograph Amplifier: 4-21
LM1896/2896
    Bridge Amplifier: 4-17
    Characteristics: 4-5, 4-6
    Low Voltage Stereo Amplifier: 4-11
LM2000/2001
    Characteristics: 4-6
    Circuit Description: 4-62
    Compensation: 4-63
    Complementary Output Stage: 4-64
    Inverting Amplifier: 4-63
LM3089
    AFC: 3-12
    AGC: 3-13
    Applications: 3-11, 3-17
    Circuit Description: 3-8
    General: 3-8
    Mute Control: 3-12
    PC Layout: 3-10
    Quad Coil Calculations: 3-11
    S/N: 3-13
LM3189
    AGC Circuit Operation: 3-14
    Applications: 3-13
    I.F. Amplifier: 3-14
    Muting: 3-14
LM3820
    AM Radio: 3-6, 3-7
    Auto Radio: 3-7
    Characteristics: 3-5
    Circuit Description: 3-4
    Configurations: 3-5
    General: 3-4
    Impedance Matching: 3-5
LM3915
    Bandwidth Display Driver: 5-15
LM4500A
    Blend Circuit Operation: 3-19
    Oscillator Waveforms: 3-19
LM13600: 5-13
Load Dumps: 4-48
Logarithmic Potentiometer: 2-46
Loop Gain: 2-1, 4-1
Loudness Control: 2-49, 4-19, 4-36

Magnetic Phono Cartridge Noise Analysis: 6-13
Masking: 2-9, 5-13
MOL (Maximum Output Level): 2-37
Meter Circuit: 2-40
Microphone Mixer: 2-65
Microphone Preamplifiers
    CMRR of: 2-45
    LF356: 2-45
    LF357: 2-44
    LM381A: 2-43, 2-44
    LM387A: 2-43, 2-44
    Low Noise, Transformerless, Balanced: 2-45
    Tape Recorder: 2-37
    Transformer Input, Balanced: 2-44
    Transformerless, Balanced: 2-45
    Transformerless, Unbalanced: 2-43
Microphones: 2-43
Midrange Tone Control: 2-55
Mixer (see Microphone Mixer)
MM5837: 2-62, 2-64
Motorboating: 2-2
Motor Driver: 4-18
Multiple Bypassing: 2-2
Muting
    Amplifiers: 4-29, 4-41
    Deviation: 3-14

NAB (Tape) Equalization: 2-30
Noise
    Bandwidth: 2-3
    Cartridge: 6-13

Constant Spectral Density: 2-3
Crest Factor: 2-8, 6-21
Current: 2-4
Differential Pair: 2-8
Effect of Ideal Feedback on: 2-4
Effect of Practical Feedback on: 2-5
Excess: 2-3
Feedback Resistors: 6-17
Figure: 6-24
Flicker: 2-4, 6-24
Generators: 2-4
Index of Resistors: 2-3
Measurement Techniques: 2-8, 2-9
Modelling: 2-4
Muting: 3-14
Non-Inverting vs. Inverting Amplifiers: 2-7
Non-Complementary Noise Reduction: 5-13
Phono Disc: 2-23
Pink: 2-62, 6-24
Popcorn: 2-4, 6-24
Resistor Thermal Noise: 2-3, 6-24
RF: 2-7
Shot: 2-3, 6-24
Signal-to-Noise Ratio: 2-7, 6-25
Thermal: 2-3, 6-24
Total Equivalent Input Noise Voltage: 1-2, 2-4
Voltage: 2-4
White: 2-3, 2-62, 6-24
1/f: 2-3, 2-4, 6-24
Non-Inverting AC Amplifier: 6-12

Octave Equalizer: 2-59
Op Amps (see Amplifiers)
Open Loop Gain: 1-2, 2-1
Oscillations, Circuit (see layout, Ground Loops, Supply Bypassing, or Stabilization)
Oscillator: 4-34, 4-39
Oscillator, Power: 4-17
Output Referred Ripple Rejection: 1-2
Overmodulation (Phono): 2-23

Panning: 2-66
Passive Crossover: 5-1
Phase Shifter: 5-10
Phono Cartridges
  Ceramic: 2-25, 4-34, 4-38
  Crystal: 2-25, 4-34
  Magnetic: 2-25
  Noise: 2-25, 6-13
  Typical Output Level: 2-26, 4-34
Phono Disc
  Dynamic Range: 2-23
  Equalization: 2-23
  Noise: 2-23
  Recording Process: 2-23
  S/N: 2-23
Phono Equalization (see RIAA Equalization)
Phono Power Supplies: 4-36
Phono Preamplifiers
  General: 2-23
  Inverse RIAA Response Generator: 2-28
  LM381: 2-25, 2-27
  LM382: 2-27

LM387: 2-25
LM1303: 2-29

Pickup (see Acoustic Pickup Preamp)
Piezo-Ceramic Contact Pickup: 5-12
Pink Noise: 2-62, 6-24
Pink Noise Generator: 2-62
Playback Equalization (Phono): 2-23
Playback Head Response: 2-29, 2-31, 2-36
Popcorn Noise: 2-4, 6-24
Power Amplifiers: 4-5, 4-6
Power Bandwidth: 6-20
Power Dissipation
  Application of: 4-49
  Bridge Amps: 4-50
  Calculation of: 4-49
  Class B Operation: 4-48
  Derivation of: 4-49
  Effect of Speaker Loads: 4-54
  Maximum: 4-49
  Reactive Loads: 4-55
Power Supply Bypassing: 2-2
Power Supply Design
  Characteristics: 6-2
  Diode Specification: 6-5
  Filter Design: 6-3
  Filter Selection: 6-1
  Load Requirements: 6-1
  Transformer Specification: 6-5
  Transient Protection: 6-7
  Voltage Doublers: 6-8
Power Supplies
  Phonographs: 4-36
  Stereo Power Amplifier: 4-58
Preamplifiers (see Microphone, Phono, or Tape)
Preamplifiers, IC: 2-12
Proportional Speed Controller: 4-18
Protection Circuits: 4-3

Quality: 6-18

Radiation: 4-65
Reactive Loads (see Power Dissipation)
Reliability: 6-18
Reverberation
  Driver and Recovery Amplifiers: 5-7
  General: 5-7
  Stereo: 5-8
  Stereo Enhancement: 5-9
RF Interference: 2-11, 4-32
RF Noise Voltage: 2-7
RIAA (Phono) Equalization: 2-23, 4-38
RIAA Standard Response Table: 2-25
Ripple Factor: 6-1
Ripple Rejection: 1-2
Rumble Filter: 2-56

S Curve: 3-14
Safe Operating Area (S.O.A.): 4-54
Scanners (see FM Scanners)
SCA: 6-21, 6-26
Scratch Filter: 2-58
Second Breakdown: 4-54

Self-Demagnetization: 2-30, 6-22
Sensitivity: 6-25
Series Shunt Feedback (see Feedback)
Shot Noise: 2-3, 6-24
Shunt-Shunt Feedback (see Feedback)
Signal-to-Noise of Phono Disc: 2-23
Signal-to-Noise Ratio: 2-7
Sine Wave Oscillator: 4-34
Single-Point Grounding (see Ground Loops)
Single Supply Biasing of Op Amps: 6-13
Siren: 4-29, 4-39
Slew Rate: 1-1, 1-2, 4-2
Speaker Crossover Networks (see Active Crossover Networks)
Speaker Loads (see Power Dissipation)
Speech Filter: 2-57
Speed Controller, Proportional: 4-18
Square Wave Oscillator: 4-34
Stabilization of Amplifiers: 2-2
Staver Heat Sink: 4-23
Stereo IC Power Amplifiers: 4-5
Stereo IC Preamps (see Preamplifiers)
Stereo Multiplex (see FM Radio, Stereo)
Summing Amplifier: 6-12
Supply Bypassing: 2-2
Supply Rejection (see Ripple Rejection)
Supply Voltage: 1-2
Sweep Generator: 5-11
Switching
    Active: 2-68
    Mechanical: 2-68

Tape Bias Current: 2-28
Tape Equalization (see NAB Equalization)
Tape Preamplifiers
    Fast Turn-On NAB Playback: 2-34
    LM381: 2-32, 2-34, 2-35
    LM382: 2-35
    LM387: 2-33
    LM387A: 2-33
    LM389: 4-38
    LM1303: 2-36
    LM1818: 2-38

Playback: 2-33
Record: 2-32
Tape Record Amplifier Response: 2-34
Tape Recorder: 2-42, 4-38
Tape Record Head Response: 2-31, 2-36
Thermal Noise: 2-3, 6-24
Thermal Resistance: 4-65
Thermal Shutdown: 4-4, 4-53
Thickness Loss (Tape): 2-30
Third Harmonic Cancellation: 3-19
Threshold of Hearing: 2-9
Tone Controls
    Active: 2-50, 4-39, 5-12
    Passive: 2-46, 4-19, 4-21, 4-25
Total Harmonic Distortion: 1-2, 6-23
Transconductance: 4-1, 5-13
Transient Distortion: 4-61
Transient Protection: 6-7
Tremolo: 4-40, 5-11
TV Sound IF: 3-7
Two Channel Panning: 2-66
Two-Phase Motor Drive: 4-18
Two-Way Radio IF: 3-7

Unbalanced Mic Preamp (see Mic Preamps)
Uncompensated Op Amp: 1-2

Variable Gain AC Amplifier: 6-13
Variable Low Pass Filter: 5-13
V_{BE} Multiplier: 4-52
ΔV_{BE} Multiplier: 4-9
Voltage-Controlled Amplifier: 4-40
Voltage Doublers: 6-8
Voltage-to-Current Converter: 4-28
V.U. Meter: 2-40, 6-26

Walkie Talkie Power Amp: 4-44
Weighting Filters: 2-9
White Noise: 2-3, 2-62, 6-24
White Noise Generator: 2-62, 4-40
Wien Bridge Oscillator: 4-34
Wien Bridge Power Oscillator: 4-17
Wye-Delta Transformation: 2-45, 6-11

notes

notes

notes

notes

notes